The Mountain Mystery

Published by

Summit Science Publishing

Calgary, Alberta, Canada

First Published in 2014

10 9 8 7 6 5 4 3 2

LIBRARY OF CONGRESS CATALOGING-IN-PUBLICATION DATA

Miksha, Ron, 1954–

 The Mountain Mystery

 ISBN 978-1-497-56238-7

 I. Earth

 QE501.M 2014

Copyright © 2014 Ron Miksha

All rights reserved

contact: www.miksha.com

The Mountain Mystery

Introduction

I was nine years old when the *Herald* brought news to our home that the continents are moving. There was a picture, a distorted map of North America and Europe with a jagged line between the two. The paper mentioned a German scientist and some people in New York City. I told my father. I probably read the story aloud. (I was a rather annoying child.) When I finished, my father told me it was all nonsense. It was obvious to him that too much education makes a man silly and he told me as much. He himself had crossed the continent. It was big, it was heavy, and it wasn't going anywhere. "These New York scientists study and study and they lose all their common sense," he told me.

My father was a farmer and a carpenter. He was good at what he did and common sense had served him well. Fifty years ago, he wasn't the only person to dismiss the foolish notion of continental drift. Nearly every geologist in the world had more sense than this new group of scientists who thought the world's oceans and continents wander about, split apart, collide together.

Common sense had served mankind well. Humans became a successful species, eliminating competitors and sometimes outwitting the environment. Common sense assured our ancestors that the world is flat and the sun is in daily motion across our sky. Just half a century ago, mobile continents were an absurdity and most geologists would have told you so. And yet, those same scientists had no way to explain one of the Earth's most beautiful features. The mountains of the world were a mystery. Volcanoes were reasonably well understood – magma emerges from somewhere below, lava spills out, cones grow. But most mountains are not volcanoes. Neither the Alps nor Himalayas have ever erupted; most of the peaks of the Andes and Rockies have lived quiet lives.

Just fifty years ago, scientists argued bitterly about the fundamental forces that brought mountains into existence. Some were quite certain the Earth is cooling and shrinking, leaving scabs of mountains in the process. Other scientists were equally willing to insist the planet is growing and expanding, widening the oceans and thrusting the mountains upwards. A daring few suggested the very crust of the Earth is alive with motion and the surface is transformed by jostling, shaking, cracking, and crushing – violent activities which build mountains. No one knew with certainty. Fifty years ago, the Earth's mountains were an unresolved mystery. They are understood much better today. *The Mountain Mystery* is the story of how we came to know the truth about mountains.

Chapters

1. **The Fish on the Mountain** 1
2. **A Heart of Iron** 13
3. **The Sacred Earth** 33
4. **Living with Fire** 45
5. **First Rocks, then Mountains** 57
6. **Everything Changes** 71
7. **The Apple Cools** 81
8. **Time for a Shake** 93
9. **Clever New Ideas** 103
10. **Hollow Mountains** 119
11. **The Ice Man's Solution** 131
12. **Lonely Voices** 153
13. **Slipping and Sliding** 163
14. **Looking into the Earth** 179
15. **The Revolution Begins** 195
16. **Poetry in Motion** 211
17. **Written in Stone** 225
18. **Rough Edges** 247
19. **Moving Back** 263
20. **Moving On** 273

1

The Fish on the Mountain

When a child is sent up a mountain, something more might return. This is the story of how we – all of us – learned how a fish of stone became part of an ancient mountain. The story begins almost three thousand years ago with a shepherd. Perhaps ten, maybe twelve, years old. It was his task to herd a few skinny sheep – short-legged animals, grey and mottled – up the steep slope north of his home. He led them away from the sea, up an ancient trail known by his father, his grandfather, and many others before them. It had been used for generations to take the sheep of the family of Palaios from their village to the meadows where tufts of green grass survived between flat rocks. Each spring, for about three months, cool damp air blew in from the sea far below, keeping the weeds and sparse grass on the ridge fresh. Then dry summer winds arrived and those sheep which were not sold or eaten were kept closer to the house.

Palaios returned each day in the early evening, after watching over the grazing animals for hours, guarding them from falling prey to other animals, keeping them from wandering into the larger herds belonging to other families. Palaios was never exhausted from the hike. These hours of his childhood passed quickly. It was time spent trapping bugs, slinging stones, and whittling sticks into spears and whistles, carving the soft sprigs with a sharp blade his uncle had given him. As he grew into an older child, his curiosity matured. One day he spotted something among the rocks he had not noticed before. It had the same grey colour as most of the stones in the meadow, but it was a rock like none he had ever seen. He carried it home.

Returned from the mountain behind his family's house, Palaios led his ragged sheep into a corral of stone and wood, then pulled the gate tight and wrapped a cord high above the reach of the goats and sheep. The family's animals – including the pair of goats that never made the daily trek with him – were safe for the night, in the pen along the back wall of the family home. He walked to the front, through the open door, towards the light of a fire.

"I have something," he announced to his grandmother, father, mother, older sister, and young brothers.

The Mountain Mystery

From his sack, Palaios removed the soft wood he had been rather poorly carving into a whistle. He allowed their interest to build. Then several grey pebbles tumbled from his hands to the table. Finally, he held his rock. He had cleaned it, had rubbed dirt from the broad flat stone overflowing his palms.

He placed the rock on the table. "A fish," he said.

And there it was. A fish. Perfect in every way. Head, tail, fins. Everyone could see it was a fish, or had once been a fish. Within the rock was the clear pattern of fish bones. A skeleton – some scales, ribs, bones – embedded in stone.

"If you want me to cook it for you, you're too late," his sister said. "We already cooked for tonight."

Palaios dismissed his sister with a smile. "How did this fish die so far up the mountain?" he asked his parents. His father, a generation earlier, had seen pieces of similar bones, also etched into bits of grey rock. But he had never seen such a beautifully preserved animal. The fossils which the older man had found – he had not bothered to keep them – were small broken shards. It was not so clear they were once a fish. But Palaios's fossil was perfect. It was, indeed, a fish.

The mother reminded her son that the gods had crossed the same hills in the time before man. Perhaps they had left the bones, from their lunch. She glanced at the stone, then away. The lunch of the gods did not belong in her house.

Palaios was not as certain. He did not tell his parents, but there were other pieces, not so perfect, that looked like broken shells from the beach, but they were also part of the grey rocks now. Would the titans carry those, too? What had brought the fish and the shells far up the steep cliff, the height of a thousand men, above the sea? He would keep his fish-in-stone until long after his childhood of whittling and collecting had ended.

What the Greeks Knew

It would take thousands of years for people to finally know why fish made of stone sometimes appear on mountains. In fact, we have only now begun to understand the mountains, even those without fish bones. Fifty years ago, geologists may have told you the Earth is cooling, shrinking. The mountains are

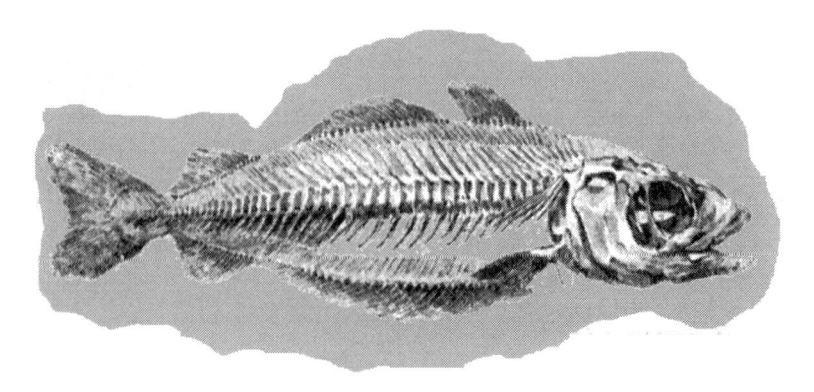

The Fish on the Mountain

the residue of the contracting planet. Or they may have described complicated schemes in which eroded bits of old mountains washed into deep coastal ravines, became compressed, heated, and melted, then rebounded as new mountains. Still others imagined the Earth expanding, thrusting up mountains in the process. There was often much confidence in their differing ideas, but little certainty. Geologists have always taken great delight in arguing against the contrary theories of their colleagues.

We are about to retrace the inquisitive investigations and the great discoveries which scientists made during the centuries after Palaios. This is a convoluted trail through laboratories where magnetism and radiation were discovered; our travels will include trips to the poles where the Earth's magnetic field generates spectacular auroras; it also includes visits to hot equatorial deserts to examine the remains of glaciers. At times we may feel as baffled as the intrepid scientists as we struggle with strange data from the Earth - gravity, magnetic, heat, and seismic signals from deep within our planet. And surprisingly, we will need to sink to the deepest parts of the ocean to really understand how our tallest mountains formed.

For a very long time, the origin of mountains was problematic. Marine fossils found on mountains were even more puzzling. The philosopher Anaximander[i], who lived along the fossil-rich north Turkish coast at the time of Palaios, wrote that the first life on Earth were fish which arose from the sea, with plants and animals later spawning from thick mud near the shores. Humans eventually evolved from fish, according to Anaximander's poem, *On Nature*. The sparest fragments of his poetry survive – he is mostly remembered through the enthusiastic discussions of his verse by his disciples.

Disciple and fellow fossil collector Xenophanes[ii] concluded that the hills where shells and fossilized fish were found had once been underwater. A bit of a humourist and a critic of the prevailing idolization of super-hero gods, Xenophanes lived most of his hundred-year life in exile. His oft-repeated adage is his reflection on religion: "If horses and cattle had hands and could draw, they would create gods that looked like

Xenophanes

---

i   Anaximander (610-546 BCE), Greek philosopher; lived within modern Turkey's border.
ii Xenophanes (570-475 BCE), Greek philosopher and religious critic; lived within the borders of modern Turkey, travelled widely in exile; probably died in Sicily.

The Mountain Mystery

horses and cattle." It was likely reaction to this sort of remark that led him to settle abroad. But Xenophanes was the first person to use fossils to help explain the changes in the Earth's geology, claiming clam shells he found on the hills of Attica were left by creatures living centuries past, when the area was flooded by seas. He was the first to propose that seas episodically covered the land, bringing soft mud that dried and hardened into rocks. Xenophanes rejected the role of deities entirely, teaching that all natural phenomena could be explained through logic and science. It was the start of a journey that would last to our own generation – a search for natural explanations for the mysteries of the Earth. As Palaios suspected, the titans of old had not dropped fish from their lunch pail.

A few years after Xenophanes, the historian Herodotus[i] described fossils he found along the north African coast. He also thought they formed at a much earlier time, when Egypt had been submerged under the sea. But Herodotus told stories of gigantic winged serpents, which he said had been killed by ibises. "In the spring long ago, winged snakes came flying from Arabia towards Egypt, but were met in a gorge by birds called ibises, who destroyed them all. It is on this account the Egyptians hold the ibis in so much reverence."[1] I have not seen his explanation of how the long-legged marsh birds killed huge flying serpents, but Herodotus, honoured as the west's first historian, did not usually bother to explain his often fanciful reports. However, the "multitudes of backbones and ribs of serpents" he saw may have been discovered recently. They seem to have belonged to large, tusked herbivores, not defensive birds or invading snakes.[2]

It wasn't uncommon for such fossils to enter history or legend. For example, the tale of the Cyclops likely began with a fossilized elephant's head. When Homer[ii] told the saga of the homeward journey of Odysseus after the battle of Troy, he led the warrior to an encounter with a member of the Cyclops race of one-eyeds. The legend of monocular-visioned giants pre-dates Homer by centuries, and likely had its origins in fossils found on the islands of Cyprus, Crete, and Sicily.

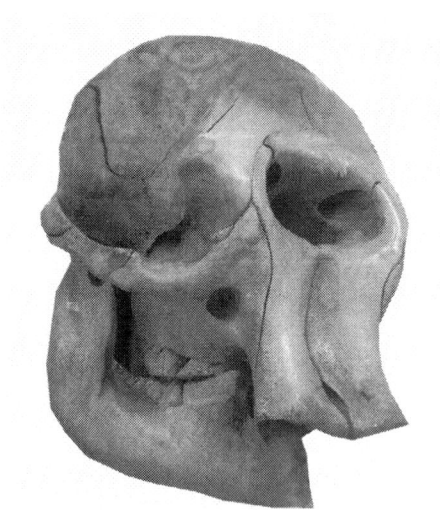

Pygmy elephant, the Cyclops

A hundred years ago, Othenio

---

i   Herodotus (484-425 BCE), Greek historian; originally from within the borders of modern Turkey, travelled to Egypt, Persia, and Babylon.
ii Homer (fl. 700 BCE), Greek poet from coastal Anatolia, today modern Turkey.

4

The Fish on the Mountain

Abel[i] conjectured that prehistoric dwarf elephant skulls (which are twice the size of human skulls) would have puzzled the folks living on those Mediterranean islands. Pygmy elephant skulls tend to have a large hole near the centre, where the elephant's trunk had been. If a person had never seen an elephant, it would be more sensible to imagine a giant one-eyed human than a big-eared, long-nosed giant grey mouse as the original owner of the head. Abel's idea, if correct, highlights a central tenet in the way we unravel the mysteries of science – we often start with explanations based on things we are most familiar with, then make corrections as more information is gathered.

Skulls of extinct small elephants are still found on Mediterranean islands. These animals arrived during episodes in the Pleistocene[ii] when the Mediterranean was dry, colonizing the intermittently accessible islands. Later, the seas rose, isolating the creatures. Eventually, of course, the elephants disappeared, probably eaten by predator humans. The elephants' cycloptic skulls were still there when more recent people settled and created legends about them.

Fossils puzzled the ancient philosophers, Greek and Roman. As we have seen, Xenophanes concluded that the land-locked remains of sea shells and fish had originated in an ocean which somehow rose as mountains. Creating reasonable narratives explaining uplifted rocks is challenging, but one solution was obvious. Babylon, Alexandria, Athens, and Rome all had citizens who had experienced earthquakes and volcanoes – either from the discomfort of their own homes, or on excursions to foreign lands. It was typical in the ancient world to imagine violent Earth activities raised the land and its dead to new heights. Less commonly, floods were blamed. In some legends, the seas briefly inundated the entire globe and carried fish carcasses onto the mountains. Thus fish arrived upon a mountain either by an elevation of the seafloor or by floods that carried the creatures. Among the Greeks, many assumed uplift had occurred; but tales of a great flood eventually dominated explanations of mountain fossils – and much more.

Gilgamesh Delivers Fossils

For centuries, the biblical story of Noah would be considered the most important event that had shaped the Earth's surface. The tale influenced scientific explanations until very recent years. For that reason, it is worth considering the story in detail. Stories of an epic flood are perhaps the most culturally pervasive in the world. The Mayans of Central America, Masai of Africa, Maidu of North America, Maori of New Zealand, and 300 other prehistoric groups had myths of a great deluge.[3] Irish legends told of Queen Cesair, who floated with her court

---

i   Othenio Abel (1875-1946), Austrian palaeontologist; a founder of palaeobiology.
ii The Pleistocene Epoch began about 2.5 million years ago and ended 11,700 years ago. This epoch is the most recent period of glaciation, ending when the last Ice Age ended. During periods of intense glaciation, sea level fell throughout the world.

until the floods receded. Spanish priests, when they arrived in the Americas, were so shocked by non-Biblical flood legends among natives that they blamed the devil for corrupting the people with apocrypha. These stories all told of horrific floods that destroyed almost everything known to man and reflected irritation of the gods with human immorality.

The most culturally pervasive legend is the Epic of Gilgamesh, the inspiration for the story of Noah. Gilgamesh was a Sumerian king of Uruk, an ancient land within today's Iraqi borders. Gilgamesh lived around 2500 BCE, which is 2,000 years before the story of Noah was added to the works which would become the Bible. The historical Gilgamesh is remembered for telling the story of a very old man who survived a world-wide flood. The gods sent a messenger named Ea who told the old man to build a boat. Ea gave exact dimensions (120 cubits high with seven decks), and told the old man to seal the joints with pitch and bitumen. The man's entire family boarded, together with his servants and representatives of "all the animals of the land." After the flood, a goddess named Ishtar felt such remorse that she promised that her colourful necklace would remind her that never again would she allow the gods to destroy the world. Meanwhile, the old man in the boat found himself grounded on a mountainside where he released a dove, a swallow, and a raven.[4]

A few centuries later, the Akkadians told a myth of gods so irritated with humans and their excessive revelry that they decided to drown the whole lot. But Enki, a good god, warns one human to build a double-decker boat, sealed with tar and loaded with family and animals so he could escape the flood.[5] The deluge destroyed the world, then drained away. For the Hebrews, it was neither a god

Gustave Doré: *The Deluge*, 1850

The Fish on the Mountain

named Ea nor Enki, but an angel named Enoch who sent a warning to Noah. At age 600, Noah saved his family and two of each of all the Earth's creatures.

The Gilgamesh tale and the other subsequent flood legends were based on a real flood of nearly biblical proportions. The cause of the deluge may have been the sudden draining of the largest lake ever known. Lake Agassiz is gone now, but it once covered much of central Canada and a vast part of the northern American plains. At the end of the last Ice Age, glaciers two kilometres thick began melting into rivulets, streams, and rivers, trapping enormous pools of water behind a series of ice dams near Hudson Bay. Before the largest of the dams broke, 8,200 years ago, Agassiz held more water than all the lakes in the world today. The ice dam spectacularly failed and its water drained into the ocean within days. Meanwhile, a smaller, but similar lake, Lake Missoula, was 700 metres deep when its ice dam broke. Although smaller than Agassiz, it nevertheless held more water than lakes Erie and Ontario combined and covered vast sections of Oregon, Washington, and Idaho. In just 48 hours, Lake Missoula drained through the Columbia River Gorge with waters rushing 150 kilometres per hour. When the waters of Lakes Agassiz and Missoula reached the oceans, the seas rose 10 feet, or 3 metres, flooding the world's coasts. In perhaps a month, an area the size of Great Britain was inundated on the Arabian peninsula. The rising water created the Persian Gulf. The coast of Africa was flooded. And the Black Sea rose, drowning a settlement recently discovered by Robert Ballard, the oceanographer who also discovered the sunken *Titanic*.[6] Most of the world's people lived along coastal plains which disappeared in mere days. Every part of the Earth was affected. The shorelines of the world were redrawn.

These cataclysmic floods drowned primitive villages which dotted coasts around the globe. A similar rise in sea level today would again drown coastal communities. Galveston, New Orleans, and Miami would be gone – so would a quarter of Boston and New York City.[7] Some island nations would disappear entirely. Eight thousand years ago, survivors described the event through oral histories that became epic legends, then became accepted as real events.[8] Until recently, the occurrence of mountains and fossils such as the Palaios fish were explained in terms of a world-wide 40-day flood that deposited hapless creatures in unlikely places.

Aristotle, Reason, and Observation

In ancient Greece, the philosophers of the day were not blaming a punishing deluge for land-locked marine fossils. They mostly concluded the fossils had been sea creatures living in the ocean; they became stuck in mud, petrified, then elevated by volcanoes or earthquakes. Aristotle,[i] though, had other ideas. Aristotle began his study of natural history on the Anatolian coast, today's Turkey. It was an

---

i   Aristotle (384-322 BCE), Greek philosopher, teacher, and scientist.

area where he could easily stumble upon interesting fossils during his daily walks.

Two centuries after Xenophanes's enlightened explanation of fossils, Aristotle set science back by teaching that fish fossils are a peculiar species that live in the ground without moving – and are spontaneously generated.[9] Aristotle based this idea on neither observation nor experimentation, but he owned quite a collection of stone fish. He had asked his ambitious student, Alexander the Great, to require hunters, fishers, and shepherds in his realm to send unusual fossils, stones, and animals to Aristotle's school in Athens. These were systematically sorted based on complexity. Aristotle excluded really odd creatures from his collection. Peculiar traits were noted, but if they didn't fit into his system, they weren't listed in his catalogue. He simply rejected outliers that didn't agree with his scheme. Aristotle was trying to find 'normal' samples and to group them into categories ranging from simple to complex – qualities which he defined in his own way.

For Aristotle, the goal of science was to systematically and hierarchically organize, to classify items into similar families and then relate those groups to each other. True scientific inquiry (at least what we have come to recognize as such) was absent. Science began with organization, not with testable hypotheses and experimentation. Classification was based on logic, but not necessarily observation; certainly not research. Aristotle had ideas about how the universe should function and nature was sometimes crudely squeezed into his models. For example, Aristotle concluded that men have more teeth than women.[10] For him, it was logical – the male is larger, needs more food, needs more teeth. Thereafter, his biology texts taught that men are toothier than women. Had he bothered to look into some mouths, he would have discovered otherwise.

Alexander receives from Aristotle

The Fish on the Mountain

Regarding mountains and topography, Aristotle had surprisingly modern ideas. From the 12th chapter of his *Meteorics*: "land and sea in particular regions does not endure forever, but it becomes sea where it was once land, and land where it was once sea. The Nile has not flowed forever. The places where seas occur were once dry, and there is a limit to their operations, but there is no limit to time. Rivers spring up and they perish; the sea leaves some lands and invades others, everything changes in the course of time." There is no limit to time; everything changes in the course of time. In these two statements, Aristotle anticipated the greatest European scholars by almost 2,000 years. Aristotle also said changes to the Earth's landscape are slow in comparison to the lengths of our lives, thus are seldom noticed. Aristotle promoted many befuddling notions in his scientific philosophy, but occasionally he got things right.

His successor Theophrastus[i] managed Aristotle's school and museum for 36 years and was strongly committed to understanding rocks and minerals. He may have been the first real geologist: identifying minerals and classifying rock personalities as magnetic, electrifying, many-sided, or simply white. It was the start of geology as a science and it set the style that would see amateur geologists sometimes compared to stamp collectors who assemble and horde collections based on scarcity, aberrations, or beauty. Aristotle contemplated nature in terms of the logical organization of the universe. He likely would have approved of Theophrastus's catalogues.

Geological change is usually not noticed in a single lifetime. Our everyday experiences argue for the eternal existence of our present scenery. It was this obvious observation about mountains, rivers, deserts, and plains that held sway for the next thousand years. The sea itself was also problematic for those who cared to think about its origin. For centuries, the oceans were explained as flooded continents; islands as partially submerged mountains. The seafloor was logically expected to be as old as continents. All these ideas are wrong. Oceans are so different from the surrounding continents that no ancient peoples – in their myths or legends – came close to explaining them correctly. It was only in the past few decades that we came to understand that ocean rocks are much, much younger, have a simpler genesis, and are made from quite different materials than continents. They are not in any respect similar to the continental rocks upon which we live.

With the ancient Greeks, we see the first efforts to understand the dead fish on the mountain. However, the science of geology, with an understanding of rock formation, mountain building, and fossilization had a long meandering journey ahead. Even the mechanics of fossilization was a great puzzle. But the occasional nearly perfect set of stone bones made even the most sensible skeptics admit the relics had been alive at some point. In reality, fossils are not so difficult to fathom. If you want to understand a fossil, you should make one.

---

i   Theophrastus (371-287 BCE), Greek philosopher and scientist, from Lesbos.

The Mountain Mystery

An Immortalized Dinner

Start with a bone. You might have some fresh dinosaur bones left from last night's dinner, if you ate chicken. Palaeontologists explain how some dinosaurs became birds a long time ago. DNA extracted from proteins in the connective tissues of *Tyrannosaurus* fossils show the animal is the ancestor of our unhappy white cluckers.[11] The chicken's ancestor – the *Tyrannosaurus rex* and some of its smaller relatives – flourished by using razor-sharp claws to catch our furry mammal ancestors cowering under nearby ferns. Thanks to those shrewd dinosaurs a hundred million years ago, mammals became smarter and swifter, if they weren't to become snacks. With dino-induced stresses, environmental adaptations, and random mutations, primitive mammals evolved into clever creatures. And eventually learned to cook the chickens that dinosaurs became.

If the discarded chicken legs you retrieved from the garbage bin are smelling less than sweet, that's helpful for the instructions that follow. It was discovered that bacteria play an important part in bone fossilization. Perhaps the critical role. Microbes help convert bone to stone, speeding up the process significantly.[12] Here is how you can convert your own discarded chicken bones into heirloom fossils.

Take an old saucepan and poke some holes into the bottom. Fill it with enough sand to bury your chicken/dinosaur bone. Pack the sand as tightly as you can by hand. Find a warm humid place to set up your experiment. Somewhere germs will thrive. Place a larger pot under your leaky sand bucket to catch the dirty wastewater that will drip from your project once it is active. Separately, find some calcium material – garden lime, Plaster of Paris, or white-wash. Mix this with water to a thickness resembling spoilt milk, then gently pour a cup of limey water on the sand every day.

You might need to wait a few hundred years to see the final result. Microbes will eat away bone bits and most of it will slowly rot. Cavities created when the ossified parts disappear will be replaced with calcium minerals dripping through the sand. The bone will turn to stone; you will have your fossil. You can stick to the theme of this book and use a fish instead of a chicken femur, but you will be less convincing when you tell people it is a dinosaur fossil.

This is also how dinosaur bones became fossils. When you see a museum dinosaur, you see stones, not bones. Ken Carpenter, at Utah State University, says some of the fossil-making is strictly a chemical process, but "the vast majority of the fossilization is due to mineral precipitation by bacteria. Bacteria feed on the organic material contained within the bones and attach their metabolic waste on various atoms or molecules, such as iron or carbonate, dissolved in ground water. The result is . . . the formation of minerals that basically turns *bone to stone*."[13]

As stone, the fossils last indefinitely. The most impressive relics – large dinosaurs – are well over 60 million years old. Surprisingly, fossilization is perhaps never complete. Palaeontologists have discovered that not every iota of

The Fish on the Mountain

blood vessel and muscle tissue become stone, not even in ancient dinosaurs. According to a 2013 study, organic dinosaur material with bits of intact DNA has been recovered from dinosaur fossils celebrating their hundred-millionth birthday.[14]

The ancient philosophers and naturalists suspected the bone stones were once living creatures. Fossils were not attributed to supernatural events. As Seneca said about such things two thousand years ago, "It will help also to keep in mind that gods cause none of these things. Neither heaven nor earth is overturned by the wrath of divinities. These phenomena have causes of their own."[15] Unfortunately, for the next thousand years, superstition, alchemy, and witchcraft often replaced science and logic in explaining fossils – and much else. At least in Europe. But during Europe's Dark Ages, scientists elsewhere sometimes understood things better.

The Slow Awakening

In 1027, a Persian philosopher described fossils as living things that became "mineralized and petrified."[16] Avicenna[i] wrote this in his *Book of Healing*, which is an encyclopedia, not a medical text. He also wrote a real book of healing, *The Canon of Medicine*, which became the standard medical guide throughout the Middle East and Europe for hundreds of years. Originally in Arabic, the book was translated into Hebrew, Persian, Chinese, Latin, German, French, and English.

Living in a tectonically active region, Avicenna also contemplated geology. The Ismaili scholar thought mountains were caused by "upheavals of the crust, such as might occur during a violent earthquake." He realized incredible spans of time are involved in shaping features of the Earth.[17] Avicenna also recognized that rock layers develop with younger deposits above older ones. Digging down through rocks, to him, was a voyage back through indeterminate time. He was right, but it would take 600 years before Europeans accepted this simple fact. When Danish Bishop Nicolas Steno presented the idea in 1669, it was greeted with skepticism, and then considered heresy.

In Avicenna's Islamic realm, scientists and philosophers developed algebra with a system that included the number zero, something that was missing among European mathematicians. Arabs mixed chemicals into new alloys, inks, and dyes. Their surgeons understood blood circulation centuries before its rediscovery in Europe. Further east, Chinese had rockets, gun powder, the magnetic compass, and the printing press. Meanwhile, European scholarship languished a thousand years, stifled by invasions, tribal warfare, and the repressive consolidation of power by the Catholic Church. But the Europeans eventually awoke from their dismal Dark Ages, furiously adopting eastern discoveries and developing their

---

i   Abu Ali al-Husayn ibn 'Abd Allah ibn Sina (980-1037), known as Ibn Sina but westernized as Avicenna, a Persian physician, philosopher, and scientist.

The Mountain Mystery

own.

European progress was rapid when independent city-states began to see their wealth grow by supporting science, technology, and exploration. An example of radical creativity is, of course, Leonardo da Vinci.[i] Supported by Milan's nobility, Leonardo had time to write, draw, and invent. He wrote in his notes that he frequented nearby hills and explored a cave on Monte Rosa where he found massive fossils. His contemporaries believed those fossils were in their lofty location because they were transported by the Flood, or (in homage to Aristotle's teachings) because they represented special organisms that lived, grew, and died within mountain mud and gravel. Leonardo dismissed both notions.[18]

"Here, above the plains where flocks of birds are flying today, fish once swam in giant schools," he wrote.[19] Leonardo recognized that the surface of the Earth changed over time, with land where sea waves once rippled. He wrote that the most powerful natural force is the movement of water in rivers. Water transforms the landscape, but he also noted that this takes a very long time. He described erosion as a slow but relentless natural process, responsible for much of the planet's topography. In his cryptic notebook, he wrote, "Why are the bones of huge fish, oysters, corals, shells, and sea-snails found on top of high mountains on the coast as well as in shallow seas? A sect of ignorant persons assert that nature or heaven has them in these places by heavenly means."[20] Leonardo rejected the biblical story of a great flood,[21] observing that there would have been nowhere for the receding water to go and that any intense rapid deluge would jumble fossils and shells within the mud. The rushing water would not leave the tidy sequences he saw, with different fossil types in different layers.

While working on canal-building projects, Leonardo da Vinci had seen thousands of shells and fossil fish buried deeply, fully encrusted in stone. As a hydrologist, he knew the simple but profound fact that shells sink in currents of water, they have no way of floating to the top of a mountain in a rising deluge. Leonardo pointed out that shells and fish bones are not only found at the highest peaks, but partly up the mountains and even deep inside them. All of which, he felt, meant something other than a great flood had placed fossils on mountains.

Italian professor Gian Battista Vai has recently pointed out that da Vinci discovered the basic principles of modern geology. Da Vinci recognized that strata, or individual layers, are originally formed horizontal and continuous. He noted older layers lie under younger ones and all such layers may be tilted by later geological activity.[22] Leonardo saw all this 150 years before the great geologist Nicholas Steno[23], who is usually credited with these geological discoveries. But like so many of da Vinci's great ideas, they were scrawled in mirror-image script into obscure notebooks and forgotten. They had to be reinvented by later generations.

---

i   Leonardo di ser Piero da Vinci (1452-1519), Italian inventor, artist, scientist.

2

A Heart of Iron

For hundreds of years after Leonardo da Vinci, Europe's authority for the Earth's geology and the planet's age remained the Bible. Irish Archbishop James Ussher[i] counted the number of years from Creation, finding an irrefutable juvenescence for the planet. Ussher performed an exact scholarly calculation. Creation had occurred in the early evening of October 22, 4004 BCE. The date was repeated frequently and printed into the side bars of the King James Bible, acquiring a biblical authority of its own. For two hundred years, it was the most cited and most widely accepted age of the Earth – 5,654 years old at the time Ussher unveiled his revelation. Ussher arrived at his date through a complicated exercise. He used the earliest known non-biblical historical event, which related to Babylonian King Nebuchadnezzar, and then worked backwards through the Masoretic Torah with its unbroken chain of lifespans, begats, and events. He convinced himself of his accuracy when the counting ended and he had arrived exactly 4,000 years before the birth of Jesus, whom Bishop Ussher thought was born in the year 4 BCE. Hence, Creation at dinner time, October 22, 4004 BCE.

It occurred to Ussher and his contemporaries that God likely intended to end the world on a mathematically appropriate date, likely 6,000 years after starting his experiment. Finding an accurate beginning date was significant for preparing for the impending end, or rapture. Other theologians agreed in general, but squabbled about the exact date of Creation. But for most, there was something reassuring in approaching a seventh millennium – it was interpreted as a sign of completion for the Earth and a celebration of the lofty heights western civilization had attained by the seventeenth century. However, the neoteric science of geology was indicating signs of an older Earth – although there was little evidence to offer other than observations of erosion and some stonified fish on mountains. Nevertheless, those who studied the rocks conjectured the planet was older than Ussher's few millennia. Most scientists of the era didn't know much about the Earth – its origin, composition, mass, or age – but many were convinced the natural transformations of the planet needed considerable time.

---

i   James Ussher (1581-1656), stridently anti-Catholic Irish theologian, remembered mainly for his chronology that defined the age of the Earth for generations.

The Mountain Mystery

Ussher's world was changing. His young, Earth-centred universe began to unravel when Tycho Brahe and Johannes Kepler proved Copernicus' claim that the Earth travels around the Sun; Galileo's telescope soon gave visual confirmation that they were right. A modern approach to science was emerging, especially within the fields of physics, geology, and geography. Empirical and experimental methods slowly began to unseat Aristotle's philosophy of nature.

The first person to create a modern atlas of the world, Abraham Ortels,[i] made a remarkable observation about the Earth's continents. The Flemish geographer noticed the similarity of the coasts of Africa and South America and imagined that the continents were once joined together before drifting to their present positions. This idea would eventually play a central role in understanding geological processes, especially mountain building. In his work *Thesaurus Geographicus*, Ortels suggested that the Americas were "torn away from Europe and Africa . . . by earthquakes and floods. The vestiges of the rupture reveal themselves, if someone brings forward a map of the world and considers carefully the coasts of the three continents." But it would take four hundred years before Ortel's remarkable conjecture would be confirmed.

After rather humble beginnings and a late start, Europeans were rapidly adopting and creating technology to explore the world. The compass was one such clever development. They didn't invent it, but they used the tool to

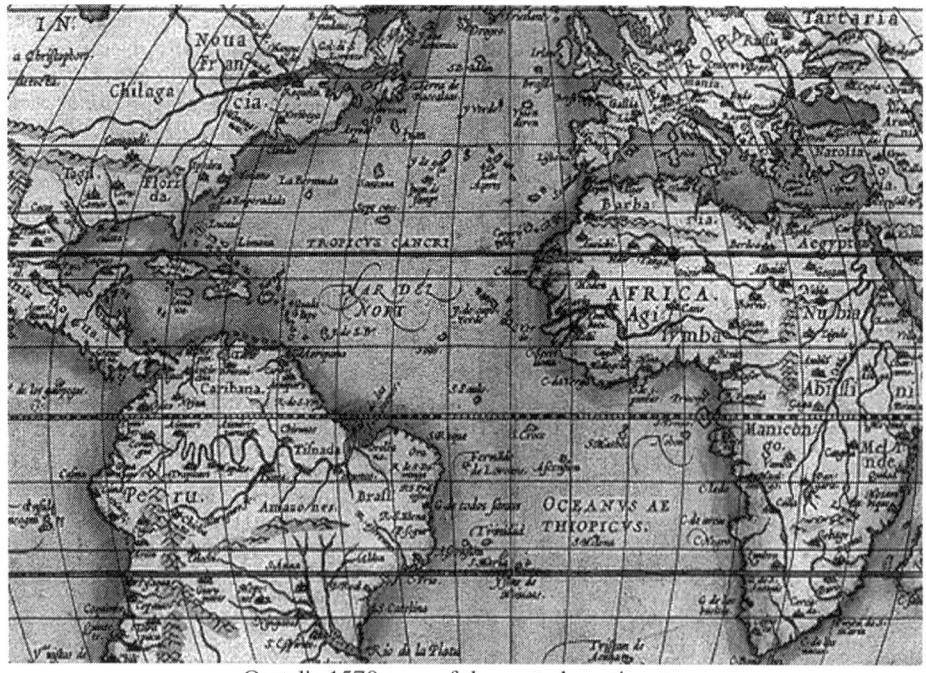

Oertel's 1570 map of the parted continents

---

[i] Abraham Ortels, also known as Ortelius (1527-1598), Dutch cartographer.

A Heart of Iron

circumnavigate the globe and to spread their empires into distant lands. They realized the Earth is an unmistakable sphere and remarkably large and varied. Soon the sixteenth-century men of adventure would discover a startling curiosity buried deep within the planet that made their compasses work.

Pointing a New Way

The Earth was being explored, mapped, and viewed in ways never before imagined. The planet was a ball, spherical, not the flat disc described in the Bible. Navigators appreciated sailing west and arriving in the east. It is a myth that Columbus was proving the Earth round while his crew feared falling off its edge if they sailed too far. In reality, almost no sailor at the time believed this. Nearly two thousand years before Columbus, the head librarian in Alexandria, Eratosthenes,[i] had calculated the circumference of the planet as 252,000 stades (a stade being the length of a Greek sports stadium), which works out to almost the same measurement known today. The sailors with Columbus were far more concerned that the spherical planet is so large they would run out of food and water before reaching India. Provisions did run low as they approached the Americas, which is much less than halfway to Asia. But there was no trepidation that they would approach a deadly waterfall at the world's end. Confident the planet is a sphere, comfortable most of it was now mapped, by 1600 few people were surprised to also learn that at the Earth's centre was a heart of iron that caused the planet to act like a giant magnet.

The knowledge that some rocks mysteriously attract others is centuries old. Lodestones[ii] had been discovered by prehistoric people. For them, a lodestone was a mysterious piece of rock with other rocks clinging to it. These natural stones were picked from the ground amid iron rocks. Greek miners found lodestones in iron ore deposits while extracting metal for spears and shields. They noticed magnetic attraction but couldn't explain it. For centuries, magnetism remained an interesting, but not particularly useful, novelty to the Europeans.

They didn't use the metal for compasses or gadgets the way the Chinese did. Over two thousand years ago, Asian fortune tellers dropped the magic rocks on colourfully annotated boards to foresee the future paths of their clients' lives. Eventually they began to realize the stones were more accurate at giving spatial directions than predicting spiritual ones. In time, Chinese navigators adopted the lodestones of their magicians and fortune tellers to navigate the seas.

---

i   Eratosthenes (276-194 BCE), Greek geographer and mathematician. His friends called him Beta, or "Number Two," because he was considered second best at everything in the world. Eratosthenes invented (and named) geography, laid the basis for western musical scales, and created the algorithm to identify prime numbers (Eratosthene's Sieve) still used today.
ii Lodestones are highly magnetic naturally occurring rocks containing magnetite. They attract iron, nickel, and other rocks containing those elements. *Lode* is derived from a Middle English word meaning to lead, or show the way.

The Mountain Mystery

The ancient Chinese manufactured compasses by carefully sculpting and polishing natural lodestones into magnetic spoons. They designed these so the handle always pointed south, which they first located from the sun's position. They could balance the spoon's bowl on a finger, wait a moment, and the handle would turn southward. Later, the compass-spoon was centred on either a plate[24] or an ornate block of wood painted with constellations representing the eight cardinal directions. Experts would slowly rotate this luopan board until the handle of the spoon (still persistently pointing south) matched the south constellation painted on the rim of the luopan. Surveyors, called geomancers, carried the gadget to building sites so they could align their constructions in ways acceptable to the traditions of Feng Shui[i].

Fashioning spoons from raw lodestones was slow work and the stones were rare enough to be expensive. So Chinese artisans, especially during the Qin and Han[ii] dynasties, mined and refined iron ore, discovering that an object cast from iron that contained magnetite became magnetized after removal from a hot furnace. Rather than spoons, artisans began creating needle-thin metal slivers, often shaped as fish. "We cut a very thin piece of iron into the shape of a fish. We make it red hot in a coal fire, and then retrieve it with tongs. The tail remaining oriented towards the north, we quench it in water."[25] The Chinese were using the Earth's natural magnetism to embed a pattern – a magnetic memory – into the iron. How this actually worked would be explained by Marie Curie in France over a thousand years later. Meanwhile, the fish-compass was thin enough to float in a bowl of water, like a small raft, and drifted until its mouth gaped southward. For Chinese, the direction south was the preferred orientation for magnets; Europeans aligned their compasses northwards.

For economics, simplicity, and accuracy, spoons and fish were replaced by thin needles which Chinese artisans similarly arranged during cooling so they would capture the Earth's north-south pattern. They also discovered that a needle acquired magnetism by stroking it consistently in the same direction against a lodestone. The result was not as permanent, but was even less expensive to produce. By the eighth century, Chinese navigators were guiding ships on cloudy nights with accurate inexpensive compasses. By the year 1000, Chinese compasses had been in use for hundreds of years and were helping military and trade excursions throughout the Far East. Years later, those magical, directional

---

i   *Feng shui* is a Chinese philosophy which attempts to harmonize human activities with the environment, largely through spatial orientation. In its earliest forms, dating over 6,000 years, it may have been as simple as aligning windows to face south for winter exposure. It grew to incorporate notions of Qi, or energy flow, believed to dissipate with wind and accumulate with water – feng-shui literally translates as wind-water. Compasses have been used for thousands of years as part of the ceremony for proper alignment of houses, tombs, and public buildings.

ii The Qin (221-207 BCE) and Han (202 BCE – 220 AD) Dynasties were times of explosive creativity. It was then that the Chinese built the Great Wall, made the Terra Cotta Army, and invented iron plows, paper, the hot air balloon, porcelain, and the compass.

A Heart of Iron

needles came into use in Europe.

During an era when Europeans were suspicious of science, the lodestone was viewed through the alchemist's eye as a thing of mystery, superstition, and power. Magnetism's practical application as a compass wasn't generally known in the west, however a reference in the Icelandic *Chronicle of Are Frode*, written at the end of the eleventh century, suggests that the Vikings may have independently invented the compass. The *Are Frode* briefly relates how lodestones helped Viking sailors cross the North Atlantic. But the compass still wasn't used much in Europe. Finally, in 1269, Petrus Peregrinus[i] gave one of the first extensive European descriptions of a working compass and built a modern one with a needle on a freely spinning pivot point.[26] Peregrinus described how a magnetic rock could draw others to it, with the north-pointing end of a lodestone always attracted the south-pointing end of another.[27] Peregrinus put the idea of magnetic attraction and repulsion to marvellous use by designing a perpetual motion machine. A series of gears and wheels were expected to rotate around a fixed lodestone, with teeth alternately pushed and pulled by magnetism, forcing the wheel to turn. Forever. At least, in his drawing – there is no record Peregrinus (or anyone else) ever actually constructed the device.

The Europeans lagged in adopting the compass. For thousands of years, Europe's sailors and navigators didn't need it. They simply avoided venturing too far from land. Other than the Vikings (who, it seems, occasionally carried a compass) Europeans weren't exploring new lands until quite late in the fifteenth century.

The Attractive Earth

Magnetic lodestones were found around the world – the Greeks, Chinese, and Central American Olmecs refined the ore two thousand years ago. But not every piece of magnetite ore acts like a strong magnet, attracting iron to it. As recently as a few decades ago, we didn't really know how lodestones became magnets, but a new theory supports an old speculation. It appears that whenever lightning hits the right sort of magnetite, the mineral melts briefly, then cools, retaining a magnetic imprint from the lightning's electromagnetic field, complete with North and South, branded into it.[28] This can be used as a compass. Unfortunately, *True North* which is the centre of the planet's rotation, and *Magnetic North,* which the compass identifies, rarely coincide, but are somewhat displaced. The magnetic field generated by the Earth is off-centre and wanders aimlessly near the pole, for reasons no one yet knows. This creates a problem for navigators seeking precise directions. All of this is important to our story of how scientists finally figured out the origin of mountains. Magnetism, we will see, returns again and again as we

---

i   Petrus Peregrinus de Maricourt, or "Peter the Pilgrim of Maricourt" (fl 1269), French scholar and physicist.

draw closer and closer to the solution – and in fact magnetism eventually proved that continents move, collide, and thrust fish fossils atop mountains.

European compass makers were among the first to unravel the mysteries of magnetism. Foremost among them was Robert Norman[i], a sixteenth-century craftsman who published a collection of his original experiments and observations on magnetism. In his catchy *The Newe Attractive,* he evaluated various lodestones ("the first and best Sorte of these Stones come out of East India, from the Coast of China and Bengalia, and is of the Colour of Iron."); he explained how to make a good quality compass; he described the five regional types of compass available (including the Levant, the Danske, and the English); and, he wrote at length about the deviations, or magnetic wanderings, that occur from place to place – something he witnessed in his twenty years at sea. Norman also considered the phenomenon of the compass needle bending down towards the Earth.[29] Not only will a compass spin to face (more-or-less) north, it also simultaneously bows its nose down towards the Earth. Norman was the first to write about this important aspect, which we now call *inclination*, or dip.

Norman wrote, "Having made many and diverse compasses, I found continually, that after I had touched the irons with the Stone [after he magnetized needles by rubbing them with a lodestone], that presently the north point bent downwards under the horizon some quantity." The needle "now required some small piece of wax in the South part to make it level again." He didn't realize that he had discovered magnetic dip, the tendency of a compass needle to align with the angle that the global magnetic field enters the Earth. In his book – both a scientific study and an experimental guide – he described how he discovered this strange property of the compass.

Robert Norman admitted he regularly counterbalanced his needles with bits of beeswax, assuming the needles he used weren't perfectly balanced from the start. But when Norman was given a perfect needle of a fixed length with instructions to use it on a compass at its original length, he ran into trouble. He polished the wire and it balanced perfectly on a pin. But when he "touched it with a stone" and placed it back on the pin, the north end dipped. Against orders, he neatly clipped off a piece from the north end, hoping to lighten it and restore it to level. But it drooped once again. So he again cut the needle to counterbalance it, and then again, "in the end I cut it too short, and so spoiled the needle wherein I had taken so much pains." Norman wrote, "Hereby being *stricken in some choller* [a Middle English phrase for *ticked off*], I applied myself to seeke further into this effect, asking learned and expert friends acquainted in this matter."

The result of this compass-maker's frustration was a series of careful experiments. In his book, Robert Norman gave precise details so other experimenters could recreate his steps, as if following a recipe. Norman found that at his workshop in London, the inclination of the north point of the compass

---

[i] Robert Norman (fl 1580), English compass-maker.

A Heart of Iron

was consistently 71 degrees 50 minutes. Norman suggested that other experimenters might get different dips at other locations. He was right, they do. Some argued that the lodestone had added some extra invisible weight to the north end of the needle, some unseen mass had drifted into the nose of the compass, transferred from the lodestone during the stroking process. To defuse critics, Norman weighed each of his many needles on a sensitive scale. He found no such mass had transferred. Compass needles weighed the same before and after stroking. But the north end always acted as if it had become heavier.

Many of his conclusions still stand, four hundred years later. His lucid descriptions of experimental procedure are prescient of the modern scientific method. But one well-designed experiment led him to draw a conclusion that was seriously flawed. In this experiment, he tried to determine the source of the magnetic force attracting and declining his compass needles. Some believed the source was a star or some other celestial body while others favoured a magnetic island somewhere in the northern seas.

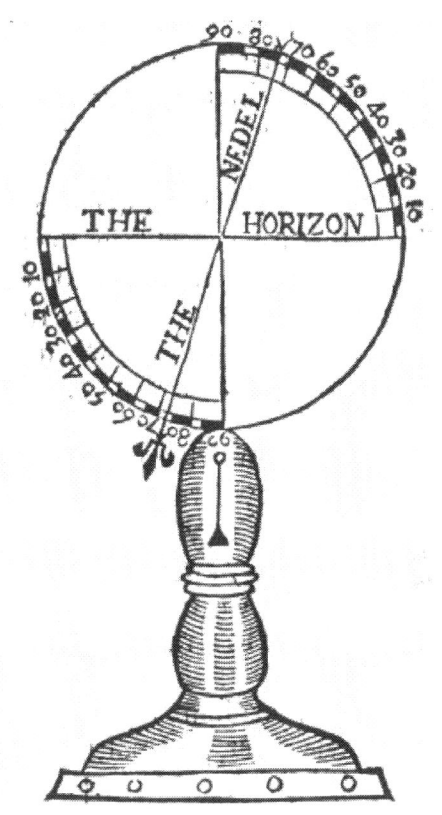

Norman's magnetic dip tool, 1581

Robert Norman disagreed. From one of his experiments he concluded that the magnetic force was not coming from anywhere in particular. His clever experiment was this: Norman sank a magnetized needle in water after driving the needle through a bit of cork that perfectly balanced it between either sinking to the bottom of his glass tank or floating to the top. It tilted down, as had all his needles, and it pointed north, but it neither sank nor rose. From this, Robert Norman surmised that an infinitely distant north magnetic pull was balanced by an infinitely distant south pull, both in straight lines, one passing downwards through the ground, the other upwards and into the heavens.

He concluded that magnetism is not caused by "any attraction in Heaven or Earth." It was brilliant reasoning, but Robert Norman was wrong in his conclusion, as Sir William Gilbert[i] would show a few decades later. Drawing

---
i   Sir William Gilbert (1544-1603), English physician and scientist.

upon the work of Norman, and other experimenters, Gilbert was able to demonstrate something entirely new about the Earth's magnetic properties. He would prove Norman wrong about the idea of an aether-like pervasive magnetism. Gilbert placed the source of magnetism squarely at the centre of the Earth. His work was the first great geophysical discovery about our planet. But it would quickly create a bitter enemy with the one scientist in England Gilbert needed as an ally.

The Queen's Doctor

Quite by accident, Sir William Gilbert found that a magnet, cut in two, immediately became two magnets, each with a new set of opposite polarities. But his greatest achievement was discovering that the Earth itself was a giant magnet. For Gilbert, this idea neatly explained how compasses work – one end of a magnetic compass is attracted to one of the magnetic poles of the Earth. Gilbert realized that the Earth, just like a lodestone, should have two magnetic poles, one residing in each hemisphere. By 1600, Gilbert had performed enough experiments to confirm his theory and to publish his primary work, *De Magnete*. One of Gilbert's keenest followers was Galileo, who wrote: "I extremely admire and envy the author of *De Magnete*. I think him worthy of the greatest praise for the many new and true observations which he has made."[30]

The eldest of five sons, Gilbert rose from an obscure family in Essex, East England, to earn his university degree at age 20, and become a maths examiner at St John's College, Cambridge. Soon he switched to medicine, becoming a physician just before his 25th birthday. He left England for a few years and practised medicine on the continent. His own notes indicate he lived in Venice, but his travels and work in Italy and elsewhere are sketchy. One biographer, writing a hundred years ago, speculated, "Gilbert had, during his subsequent sojourn in Italy, conversed with all the learned men of his time. He had experimented on the magnet with Brother Paolo Sarpi; he had, there is reason to think, met Giordano Bruno.[i] Being a man of means and a bachelor, he spent money freely upon books, maps, instruments, minerals, and magnets. For twenty years he experimented ceaselessly, and read, and wrote and speculated, and tested his speculations by new experiments."[31]

Since Gilbert lived in Venice at the same time as Giordano Bruno, and since Gilbert seems to have been curious and gregarious, it is likely that he did meet the wayward monk. Bruno would have been an amazing person for the young doctor from Essex to befriend. Not just because Bruno supported the Copernican theory, nor because he suggested the Earth started as a molten rock on which mountains formed as the planet shrank and cooled, but because Bruno represented the best of

---

i   Filippo Giordano Bruno (1548-1600), Italian theological philosopher, author, and scientist.

A Heart of Iron

sixteenth-century reasoning confronting the worst of the doctrines of his times. If Gilbert did in fact know Bruno, his acquaintance would have presented exciting new ways to ponder the universe. Gilbert would have been encouraged to question everything. For us, knowing Bruno brings an understanding of the inertia against which Renaissance scientists and philosophers fought, and the tragic consequences of a society dogmatically protecting tradition over free thought.

Heresy without Redemption

Bruno was an Italian Dominican friar, a philosopher, mathematician, and astronomer. He envisioned a universe far larger than the Copernicus model. Bruno suggested the sun was a mere star – and the universe contained an infinite number of stars and an infinite number of inhabited worlds populated by other intelligent beings. For this, and other unorthodox views, the Roman Inquisition found him guilty of heresy. They ultimately murdered him in a disturbingly cruel way.

Born Filippo Bruno, he adopted the name Giordano when he entered the Dominican Order at a monastery in Naples, thirty kilometres from his family's village near Italy's western coast. Bruno was 17 when he began his studies in theology and metaphysics, and he became an ordained priest at 24. He was remarkably intelligent and gained considerable fame for his tricks of memory. His mnemonic gymnastics were rewarded with an audience before the pope where he performed well and ingratiated himself with the pontiff.

His early friendship with Pius V wasn't enough to prevent Bruno's eventual execution by the church, accused of heresy for promoting the notion that Christ was different and separate from God, thus questioning the doctrinal definition of the Trinity. His main crime, however, was his incessant free-thinking independence. Even during his years at the monastery, Bruno committed a host of insubordinate transgressions – from reading the banned works of Erasmus (he kept a copy hidden beside the toilet) to removing images of saints from his monastic cell. Bruno wrote an allegory he called *Ark of Noah* in which donkeys, representing monks, brayed their displeasure at their seating assignment on the ark. It was meant to symbolize the pettiness and lack of serious aspirations among his brother friars. (Much later, at his inquisition, Bruno said the then current Pope Pius V had rather liked the tale.) After eleven years as a monk in Naples, Bruno fled when warned charges of heresy were being drafted against him.

Bruno dressed as a peasant rather than a monk when he left Naples, travelled to Rome, then Genoa and Turin. He stayed in Venice long enough to publish a book on the innocuous subject of memory, written simply to earn a bit of money. But by age 30, he felt Italy was unsafe, so he moved north. Over the next 14 years, Bruno lived in Geneva, Paris, Prague, London, and Frankfurt.

He tried his hand as a playwright in Paris, creating *The Candlemaker*, a comic satire, which exposed prevailing superstitions. But mostly he wrote philosophical

The Mountain Mystery

Brother Bruno

tracts and made studies on the nature of memory. King Henry III was a patron, especially appreciating Bruno's lectures on tricks of memory. This led to Bruno's 1582 book, *The Art of Memory*, a somewhat scholarly treatise, published in Paris with the king's encouragement. Through court connections, he lived in London for two years, staying at the home of the French ambassador. Renowned as a philosopher of nature, he was invited to lecture at Oxford, but his defense of Copernicus and his opposition to Aristotelian philosophy kept him from receiving the teaching post he anticipated.[i] But perhaps it was more than his vision of science – Bruno's religious ruminations were unpalatable everywhere in Europe, even in relatively liberal England.

For Bruno, our planet was one of many worlds and doomed to eventually disappear. In *De Immenso,* Bruno wrote: "The earth, which is of the same species as the moon, is of creatable and destructible substance, therefore the worlds are able to be created and destroyed, and it is not possible that they have been eternal, since they are alterable and consisting of changing parts."[32] Bruno thought that our planet had formed out of hot mass, then cooled and shrunk according to laws that act the same everywhere in the universe. According to Bruno, the mountains formed as the Earth's surface cooled. This concept of a cooling, contracting Earth would return again and again among geologists. It was an important insight into the understanding of the evolution of the world.

While in London, in 1584, Bruno published *On the Infinite Universe*. This was likely the first book to describe an endless universe, with innumerable worlds populated by intelligent beings. The stars are like our sun, around each revolved a planet like our own. Bruno, incorporating his philosophical description of existence into his concept of the universe, claimed all matter has intelligence – every part of the universe, every rock, drop of water, plant, and animal, has a soul or is part of an all-encompassing soul. By publishing this philosophy, he

---

[i] In 1583, when Bruno lectured at Oxford, the school had on its statutes that "Bachelors and Masters who did not follow Aristotle faithfully were liable to a fine of five shillings for every point of divergence." Bruno was not in any way a follower of Aristotle.

A Heart of Iron

committed the heresy of pantheism.

Without the Oxford appointment in England, he returned to Paris, but it had swung against him and his liberal interpretation of religion, so Bruno travelled to Germany where he taught for two years. Then Prague. He was running out of places and people to trust. But during all his wandering, Bruno continued to write – *Theses on Magic; Composition of Images and Ideas; A General Account of Bonding*. These were largely concerned with thought processes and psychology – his book on bonding, for example, related to the interconnectedness (the *bonds*) of society.

Bruno was hired to tutor an aristocrat's child, but this took him back to Italy where his young student denounced him to the Venetian Inquisition. In Venice, he was tried, found guilty, then sent to Rome for a second trial. The Church imprisoned him for blasphemy, for practising magic, and for heresy. The new Pope recommended execution. "I neither ought to recant, nor will I," Bruno said.[33] So they tortured the former priest, philosopher, mathematician, and astronomer. Iron spikes were driven through his jaw, tongue, and palate. Bruno was pulled through the streets by a hooded, chanting group known as the Company of Mercy and Pity. He was stripped of his clothes, tied to a stake, and burned to death.

The execution of Bruno, in the year 1600, was but one example of hundreds of attempts by the church to coerce scientific and philosophical harmony with theological dogma. Historian Andrea Del Col estimates between 50,000 and 75,000 cases were judged by the Inquisition in Italy alone, resulting in 1250 death sentences.[34] This was in addition to hundreds of thousands of accusations of witchcraft where children and women were the principal victims.[35] Undoubtedly, many allegations arose from civil disputes and grudges, but the stifling atmosphere of rigid church-sanctioned scientific doctrine prevented much genuine unbiased inquiry into the nature of the universe. The martyrdom of Giordano Bruno, immolated at age 51, set an example which stifled free-thinking scientists and philosophers for generations. Galileo, forced to utter words of obedience to the church and recant his belief in the motion of the Earth and planets around the Sun, was fully aware of Bruno's fate. In fact, Robert Bellarmine – the chief inquisitor who had directed the torture of Bruno 16 years earlier – was now Cardinal Bellarmine and had become the pope's intellectual adviser. The cardinal spoke with Galileo, offering leniency in exchange for obedience.

Giordano Bruno was more poet, lecturer, and philosopher than scientist. He didn't work in a laboratory, testing and experimenting. However, when his philosophical wanderings brushed upon science, his insights were often uncannily accurate. We can't be sure William Gilbert met him in Venice, or later when Bruno lived for two years in London. But he certainly would have read his tracts and books, and been influenced by his thoughts. Gilbert lived in England, a safer place to express opinions and investigate science. Safer, but as Gilbert would discover, politics is at work everywhere. Gilbert's scientific pondering created trouble for him, too.

It is unclear why Gilbert developed such an attraction to magnetism. Perhaps it had been kindled in Venice when he experimented with magnets while visiting his friend and fellow scientist, Brother Paolo Sarpi[i], a free-thinker of similar ilk and age as Bruno. Sarpi's fate was far less disastrous. Like Bruno, he too, suffered excommunication. But unlike Bruno, Sarpi twice survived Vatican assassination attempts and he outlived several popes. His correspondents included the scientists Lord Cavendish, William Harvey, Francis Bacon, and, of course, William Gilbert. The community of Europe's diverse and dispersed scientists was surprisingly bonded, as Bruno would have said.

The Magnet Earth

In 1573, just shy of 30 years of age, Gilbert returned to London from Italy and began to earn a good living as a doctor. He was either very capable, or very well connected, or both, because he soon became the queen's own physician. Medicine was his occupation, magnetism was his obsession. To understand the compass, Gilbert drew heavily on Robert Norman's well-documented experiments and on his own Italian contacts, but he moved the science much further along.

As we have seen, long before Gilbert's time, it had been noted that a suspended compass needle points north, but with some error. Columbus, an excellent navigator, found he needed to recalibrate his magnetic compass several times in order to match geographic north (determined by charting the stars) as he sailed across the Atlantic.[36] Columbus discovered a spot far out at sea, nearly in the mid-Atlantic, with no variation at all between his compass and true north – geographic north exactly matched magnetic north.[37] But as he sailed farther west towards the Americas, his compass was off again, erring the opposite direction from when he left Spain. Columbus, Cabot, and other early navigators carefully recorded all those variations between geographic north and magnetic north. By 1530, maps showing the requisite compass corrections were being printed and sold to sailors.

Norman had catalogued the navigational declination, or straying, and had discovered the inclination, or dip, of magnetism into the Earth. But it was Gilbert who was able to correctly identify the source of the magnet's pull. In 1600, Gilbert published his work and conclusions in *De Magnete (On the Magnet and Magnetic Bodies and on That Great Magnet Earth)*. We will see shortly how he discovered that our planet has an iron core and why Gilbert called the Earth "That Great Magnet." Recognizing the Earth as a magnet with an iron core explained Norman's compass dip, but not the directional variations Columbus encountered.

Gilbert's *De Magnete* presaged modern scientists' methods of presenting evidence. Gilbert began by listing what was already known about magnetism –

---

i   Fra (Brother) Paolo Sarpi (1552-1623), Italian Venetian patriot, scientist, and philosopher.

A Heart of Iron

then he debunked most of it. As a physician, he dismissed many of the various medical attributes claimed for magnetism and lodestones. He recited, then rejected, claims that magnetism causes mental imbalance and depression and that powdered lodestone can reduce fever, cure tumours, and restore youth and vitality. Ground into powder and eaten, some lodestones were reported to cure diarrhea, others to cure constipation, but Gilbert suggested these opposing efficacies might be due to dirt or soil contamination from the local lodestone mine, not the magnetite itself.

To support his objections, the physician devoted a lengthy chapter of his book to "The Medicinal Power of the Iron." Gilbert explained that lodestone is a type of iron, but since it attracts iron that the body needs, powdered lodestone will induce the opposite of desired benefits. For those doctors who were using iron to treat their patients, Gilbert shared a lengthy recipe, hoping to stop physicians from killing the sick with poorly prepared doses. He gave away his secrets so that inexperienced doctors using iron "as a medicinal agent may learn to prescribe it more judiciously for the curing of patients, not, as is too often the case, to their destruction."[38] Gilbert gave just one example of when to use powdered iron: "young women of pale, muddy, blotchy complexion are by it restored to soundness and comeliness."[39] He was describing a treatment for anaemia.

Gilbert's book, of course, was primarily about magnetism. But in it, he also discussed static electricity. He originally wrote *De Magnete* in Latin, using the word *electricus* for the electrical effects he saw, thus introducing *electricity* to modern science. He studied electricity by stroking amber and attracting bits of thread and paper. Using the electroscope he invented, Gilbert was first to show electrical attraction was not limited to amber, but included a dozen other materials. He also meticulously catalogued substances that didn't carry any electrical charge – insulators such as bone, pearl, and wood. He was rather surprised when he discovered lodestone does not hold a static electric charge when rubbed the same way as amber. Gilbert was unaware that lodestone, an iron, could conduct an electric current. "The substance which above all others possesses the magnetic property of attracting iron shows no trace of electric action when rubbed in the hand."[40] Another major difference between electricity and magnetism that Gilbert described in *De Magnete* was that electric static could be walled off by placing metal, or even paper, between the statically charged item and the things it normally attracted. But he discovered magnetism ignores all barriers. Gilbert's experiments led to a belief that the forces of magnetism and electricity were unrelated. It took two centuries before scientists reunited them and discovered each can produce the other.

Gilbert found that ambient magnetism – the constant background magnetism that everywhere attracts the compass needle – was as strong indoors on the laboratory bench as outdoors in the garden. It seemed to be global, but it also had local perturbations. Gilbert separated the global magnetic field from local anomalies that confused the mariner's compass, showing these were due to interference from relatively close sources, such as magnetic minerals in nearby hills. In Gilbert's day, many assumed the compass needle's source of attraction was a large anomalous magnetic object – either the legendary magnetic island Insula Magneta or perhaps the North Star, Polaris. Once anomalies were removed, Gilbert was able to explain global magnetism in a way that revolutionized our understanding of the Earth. Through a series of careful experiments, he finally concluded magnetic attraction was emanating from within the planet itself – he discovered that the Earth is a giant magnet.

In *de Magnete*, Gilbert explored the composition of the interior of the Earth. He dismissed Aristotle's notion that all matter is composed of varying amounts of just four elements – earth, fire, air, and water. Brushing aside the idea, he said seventeenth-century followers of Aristotle were "misled by their vain dreams about elements." Gilbert then refuted a common belief that the Earth's interior is solid rock. Instead, Gilbert said, in a tone reminiscent of Bruno, "as all the other globes of the universe," the inside of the Earth is a solid homogeneous body of "pure native iron."[41] Gilbert thought iron was in the same family as the magnet, so he considered it a lively energetic substance.

Sir William Gilbert experimented for years in his laboratory, surrounded by magnets, scales, lathes, rocks, and various gadgets of his own design. Unmarried, with no financial obligations, and paid a massive salary as Queen Elizabeth's physician, he is said to have spent over five thousand pounds on lodestones and gadgets[42] – a fortune, even today. His principal tools were little models of the Earth which he fashioned from round magnetic stones. He had acquired several that were nearly spherical, like the Earth. He chiseled, lathed, and polished the spheres, calling them *terrellas*[i] – earthkins, or little Earths.

In his experiments, Gilbert suspended a tiny compass from a thread and approached an earthkin, then waited until the magnetic force of his round magnetic rock overpowered the Earth's background magnetism. The needle danced a bit, then settled. He recorded the inclination (dip) and the direction of the needle's north end. Then he sketched a line on the stone with chalk, heading north from the needle. (At least, to an army ant living on the ball and properly outfitted with a hand-held compass, it would seem to be north.) Gilbert took several readings on his experimental lodestone. He finished his work by extending the chalk lines on the rock's smooth surface until the lines completely encircled the globe, meeting at their starting points. These were great circles, like equators, drawn on the ball. Gilbert found that those circles, all starting at different points and heading off towards what a terrellian would call north, coincided twice – at the lodestone's north and south poles. And where they met,

---

[i] Gilbert is sometimes credited with inventing the terrella, but Peregrinus described making one 350 years earlier, in his *Letter on Magnetism*, and had already performed some of Gilbert's famous experiments. He had not, however, made Gilbert's greatest inductive conclusion.

A Heart of Iron

the compass needle always pointed straight into the ball. Halfway between the magnetic north and south pole, the needle's ends didn't dip at all, but suspended horizontally, parallel to the terrella's surface. It all perfectly matched observations reported by mariners trekking around the world.

The needle and terrella behaved exactly as their full-scale counterparts. The needle pointed towards the miniature magnetic pole, and inclined into it, as Robert Norman had once described compass needles bowing in submission to the Earth. Gilbert made a bold leap of inductive reasoning, concluding that since his terrellas all acted the same, and since the planet was showing the same activity on a grand scale, then the Earth is a magnet and has a core of magnetic iron under its crust of dust and stone. The Earth is a giant terrella, a lodestone, a magnet.

Gilbert was steadfast in his experimental approach to science. His studies of magnetism and electricity were highly innovative. He created his own tools and repeated his tests again and again, changing parts of the experiments to isolate variables. This was a new way of conducting science, an approach based on experimentation and observation. Gilbert advocated inductive reasoning. Since each of his terrellas mimicked the Earth in his tests, and since the terrellas were iron lodestones, he concluded that the Earth's interior was an iron lodestone.

Induction and deduction are two of the familiar paths followed to form broad conclusions through logical reasoning. When we solve a problem with deductive reasoning, we eliminate various explanations until only one likely solution remains. "Deductive" sounds like "reductive" and that's a great way to think about it. Here's an example: Imagine a bright light streaks across the night sky. Observers speculate on secret military operations, meteors, reflections of moonlight off clouds, and, of course, flying saucers. Using the deductive process you eliminate the less plausible, until a single explanation remains. Your friend at the military installation in the desert tells you it wasn't them. The moon was away for the night, so that wasn't the cause. However, it was August and the annual Leonids meteor shower was in progress. Through deduction, you have reduced the possibilities to a meteor streaking through the sky or, perhaps, an extraterrestrial visitor. Based on the speed and position of the sky-streak, and other considerations, you write your paper. If you ever want to become a famous scientist, you present the idea of the meteor, but you keep the faint possibility of aliens in your mind. You have a "Theory of the August Flash" to offer at the upcoming conference. You know it is not necessarily the correct option. With new evidence, you may one day have a different explanation. Until then, other streaks of light will be compared to your conclusion, which has become the working model to explain such phenomena. That's deductive reasoning: resolving a question by rejecting unlikely possibilities.

The other method of logic that could result in a scientific model is inductive reasoning. Induction involves increasing the applicability of a group of observations, arriving at a universal general rule. Gilbert did this by expanding his terrella observations to include the core of the Earth. Here is another example: You are a caveman scientist. It is mid-summer, the women have come home with berries and you have returned from climbing a tree and fetching a few honey combs. Sitting outside by the fire, everyone gets excited when meteors light the sky. Here they are again, just like last year. This year, though, a streak of light, a loud boom, and hole in the ground startles Ogg, who is sitting alone by the edge of the clearing. Everyone runs over to see what made the hole. It's a rock – quite hot and heavy. Immediately a second and a third meteor strike nearby, each making a bright flash and leaving a hole with a hot rock. Inductive logic instructs you to conclude that every nighttime streak of light is a hot rock. You have taken an example (or two or three, since you are incredibly lucky) and extrapolated to create a general law, the "Theory of Bright Streaks during Berry Season" which explains all such phenomena as caused by hot flying rocks.

This sort of logical reasoning was not so apparent for a thousand years of dark European history. Nor was the idea that one could propose questions for research, then investigate through experiments. William Gilbert expertly applied science to determine the source of ambient magnetism. In his day, most scientists used a non-experimental approach to study nature. This was Aristotle's approach – philosophy trumped examination. For example, Aristotle taught that heavy objects fall faster than light ones, as logic dictates. Aristotle never tested the idea. Over a thousand years passed before Gilbert's contemporary, Galileo, dropped weights from a leaning tower, testing the notion, and finding Aristotle was wrong.

Even eminent scientists like Francis Bacon, sometimes described as the first modern scientist, were locked in the old tradition. For example, Bacon wouldn't switch from an Earth-centred universe – he vigorously opposed the Copernican model. He mostly used philosophical and theological arguments for his position, rather than the physics being developed by others of his time. Modern scientists would find the domination of philosophy and religion over scientific investigation smothering. But in 1600, even bright scientists frequently began their tracts extolling the "perfection of the sphere," the "logical rightness of man's dominance," or the "fine true aspects of an arrow's forward energy" – these and many other facets of nature, in the spirit of Aristotle, were considered truths that didn't need proofs and, indeed, could not be questioned. Add to this a slate of theological dogma, science treatises banned by churches, threats of investigation by the Inquisition, and a power struggle between declining Catholicism and ascending Protestantism, and the situation for science becomes tenuous.

A Slice of Bacon

For his day, Gilbert was the epitome of experimental science and applied logic. However, *De Magnete* was not a bestseller;[43] nor was it immediately heralded for its scientific achievements. The book, printed in only a few hundred copies and

A Heart of Iron

dismissed by contemporary scientists, was only later regarded as revolutionary. Meanwhile, Sir Francis Bacon,[i] arguably the most important English scientist to glance at a copy of Gilbert's newly-published manuscript, dismissed it, describing Gilbert's *De Magnete* as a "painful and experimental work."[44]

Some of Bacon's criticisms were warranted. When one reads *De Magnete*, it does seem, as Bacon said, "Gilbert has made a philosophy out of observations of the lodestone,"[45] and "he has ascribed too many things to that force and built a ship out of a shell."[46] Certainly, Gilbert was bewitched by magnetism – in fact, he wrote "the magnetic force is animate or imitates a soul; in many respects it surpasses the human soul."[47] But the animosity of Sir Francis Bacon was profound. In the 1898 book, *A History of Electricity*, historian and engineer Park Benjamin wrote that he had never encountered anything written by Bacon which did not show "a failure to understand Gilbert's magnetic and electric discoveries, or direct charges of jealousy, malice, and injustice."[48] Bacon fully rejected Gilbert's book, dismissing nearly all of Gilbert's conclusions about the planet's internal structure.

Although the ideas of Copernicus, Kepler, Brahe, Bruno, and Galileo were readily available to him, Sir Francis Bacon never completely gave up Ptolemy's model of an Earth-centred universe with its 55 layers of mathematically defined spheres of orbit, each loaded with planets, comets, and stars circling at different speeds around the Earth. It was complicated and cumbersome, but didn't disagree with the intuitive purity of a man-centred universe. Bacon conceded that Mercury and Venus might orbit the Sun, but for him, the Sun revolved around planet Earth. For political, philosophical, and theological reasons, humanity, the focal point of Creation, remained at the heart of his universe.

Others have pointed out a fundamental difference between the philosophical science that preceded the Renaissance and the experimental science that followed by this simple allegory: An Aristotelian might argue that water is necessary for life by pointing out the lack of water on lifeless desert landscapes, and by invoking the innate and obvious purity of water. By the late Renaissance, scientists who began to reject this philosophical approach might perform experiments on different living creatures to determine whether water deprivation results in death. Measurements and experiments were the new scientific tools. With data, conclusions could be drawn.

Refinements in the scientific method have led to our modern system of posing research questions as hypotheses and building experiments to test theories. Bacon contributed to this new pattern, but he was not the first modern scientist and he fought against empiricism. Here is what he said about experimentation in *Novum Organum*, his most famous book about the philosophy of science: "The empirical school produces a dogma of a more deformed and monstrous nature than the theoretical school: It is not founded in the light of common notions, but in the

---

i Sir Francis Bacon (1561-1626), English philosopher, scientist, politician.

confined obscurity of a few experiments. Hence, this species of philosophy appears provable and almost certain to those who are daily practised in such experiments, and has thus corrupted their imagination. We have strong evidence of this in the alchemists and their dogmas, but it would be difficult to find another in this age, unless perhaps in the philosophy of Gilbert."[49] So again, Francis Bacon discredited William Gilbert, and with him, all of experimental science. Bacon's tone and unwavering animosity suggest a deep, perhaps personal, dislike or unease with the queen's physician. Certainly Bacon's opposition to both Copernicus and Gilbert belie a stubborn conservatism of thought and philosophy.

Although generally cautious, Sir Francis Bacon was nevertheless astute enough to have interesting observations about the natural world. His philosophy encouraged him to look for patterns everywhere, searching for great principles in the "real and substantial resemblances" he found in nature. Interestingly, this included the shapes of the African and American coastlines. In his *Novum Organum* Bacon wrote, "both possess a similar isthmus and capes, a circumstance not to be attributed to mere accident. The new and the old world are both broad and expanded towards the north, and narrow and pointed towards the south."[50] Bacon's description of the globe's geography imply that the southern hemisphere's continents could be aligned if the bulge in South America were tucked into the armpit of Africa – suggesting they were once joined. We can't be sure Bacon meant the continents had drifted apart, but some later geologists saw Bacon's cryptic lines as an endorsement for the theory of mobile continents. Bacon further observed that the fit of the southern continents was like the petals of an opening flower – by this he may have been suggesting the Earth is expanding in size, as other geologists eventually interpreted his description.[51] Perhaps, for Bacon, the Earth was expanding – the growing oceans were the gaps between a flower's petals. For Bruno, it was cooling and shrinking. These two themes would long outlast the seventeenth century, each having their chance to enchant future scientists.

Bacon led a clearly dichotomous life – as England's Attorney General and later Lord Chancellor, he was a good politician and rose far, but his career ended in disgrace and included time in the Tower of London. He admitted taking bribes, resigned his lofty position, and briefly went to prison. As a moralist, he envisaged a science fiction utopia, an island called Ben Salem, with pure and chaste inhabitants. But at age 45 he married the 14-year-old daughter of a politically well-placed colleague. As a scientist, as we have seen, he was a vociferous critic of Gilbert's work and dismissed the latter's book as careless experimentation followed by wild speculation. Despite this, we can't ignore the dedication of a scientist who died from pneumonia contracted while observing how meat is preserved through freezing, even if, for him, the sun revolved around the Earth.

In contrast, in the final section of *De Magnete*, Gilbert unequivocally supports Copernicus's heliocentric model. This was before either Kepler or Galileo confirmed it, making him among the first to agree the Earth is not the centre of the universe. He believed a magnetic force propelled the planets around the sun. He was wrong in this, but Gilbert's book concludes with a final rebuke of the establishment and full support of the new model of the solar system.

Largely because of Bacon, it took a generation before scientists accepted that the global magnetic field was generated in an iron-cored Earth. It didn't help that Gilbert pointed out his critics were "old-womanishly dreaming things that were not,"[52] and were "deluded by vain opinion."[53] None of this helped make *De Magnete* popular among the contemporary naturalists to whom the insults were hurled. It also didn't help that William Gilbert died soon after his book was published. Nor did it help his legacy that all his books, charts, globes, magnets, lodestones, lumps of amber, letters, and hand-built instruments went to the Royal College of Physicians where everything was soon destroyed by the Great London Fire.[54]

The Wandering Pole

Gilbert showed that the Earth's magnetic field comes from within the planet and its direction and dip change with location. Earlier, Robert Norman and other compass makers noticed that although the compass points in the general direction of the North Pole, it wasn't exactly correct. It seemed to them that the two – magnetic north and geographic north – were similar, probably related, but clearly not identical. Gilbert had accepted this distortion, blaming it on localized ore deposits, perhaps hidden in distant mountains. But Gilbert, with his solid iron terrellas and his solid iron Earth, said there were no conditions that would cause the magnetic north pole to stray. In this he was quite mistaken.

Surprisingly, the north magnetic pole neither aligns with true north nor does it stand at rest in its off-centre locale. At an observatory in Paris, the various locations of the wandering magnetic pole have been recorded faithfully, monthly, since 1540. This is the longest continuous record of geophysical data on Earth. It shows that the magnetic north pole wanders around considerably. Enough to make magnetic variation maps purchased by sailors useless in a decade or two.

When the Parisian scientists first began recording the difference between true north and magnetic north, they found their compasses pointed 6° east of geographic north. In other words, the magnetic pole was in western Siberia rather than the north pole. By 1600, it had drifted still farther, up to 8° east, but then the declination reversed and the pole journeyed back, approaching true north. In 1660, compasses in Paris pointed precisely at geographic north. For a few months. Then, frustratingly, compasses became progressively off the mark, ending up a thousand kilometres west of north in 1815, before slowly correcting again. Today, a compass at the Paris Observatory points almost exactly towards rotational north once more. It took 350 years, and champagne might be spilled marking the

The Mountain Mystery

prodigal force's return, but it will soon wander off to the east again. Meanwhile, the geographic north pole, that is the hub of the central axis around which the Earth spins, is not moving in this rapid irregular pattern. If it were, the planet would not rotate smoothly, but would vibrate like an unbalanced washing machine. Thus it was clear, even hundreds of years ago, that the magnetic pole wanders while the rotational pole is almost stationary.

Shortly after William Gilbert, Edmond Halley[i] had an idea that he thought explained the Parisian magnetic drift. Halley suggested extreme internal pressure gives the Earth a solid iron heart. The crust we stand upon is also obviously solid. But Halley brilliantly surmised that some type of fluid separates the two solids.[55] He believed the fluid region of the inner Earth produces the magnetic field, but due to sluggish rotation of that liquid, the magnetic pole is drifting.

Halley was partially correct; he was certainly right when he identified distinct inner-earth layers. He had a knack for clear reasoning in other areas of science, too. He said solar heat was the cause of wind. He discovered the relationship between barometric pressure and elevation. Halley edited his friend Newton's great study of physics, *Principia,* then printed it when the Royal Society failed to find the cash. And, of course, Halley predicted the return of the comet which others named in his honour.

Halley was England's Royal Astronomer but he didn't receive the job he really wanted, Oxford Professor of Astronomy. The Archbishop of Canterbury vetoed the appointment due to Halley's well-known atheism. It would have been easier for him to work within the church as many of the great scientists of his time did. However, even that could guarantee neither acceptance nor contentment, as we will soon see in the brief and painful life of the geologist Saint Nicolas Steno.

---

i   Edmond Halley (1656-1742), English astronomer, geophysicist, mathematician.

3

The Sacred Earth

Having noted the precarious situation for scientists in an era when theocracy ruled, it is worth remembering that a number of the best scientists were members of the clergy. Nicolas Steno[i], for example. Steno, born into a Lutheran Danish family, converted to Catholicism just before his 30th birthday. He eventually became a bishop in the Catholic Church. Pope John Paul II beatified him as Saint Steno in 1988, partly for his example of selflessly serving the poor. But he is better known to science as a founder of modern geology. Much of what he discovered over three hundred years ago set the foundation for future geologists.

Steno moved from Copenhagen to Florence to work as a physician, museum curator, and artifact collector for the Grand Duke of Tuscany. It was 1665, and the young scholar from Denmark had already studied medicine in Holland, dissected animals in Italy, and become an expert human anatomist. He specialized in the lymph gland system, but also found that tear drops form in the eyelids and are not the brain drippings other physicians detected. (Physicians of the time felt this was not an unreasonable idea – they claimed that women and the weak were more prone to tears, resulting in loss of brain material and subsequent poor sense of judgement and diminished mental abilities.) Steno also created a geometric model of human muscles – he demonstrated that contracting muscles change shape, not volume. By studying cows' hearts, he realized the heart is a muscle, not a "centre of warmth." Phenomenal as these discoveries were for a young man in his twenties, Steno's work in geology was truly revolutionary.

During 1667, working in Tuscany for the Grand Duke, Steno dissected a shark's head. Examining its teeth, he was reminded of stones he had found inside local rocks. Those stones were called *glossopetrae*, or tongue stones. Their origin was in dispute. Some claimed they were gems, fallen from the moon. Others believed Aristotle's explanation – these special stones were an integral part of the soil, had simply always been there. But Steno recognized the fossilized shark teeth for what they were. And, against conventional wisdom, he speculated that petrified bones entombed between layers of rocks were also once living creatures.

---

i   Nicolas Steno, also known as Nils Steensen and Stenonis (1638-1686), Danish Catholic bishop, geologist, medical scientist, and philanthropist.

The Mountain Mystery

Steno examined crystals, dispelled the notion they were magical, and pointed out they are simply rock. He noticed the angles between the various faces on a crystal are always the same for each crystal type, regardless of where it was found or the size it achieved. Geologists still call this basic principle Steno's Law.

Three indispensable rules of modern geology were also discovered by Steno. These all deal with how rocks are arranged in layers around the world. Although Avicenna, the great Persian physician, had noted the same universal laws of geology six hundred years earlier, it was Steno who independently considered their implications and made them widely known to Europeans. Steno explained geology in these terms:

> All rocks form as flat layers;
> Those flat layers could extend forever; and,
> Older layers get buried by younger layers.

Today, these rules are obvious and so elementary that they are introduced to children in grammar schools. But they were not such self-evident laws before Steno described them. With these rules, coal seams are traced across river beds, iron ore is followed across crests and ridges, and oil fields are expanded. Significantly, Steno's observations implied that various floods had created some of the rock layers and the expanse of geological time involved was enormous.

In 1669, Steno wrote that layers, or strata, of the Earth's crust had been deposited by former seas, then altered by erosion, earthquakes, and volcanic eruptions. This challenged the orthodox position that the Earth had remained unchanged since Creation, except for one single major event, Noah's Flood. Steno realized rock layers formed intermittently from sediments in extensive seas. The layers were as broad and flat as the oceans. Geologists call this the Principle of Original Horizontality – rocks form as horizontal planar layers.

It was significant that Steno voiced this observation because rock layers almost everywhere are presently twisted, and not flat at all. To Steno this meant that the Earth was not a fixed dead planet, but has had some powerful forces at work bending, folding, wrenching, contorting, and twisting those originally horizontal rock layers. Steno wrote, "All present mountains did not exist at the beginning of time – mountains can be overthrown, peaks raised and lowered; the Earth can be opened and closed again." Some will not accept that these things occur as natural phenomena, said Steno, and they "escape the name of science, considering these phenomena in terms of myths."[56] The laws Steno devised meant that the fossils which he described as evidence of ancient life also corresponded to gradual changes with time. To him, it explained why simpler, older fossils consistently appeared in lower strata, buried by rock layers containing newer and more complicated fossils. The future saint's discovery set in motion ideas that contributed greatly to Darwin's description of evolution.

The Sacred Earth

Such flashes of insight did not come easily to Steno. He wrote, "I studied day by day the details [of sedimentation and shells] and found myself again and again brought back to the starting place when I thought I was nearest my goal. I might compare those doubts to the head of the Lernean-Hydra for when one was rid of, numerous others were born, and I saw I was wandering about in a sort of a labyrinth where the nearer one approaches the exit, the wider the circuits one treads."[57] Science is still described this way by the best scientists.

Nicolas Steno, the anatomist, physician, and revolutionary geologist, had broad interests. His curious, questioning mind led him to leave the Lutheran faith of his family. Influenced by colleagues in Florence and a conversation with Pope Alexander VII, Steno converted to Catholicism. He was soon an ordained priest.

Steno demonstrated such zeal as a proselytizing convert that in just two years Pope Innocent XI made Steno a bishop. His life and beliefs were evolving ever more quickly. Within a year, Steno became a leader of the Counter-Reformation and recommended burning books by philosopher Spinoza[58] and by progressive scientists of the day. He entered such a stage of self-sacrifice and commitment to the poor that he sold everything – including the small silver cross he had worn for years – to provide for others' welfare. He dressed in rags and ended up shuffled off to a small mission in northern Germany where his self-sacrificial example was less of an embarrassment to other members of the clergy. In Germany, without friends and family for support, he seems to have fallen into a mental state that led him to starve himself in what appears an act of personal penance. Bishop Steno died, emaciated, stomach swollen, and in severe pain, at age 48. But his scientific discoveries, accomplished in the ten years before becoming a priest, endure through today, as does his example of service to the poor.

A Closer Look

Another versatile genius, an Englishman named Robert Hooke[i], was a contemporary of Steno. Hooke's most famous work was with the microscope, but he was among the first to build telescopes and vacuum pumps; he deduced the wave theory of light, was a geometry professor and architect. As a physicist, Hooke figured out how a metal spring's energy changes with stretching – creating the relationship we now call Hooke's Law. As a surveyor, he assessed the smouldering ashes of the Great London Fire of 1666. The fire destroyed thousands of homes and public buildings, including the museum that held all of William Gilbert's original manuscripts. As an architect, Robert Hooke worked with designer Christopher Wren to rebuild London.

Galileo preceded Hooke, but his single-lensed tool, the *occhiolino*, as Galileo called the microscope, was little more than a magnifying glass. Dutch eyeglass-makers built the first compound microscopes and it was their design Hooke

---

i Robert Hooke (1635-1703), English scientist, naturalist, and inventor.

copied when he crafted his own. With it, Hooke began comparing fossils with living creatures using a greatly amplified view.

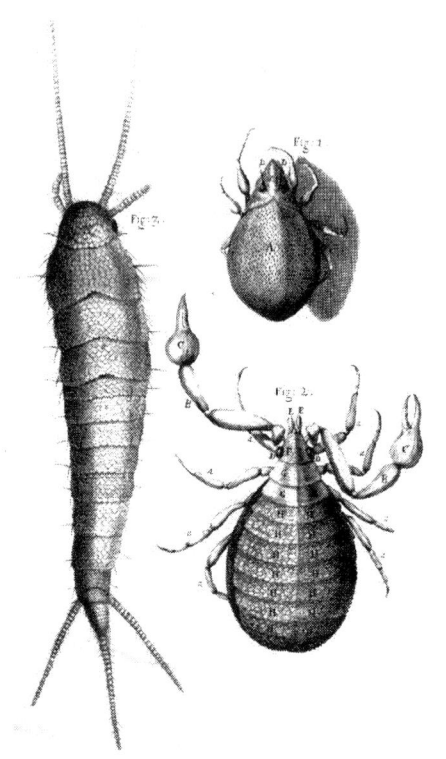

Hooke described tiny biological units he called *cells,* a label which has stuck. From observations of fossils and mites, Hooke was one of the first to suggest biological evolution. In *Micrographia,* he described evolutionary changes due to climate and soil as especially apparent in insects and mites, just as today's scientists point to even tinier bacteria and viruses as the best observable examples of evolution.

After describing the effects of environment on biology, Hooke was careful to add that his ruminations about evolution were mere theories. "Though I propound this as probable, I have not yet been certify'd by Observations to conclude anything, either positively or negatively, concerning it. Perhaps some more lucky diligence may please the curious Inquirer with the discovery of this to be a truth, which I now conjecture."[59] The lucky inquirers were Wallace and Darwin but they would not arrive on the scene for nearly 200 years.

Hooke's 1665 drawings of mites inspired his ideas of evolution.

Having studied the cells of living creatures, Hooke turned his microscope to dead things – fossils. About the time Steno was concluding fossils were not rocks but were ancient petrified lifeforms, Hooke realized the same. Because of similarities between living cells and fossils as viewed through his microscope, Hooke concluded that petrified wood found in farmers' fields, as well as old shells embedded in rocks, were the remains of living creatures, long since dead, decayed, and exposed to mineralized water. He even suggested that some of the more peculiar fossils belonged to ancient extinct creatures. This drew the ire of contemporaries who believed extinction was theologically untenable, but Hooke apparently suffered no repercussions for his musings.

Hooke believed that twisted, tilted, and vertical strata were the consequence of earthquakes. He stated this opinion at a meeting of the Royal Society in London, two years before Steno published the same thoughts, and Hooke inferred that

The Sacred Earth

shells he had observed in a cliff on the Isle of Wight had been lifted above the sea by ancient earthquakes.

Dramatic advances in understanding the form, nature, and evolution of the planet were taking place. Steno, Hooke, and Gilbert all suggested large-scale events had taken long periods of time. They each said processes at work in modern times were also active much earlier in Earth's history. However, tradition and theology continued to influence European views of the Earth's geology and its history. Most of their contemporaries believed geological changes were supernaturally directed.

The Sacred Theory of the Earth

In the late Renaissance, it was rather fashionable to combine the nascent study of geology with religious scholarship. Men with money and time for leisure speculated at great length about the many changes the Earth had endured. Usually they were careful to keep their deliberations within locally accepted theological bounds. Within a few decades bracketing 1700, several million words of descriptive cosmogony[i] were typeset from the scrawled notes of the natural philosophers François Placet, Isaac La Peyrère, John Beaumont, William Whiston, John Woodward, and others. Thomas Burnet,[ii] for example, published over 3,000 pages, trying to reconcile the Bible with the Earth's geological history.

Burnet was not a scientist, he was the English king's private chaplain. But like many professionals of his time, geology was his passion. He and others filled cabinets with misshapen mummified animals, fossils, and peculiarly formed rocks. From his armchair observations – Burnet did not hike the countryside looking at geologic processes – he wrote a lengthy natural history of planet Earth as a four-volume series: *The Sacred Theory of the Earth.*[iii]

The Sacred Theory spanned four volumes: *Concerning the Deluge* explained the impact of the Great Flood; *Concerning Paradise* explored the Earth's original perfection; *Burning of the World* described the divinely planned incineration of Earth; and *New Heavens and New Earth* predicted a future paradise. Burnet did his best to bend but not break the words of the Bible and merge geology as he understood it. His books exude confidence and employ scripture almost as frequently as observations of nature to explain the history of the universe.

Burnet's most important work, *Sacred Theory of the Earth,* was published in Latin in 1681 as *Telluris Theoria Sacra*, then translated into English three years

---

i Cosmogony refers to any scientific theory that attempts to explain how the cosmos came into existence. The current prevailing cosmogony is called the Big Bang Theory.

ii Thomas Burnet (1635-1715), English theologian, chaplain to the king.

iii In full, Burnet's book was entitled *The Sacred Theory of the Earth: Containing an Account of the Origin of the Earth, And of all the General Changes which hath already undergone, or is to undergo, till the Consummation of All Things.*

later. The English version was extraordinarily popular. Burnet describes the antediluvian Earth as a perfectly smooth sphere with a hollow inner ring where God hid the water needed to destroy his errant humans with the Biblical Flood. This implied that God knew from the start that his creation would go badly and had prepared for its destruction by keeping the necessary water at the ready, a thought that some theologians of the time faulted. Prior to the Flood, Burnet said the Earth was a paradise. Perfect spring weather dominated the globe. Rivers flowed from the poles to the Equator and things were lovely – until the Flood, when the water burst out, the hollow ring collapsed, and mountains and oceans were carved onto the Earth for the first time. Burnet claimed the break-up of "the fountains of the deep" was the cause of the Deluge in a literal world-wide event. He believed there was not enough water in the Earth's present oceans to have caused the Flood, so he disposed with their existence until after the event. Burnet even walks us through his calculations of the amount of water needed to engulf the planet.

Burnet did not know the immense height of the Himalayas, so he guessed that the world's tallest mountains reached only 2,400 metres, or one and a half miles in the units he used. He immediately approximated that as one mile. Burnet estimated that it would take an extra "eight oceans" to cover those mountains – a single "ocean" being his estimated volume of all the current water on the planet "covering half the Earth, as 'tis generally believ'd." Burnet felt that it couldn't possibly rain wickedly enough in 40 days to inundate the Earth with water a mile deep. The logical conclusion was that the water must have been stored inside the hollow Earth, waiting to be released when God needed it. According to Burnet's theory, the Flood changed the shape of the Earth, and all the mountains were created during forty days of destruction. If Burnet's description of a perfectly smooth antediluvian Earth were correct – a world without mountains – it would have taken very little water to flood everything. Confusingly, Thomas Burnet chose to flood the planet as it was geographically known to him in the late seventeenth century, complete with his mile-high mountains.

Burnet can be lauded for estimating the volume of water and for finding hiding places for it within the Earth. He was building a scientific explanation consistent with his belief in a world-wrecking deluge. He had a challenging task – completely covering the entire globe, deeper than the known mountains, using water from a source other than clouds. Rain, in any amount, was problematic in Burnet's scheme. It was believed by many in Burnet's day that rainbows never occurred until after the Flood, so rain also would have been absent. Burnet wrote that an abyss opened below and rain descended from above ". . . and in these two are contained the Causes of the great Deluge, as according to Moses, so also according to Reason and Necessity."[60]

Burnet, with Reason and Necessity, calculated that a constant storm of rain

The Sacred Earth

should bring a metre of water in a day, 40 metres in 40 rainy days and nights.[i] It would take 90 times as much water, said Burnet, to "cover the Mountains of Armenia, or to reach 15 Cubits above them."[61] As we've seen, Burnet found the extra water by allowing the Abyss to open. When it did, the inner water was violently released, and the planet was knocked off its perfectly created vertical axis, forcing it to tilt as it does today. The result of the tilting was an end to constant spring-like weather, and the introduction of cold wet winters and hot dry summers. With the biblical Flood, seasons had come to the Earth, said Burnet.

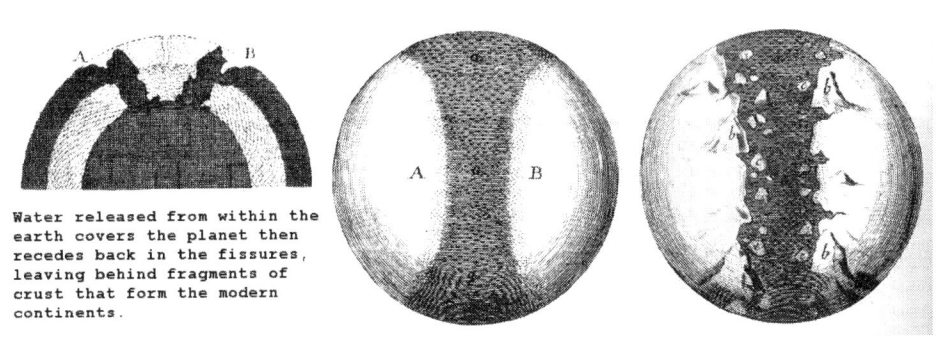

Water released from within the earth covers the planet then recedes back in the fissures, leaving behind fragments of crust that form the modern continents.

Thomas Burnet's 40 days of transformation during the Flood

Burnet wrote *The Sacred Theory of the Earth* partly as a rebuttal to Isaac La Peyrère[ii], and others, who claimed the Flood was not universal. La Peyrère was born into a Huguenot family of Marrano heritage (Christian converts of Jewish-Iberian descent), was raised as a Calvinist, but converted to Catholicism. Some have "interpreted him as a heretic, atheist, deist... and father of Zionism."[62] La Peyrère originally trained and worked as a lawyer, was a French Millenarian theologian (believing the Messiah would appear momentarily) and he invented the Pre-Adamite explanation of human origins, which avers God created Gentiles long before God created Adam, an idea that was not popular within the Catholic Church he had joined. His *Prae-Adamitae* book was publicly burned in Paris.

While visiting Catholic Spanish-controlled Netherlands, La Peyrère was briefly imprisoned. It didn't help that his writings included views that Moses did not write the Pentateuch, that no accurate copy of the Bible exists, and that the Flood was a local event. When La Peyrère was finally arrested and tried for heresy, he told the Pope that his strict Calvinist childhood had caused his heresies. Although he could not find any errors in anything he'd previously written, he said

---

i Burnet used the old Imperial measuring system and actually wrote "4 feet in one day" would be an extreme, but possible amount of rain. His calculation of potential extreme rainfall was quite reasonable. The greatest rainfall ever recorded in the USA in a day is 43 inches, very nearly Burnet's 4 feet. The world's record is nearly twice as much, occurring in 1952 on the Indian Ocean island of Réunion.

ii Isaac La Peyrère (1596-1676), French lawyer and theologian.

that he would recant all of his many heresies because the Church said he was wrong – and for no other reason. In response to his logic, boldness, and clearly feigned contrition, the pope offered the lawyer a job.

When Isaac La Peyrère died, aged over 80, one of his friends suspected La Peyrère had been an agnostic. He wrote as an epitaph:

> *Here lies La Peyrère, that good Israelite,*
> *Huguenot, Catholic, and finally Pre-Adamite.*
> *Four religions pleased him at the same time*
> *And his indifference was so uncommon*
> *That after eighty years, when he had to make a choice*
> *The Good Man departed and did not choose any of them.*[63]

Thomas Burnet fought against such opinions as La Peyrère's, defending his own rendering of the Bible and reconciling it with the new discoveries of the scientific awakening. Burnet was in many ways the great-grandfather of scientific creationism. For him, there was no conflict between scripture and science, it just took some work to show how they complemented each other. When it became too difficult, he found himself occasionally suggesting parts of the Bible are allegory. But generally he believed science confirmed the literal biblical narrative.

His piety did not keep him out of theological trouble. Burnet was particularly assailed by Bishop Herbert Croft, chaplain to King Charles I and therefore as well-positioned as Burnet, who became chaplain to King William III. Bishop Croft publicly chided Burnet's use of the Second Epistle of Peter, rather than Genesis.[64] But Peter's letter in the New Testament, said Burnet, supported the idea that flood waters had exploded out of the Earth's abyss. Burnet argued that the Scriptures' truth about the Earth's interior was being subverted by his opponents. To reconcile the Bible to science, Burnet began to interpret the Old Testament as parable, going so far as to claim the snake in the Garden of Eden was mute and did not actually have a conversation with Eve. Finally, the king's chaplain found himself dismissed from the king's court.[65] Although Burnet thought he was doing a great service – reconciling scripture with nature – it ruined him politically.

It is obvious that *The Sacred Theory of the Earth* drew significant attention. Not all of it was negative. Sir Isaac Newton admired Burnet's theological approach to geology, but told Burnet that when God created the Earth, the days were longer, thus allowing more time for geological processes to occur. Burnet responded that God didn't need longer days to do his work. Although short on science and hotly debated, Burnet's book was the most popular geological work of the seventeenth century.[66] Because the bestseller was so widely owned and read, its influence on common thought about the compatibility of religion and science was enormous.

To the modern reader, it would seem so much easier to simply shrug and point

The Sacred Earth

out that God can do anything; or to suggest that the Scriptures weren't intended as science. But according to modern palaeontologist Stephen J. Gould, Burnet assumed that the Bible is unfailingly accurate, then he searched for "a physics of natural causes" to prove it.[67] For that, Burnet's huge and interesting effort was ridiculed by geologists for the next three hundred years. Then it was forgotten.

One hundred years after publication, geologist James Hutton, in 1795, wrote of Burnet's *Sacred Theory*, "This surely cannot be considered in any other light than as a dream, formed upon a poetic fiction of a golden age."[68] Charles Lyell, a friend of Darwin and the foremost geologist of his time, also dismissed Burnet, saying "even Milton had scarcely ventured in his poem to indulge his imagination so freely as this writer."[69] But as philosopher Michael Ruse has said, ". . . at the time of Lyell it is not easy to think of anyone writing on geology in Britain who did not have some theological axe to grind."[70]

Still another hundred years later, Archibald Geike, in *Founders of Geology*, wrote that Burnet's was among the "monstrous doctrines" that infested science: "Nowhere did speculation run so completely riot as in England with regard to theories of the origin and structure of our globe."[71] However, to be charitable to Burnet and his thousands of pages of effort, we should say that his books remain readable even today, and as geologist John Laurance points out in *Geology in 1835*, the work has great merit as a literary composition which describes the world created with a thin crust which later breaks up to produce mountains.[72] This makes Burnet's *The Sacred Theory of the Earth* one the earliest attempts to explain the origin of mountains as derived from broken bits of crustal material.

More Sacred Theories

Another attempt to wed theology and geology was produced a generation after Burnet by a physician who started as a linen-seller, became Cambridge professor of medicine without ever attending university, founded a department of geology, and fought a duel over his theory about the spread of smallpox. John Woodward[i] was born in a village in the East Midlands county of Derbyshire. He arrived in London at age 16 to work as a shop apprentice, but became attached to Peter Barwick, personal physician to King Charles II. Woodward never studied medicine, but was trained by his mentor, who lodged him for four years in his home. At age 27, he won the prestigious appointment of Gresham lecturer in medicine, giving free lectures to the unschooled public.

Woodward was a leading English supporter of experimentation and observation in science. His greatest triumphs were in the field of botany – he grew plants hydroponically (in water, not soil) in order to understand how plants eat and breathe, showing they derive nourishment not from water but from minerals within water. His observations in geology were equally astute. He was among the

---

i   John Woodward (1665-1728), English physician, botanist, and geologist.

growing group which insisted fossils represent dead organic beings. However, like Burnet, he explained much geology in terms of the Flood, leaving himself open to ridicule from later geologists. Woodward imagined a "principle of cohesion" held matter together. He said this strange property had been temporarily suspended during the Deluge. This, said Woodward, allowed rocks to become a pasty material easily penetrated by marine shells and fish bones. When the waters reached their apex, fossils and sediments settled out in order of their density, thus creating layered rocks with fossils embedded. He put these thoughts into *An Essay Toward a Natural History of the Earth*, in 1695. However, had he experimented, he would have discovered that fossils are not sorted by density.

The frugal former draper's apprentice amassed a fortune. Woodward donated a great deal of his money to Cambridge for a permanent professorship of geology. He also gave the school his life's collection of rocks – including 9,400 fossils and minerals. They are still on display in their original walnut cabinets.[73] Woodward bequeathed his fossils (the oldest such collection in the world) plus money for a curator and a lecturer. The terms of his will required that the lecturer must be in the collections room "from the hour of nine of the clock in the morning until eleven, and again from the hour of two in the afternoon until four, three days a week, to show the Fossils, *gratis*, to all such curious and intelligent persons as shall desire a view of them for their information and instruction."[74] Three hundred years later, the Woodwardian lecturer tradition continues.

Just one year after Woodward's *Essay of the Earth* and fifteen years after Burnet's *Sacred Theory of the Earth*, came William Whiston's[i] *A New Theory of the Earth*.[ii] Rather than hiding Noah's flood water in a concentric inner ring, Whiston had the necessary water delivered on the tail of a comet. Whiston correctly identified that comets contain water and his 1696 drawing of a comet's orbit is among the earliest printed diagrams showing the Earth revolving around the Sun. Except for perfect circles, rather than ellipses for the planets' paths, his sketch of the solar system could almost sneak into a modern Kansas high school textbook.

Whiston also maintained that the Earth itself originated from the atmosphere of a comet and most changes in Earth's geological history could be attributed to the action of comets. He rejected Burnet's approach. For Whiston, the Earth was not decaying after a perfect creation. Instead, he believed that God's Creation was order out of chaos. The Earth was in harmony with natural events – even cataclysmic ones caused by near misses with comets. God created the Earth out of the atmosphere of a comet, and then engulfed it in a Great Flood with the tail of

---

i William Whiston (1667-1752), English mathematician, philosopher, theologian.

ii Whiston's book's full title is *A New Theory of the Earth: From its Origin to the Consummation of all Things wherein The Creation of the World in Six Days, The Universal Deluge, and the General Conflagration, as laid down in the Holy Scriptures, are shown to be perfectly Agreeable to Reason and Philosophy*.

The Sacred Earth

another.[i] His book, with its nod to a world designed by God, set in motion, and allowed to run on fixed laws, was well received by Isaac Newton, John Locke, and other notable contemporaries.

Whiston was undoubtedly brilliant. He followed Newton as Chair of Mathematics at Cambridge and he produced an extraordinarily popular translation of Josephus's history of the Jews – after the Bible, Whiston's version was the most widely-owned book in England. Despite his considerable efforts, Whiston's contributions to geology were slim. However, he suggested a geological history that lasted much longer than six thousand years, and he showed that geological changes were due to natural processes. Others had already suggested both ideas, but Whiston's popular promotion of them helped the advancement of science.

Burnet, Woodward, and Whiston were not alone in their sacred theories of the Earth. There were many others. In their discussions, it seems learned Englishmen were arguing about dancing angels, with science as their pin. Most of the discussions were unverifiable philosophical musings. A lot of energy was exerted by quite a few scholars whose time might have been better served farming or designing bridges than trying to reconcile biblical stories against the emerging sciences of physics, biology and geology. Perhaps they could have used the character of a German monk as a guide.

One of the world's greatest polymaths, Jesuit priest Athanasius Kircher[ii], has been described as the last Renaissance man, perhaps the last person to have deep knowledge of everything known in the world. Because he died in 1680, there was obviously a bit less to know. But the Jesuit's curiosity had no bounds. He spoke German and Italian, and of course Latin. As a youngster, a rabbi taught him Hebrew. Kircher attempted translations of Egyptian hieroglyphics, using his knowledge of Coptic languages and conjecturing, correctly, that the various Semitic languages are related. He wrote an encyclopedia about China. His interests in science included physics and mechanics (he invented the megaphone and a magnetic clock), he proposed the evolution of species two hundred years before Darwin. (When a deer moves to a colder climate, its descendants transform into more suitably adapted animals, such as reindeer, the priest wrote.) Kircher was likely the first person to examine microbes through the newly invented microscope, speculating the little creatures had caused the plague. Realizing this, he was one of the first to advocate the use of face masks and hygiene to reduce the spread of germs.

Sometimes his curiosity led him into dangerous situations. For example, he was lowered into the throat of Mount Vesuvius as the volcano hissed up at him. Luckily, the poisonous gases that were known to kill sheep near the summit were

---

[i] Comets enamoured Whiston. He caused a bit of a panic in London when he predicted that one of those comets would collide with the Earth on October 16, 1736, and bring an end to life on the planet.

[ii] Athanasius Kircher (1602-1680), German Jesuit scholar.

not belching that day. Kircher was obviously extremely curious, but capable of drawing conclusions that we learned later were not quite right. He thought tides were caused by subterranean waters flowing in and out of caves under the oceans. And he could not accept that fossils found inside mountain rocks were formerly living creatures because, he thought, all the world's mountains had been created long before sea life developed.

Mountains, Kircher explained, were created when the Earth was created – to serve as the Earth's skeleton, keeping the planet from collapsing. In his *Mundus subterraneus*, he claimed mountain chains formed a structural grid, at right angles to other mountain chains, and were slowly being revealed through erosion.

The big geological advances up to the end of the Renaissance were mostly concerned with sedimentary strata – the layers of rock stacked one on the other as gravel, sand, grit, mud, and silt accumulated in the seas bordering the edges of continents. For this, Nicolas Steno played the key role, although this feature of the Earth could be explained by some, as we have seen, in terms of a great biblical flood. However, there was more happening to the planet than the stacking and eroding of layers of sediments. There were also volcanoes. At about the time Father Kircher was being lowered into the gases of Vesuvius, something totally unexpected, and quite intriguing, was happening. The remains of a once-thriving tourist resort, missing for over fifteen hundred years, were being exposed on the slopes of Kircher's Mount Vesuvius. This encouraged a new look at one of the greatest forces of nature, volcanoes, and all the growth and destruction accompanying the hot rocks extruded from deep within the planet.

4

Living with Fire

Engineers were digging an underground channel to divert water from the Italian River Sarnus to a nobleman's villa. The excavation stopped when workmen slammed up against solid walls, buried in ash and rubble. No one expected this. Ancient Pompeii had been discovered. It was 1599, archaeology had not really been invented yet, so a Swiss-born builder, Domenico Fontana,[i] was sent from Rome to make sense of what was being unearthed. Fontana had designed chapels, including the Santa Maria Maggiore Basilica, but he is best known for erecting the 300-tonne Egyptian obelisk called Vaticano in St. Peter's Square – a task that demanded 900 men and 75 horses. This great architect-engineer of Rome was sent by the pope to Naples soon after strange things were uncovered near the Sarnus channel. Fontana had a problem with what he saw.

Domenico Fontana's best client was his friend, the pope. The Counter-Reformation, a backlash against science and liberalism, had a strong hold on the lives and careers of professionals working in Rome's shadow at the time, so Fontana cautiously considered his report from the construction site. Among the first murals uncovered in the tunnel was an explicit larger-than-life advertisement on the outside wall of a long-buried brothel. It was decorated with detailed images of the many pleasures patrons might, for a fee, experience inside the attached building. Embarrassed by the revealing paintings of their ancestors, Fontana's team could have destroyed what they found. But they quietly reburied the wall. Politically, it was the least troubling solution. Archaeologically, it was a gift from Fontana to future generations. Fontana discreetly documented his discovery for the Church's records and no one considered the event for a hundred years. Not even the Roman words *decurio Pompeii* on a gate sparked much interest, though scholars had been speculating about the lost city of Pompeii for generations.

The people of the prosperous Pompeii region owed much to their volcano. The weathering of Vesuvius's ancient volcanic rock, rich in potassium and able to hold water through drought, created some of the most fertile farmland in Italy. Etruscan, Greek, and Roman settlers had each colonized the rich volcanic soils along the Mediterranean Sea. All had prospered.

---

i   Domenico Fontana (1543-1607), Swiss-Italian architect and engineer.

The Mountain Mystery

Mount Vesuvius's location near the sea drew cool Mediterranean breezes in the summer while the heat from her shallow magma encouraged mineral spas, making Pompeii and nearby Herculaneum popular resorts for wealthy vacationers from Rome, two or three days' journey north. It was serene, but fifty years before the eruption that smothered Pompeii, the geographer Strabo[i] described Vesuvius as a burnt mountain.[75] That description seemed unlikely – the volcano's crater was covered with a dense forest, its slopes adorned with cultivated fields and olive orchards.[76] Villages on the volcano's slope were prosperous in the decades before their destruction. It was the time of Emperor Octavius, the General Germanicus, and Jesus. The Roman Empire occupied much of Europe, North Africa, and the Levant. To the east, Persia, Kush, and the Han Dynasty were strong nations. Science was advancing, Rome was at peace. Pompeii was rich and progressive. It was arguably the best place in the world to live.

Pompeii was rich and satisfied, thriving in that optimistic era. The soils and gentle grade of Vesuvius's slopes spawned fine vineyards, the volcano's pumice was used as paving stones for streets and as mill stones for grinding grain. Grapes, olive oil, tourism and trade contributed to the robust economy. Pompeii had at least 130 bars, pubs, and taverns. Two hundred restaurants. Thirty bakeries. Good food, beaches, an amphitheatre, bath houses, and regulated, taxed brothels were attractions. People enjoyed a relaxed secular life. "Money is Welcome" was engraved on the floor of a trading house; "Profit is Joy" announced a sign in front of a merchant's home. Bawdy graffiti and wall advertisements were everywhere, delighting and informing modern archaeologists as much as they did residents in the year 79 when it all came to an abrupt end.

The day before the city was destroyed had been a bright, hot, relaxing holiday. It was the Vulcanalia festival, though few people took their religious heritage seriously. The pious might have sacrificed something to Vulcan, god of fire. Many people started the day with the festival's tradition of lighting a candle; some may have even committed a minor personal treasure to ash. But at the time, Vulcanalia was a day for sporting competitions. It ended with a huge bonfire. Tossing live fish into the bonfire was part of the fun, but most people were not so religious. Perhaps if they had realized they were living in the shadow of violent devastation, they may have made more significant and solemn sacrifices to their volcano god.

But there was little to link the Vulcanalia customs to the mountain looming on the horizon. Other than a few ancient myths, there was no record of the mountain being dangerous. But in one legend, the Greeks told about Hercules fighting the fire god on Mount Vesuvius, "a hill which anciently vomited out fire."[77] Nearby Herculaneum was named for the battle that set Hercules against Vulcan; Pompeii, for the *pumpe* and ceremony, or victory procession, which Hercules celebrated after his victory over the fire god. But these were legends.

No one imagined the mountain could explode and obliterate their city. The last

---

i   Strabo (63 BCE – 24 AD), Greek-born Roman geographer and historian.

46

Living with Fire

eruption was two thousand years earlier. It had destroyed Bronze Age villages, but that history had passed into mythology, and was hardly believable to Pompeii's sophisticated first-century residents. By August 24, the Vulcanalian holiday was over; people were back at their routines. Then the mountain exploded. Smoke billowed from the lush crater amid rumbles louder than thunder. At midday, a tremendous explosion blew apart the cone of Vesuvius, sending a plume of ash and pumice forty kilometres into the sky. The power of that single explosion released as much thermal energy as 100,000 Hiroshima atomic bombs. At the volcano, molten rock was launched at a rate of a million tonnes each second. Farther away (but not far enough), ash began to fall like snow on Pompeii. Within minutes, the city centre was knee-deep in warm tephra that began to fuse into tuff. People fled.

Fortune Favours the Cautious

From relative safety across a bay, in the town of Misenum, the Roman admiral Pliny the Elder[i] received a message that his friend Rectina was trapped at her seaside villa near the foot of Vesuvius. He immediately launched a fleet of galleys for the evacuation of the entire coast. Pliny himself set off with a handful of aides in a light ship to rescue Rectina's group. He covered the 35 kilometres in a few hours, but as he approached the volcano, showers of hot cinders and pieces of rock pelted his boat. Pliny reached Stabiae, about three hour's walk from Pompeii but was trapped by falling debris, pumice floating in the water, and winds that kept him from rescuing Rectina. Pliny's team was stuck; they spent the night on the beach as ash accumulated. In the morning, rocks and wind still blocked their escape by sea. The group tied pillows to their heads – helmets against falling rocks – and withdrew by land. Pliny, in his mid-fifties, sick and weak, decided to stay on the shore and wait for his crew's return. He died on a white sail that his companions had set out for their admiral. They tried to make him comfortable before abandoning him. The group hiked out of the doomed spot, and returned to Misenum. Their safe return is how we know that while they were still at sea, just before reaching the spot where he would die, urged by his helmsman to turn back, Pliny responded, "Fortune favours the brave."[78] Death, too, apparently.

We know much about Admiral Pliny's heroic attempt to save his friend and about the manner in which Vesuvius erupted because at the villa of Pliny the Elder were Pliny's sister and her bookish teen-aged son, remembered by us as Pliny the Younger[ii]. Much later, the junior Pliny wrote to his friend Tacitus that his uncle was likely killed by poisonous sulphurous gases. However, gases did not affect anyone else so it has also been speculated that the admiral suffered either a stroke or an asthmatic attack.

---

i Pliny the Elder, Gaius Plinius Secundus (AD 23–79), Roman scholar, author, scientist, admiral.
ii Pliny the Younger, Gaius Plinius Caecilius Secundus (AD 61–112), lawyer, author, magistrate.

The Mountain Mystery

Back in Misenum, Pliny the Younger was taking notes on what he could see across the bay. He had been invited to join the rescue party, but he was a bit sickly, only 17, and no fool. His motto could have been fortune favours the cautious, as Pliny the Younger would live a long successful life navigating the intriguing labyrinth of Rome's power, becoming wealthy and popular in the process. He not only survived the rule of various (and frequently contradictory) emperors, but consistently rose in rank all the while. Pliny's observations of the Vesuvius eruption give a splendid account of the power of the explosion. In his letter to the historian Tacitus, Pliny the Younger described an eruption that resembled "a pine tree – it shot up to a great height in the form of a tall trunk, which spread out at the top as though into branches, occasionally brighter, occasionally darker and spotted, as it was mostly filled with earth and cinders." During the eruption, tremors forced young Pliny and his mother – the admiral's sister – out of their house and into the courtyard, lest an earthquake bring the building down on them. After another sharp tremor, they left the village entirely. Pliny noted what seemed to be a small tsunami: ". . . the sea seemed to roll back on itself, driven away from the shore."[79]

Pliny the Younger matured into a gifted Roman lawyer, author, and politician. He was an incessant letter-writer. His notes are an important source for historians because Pliny illustrated the lives of the rich and famous when Rome ruled the western world. It was his friend, the historian Tacitus, who urged Pliny to put his Vesuvius experiences in writing. Pliny's description is the world's oldest written eyewitness account of a volcano. He wrote with such detail that modern volcanologists refer to similar eruptions as *Plinian*, in his honour. A Plinian eruption is a volcanic explosion in which columns of gas and ash are blown high into the stratosphere, lifting enormous amounts of pumice. This type usually includes the loud explosive sounds and pyroclastic flows which warned, frightened, then smothered the citizens of Pompeii.

Those pyroclastic flows ended the lives of thousands and left their figures encased in solidified ash. The buried organic material of humans, pets, and trees disintegrated into gas, leaving behind hollow outlines of former lives. Dogs chained to wooden gates. Couples embracing.

Over a century after Fontana discretely reburied the place, Giuseppe Fiorelli[i], an archaeologist from nearby Naples, took over the excavation of the newly rediscovered ruins of Pompeii. He realized that the occasional voids in the ash layer were cavities left by decomposed bodies. Fiorelli conceived the idea of injecting plaster into those hollows, recognizing the ash tombs would serve as moulds. After chipping away the ash, he revealed the eerie casts of Pompeiians in their dying moments. It was a brilliant insight for the archaeologist who meticulously detailed all his work in a three-volume description of his excavation techniques, much of it written while incarcerated in prison for treason.

---

i   Giuseppe Fiorelli (1823-1896), Italian archaeologist.

Living with Fire

Fiorelli was jailed for his commitment to archaeology, which interfered with local politics. Before Fiorelli, archaeologists were not easily distinguished from grave robbers. By 1847 when Fiorelli took charge of the excavation, Pompeii had been picked apart by treasure hunters. Archaeology had been recklessly performed and artifacts were hurriedly boxed and shipped to the museums of the current rulers. When Fiorelli arrived at Pompeii, it was a mess. He began by clearing away the mounds of garbage littering the place.

The priority of previous archaeologists had been retrieving portable treasures for their financiers. Earlier workers missed the point that the buildings themselves were treasures worth preserving. Fiorelli built roofs over the exposed structures to protect them from rain and sun. He encouraged slow careful work. Before his leadership, the streets at Pompeii were scraped clear, then buildings were haphazardly dug into from their sides. Fiorelli started excavating from the top down, documenting the layers as they were unearthed. He introduced a *Journal of Excavations*, requiring constant cataloguing of everything. As far as possible, he wanted artifacts left in place. Leaving interesting antiquities *in situ* violated the demands of Fiorelli's boss, a demoted bureaucrat named Carlo Bonucci. Fiorelli was accused of disloyalty and Bonucci manoeuvred his rival into prison where he was held for three years. However, with a regional change in politics, Fiorelli was released. He returned to Pompeii and brilliantly employed the trick of pouring plaster and glue into the moulds of Vesuvius's victims, capturing the style of their clothes, the positions of their bodies, even their expressions in death.

Robert Scott Duncanson: *Pompeii and Vesuvius Excavation*, 1870

The Mountain Mystery

Even Worse

Although at least 10,000 people died, the destruction of Pompeii and Herculaneum does not represent the worst volcano-related loss of life recorded. That distinction belongs to an Indonesian volcano. In April 1815 about 90,000 people perished when Mount Tambora exploded, ejecting more rock, ash, and dust than any other volcano in recorded history. Most of the lost lives were due to an ensuing famine caused by loss of cropland buried under the fallen ash, but an estimated 12,000 vanished during the eruption. In 2004, excavations began to uncover remains under 3 metres of pyroclastic deposits, similar to those found at Pompeii, using techniques adopted from Fiorelli.

Millions of tonnes of ash were expelled from Tambora. Its ashes encircled the globe, causing havoc by shading and cooling the planet. The dust was suspended in the sky for months, concentrated in a cloud drifting over North America where it created "The Year without a Summer" in 1816. In the United States, late spring frosts killed crops. For weeks, the sky was shrouded by a red fog, making daylight dark. Ice was reported on Pennsylvania rivers and lakes in July and August. Crop failures inflated food prices. Corn increased from 12 cents a bushel to nearly a dollar. Temperatures dropped to minus thirty in New York City that winter. The cloud drifted to Europe, resulting in disaster for wheat, oats, and potato crops. Riots and looting broke out as desperate people pillaged grain houses. The veil of ash continued to block the sun. There were famines in Wales, failed monsoons in China, an ice-dam break in Switzerland. All the result of a single volcanic eruption in Indonesia. The effects continued for two years, then weather patterns stabilized.

A century later and a hemisphere away, another island eruption killed even more people in its initial blast, but the fallout on the rest of the world was less dramatic. In May 1902, the Martinique town of Saint-Pierre ("The Paris of the Caribbean") was obliterated. All 30,000 inhabitants were smothered. Nearly all. One gentleman who lived on the edge of the city and one prisoner locked deep in a dungeon avoided the incinerating pyroclastic surge from Mount Pelée. The prisoner, considered spared by God, was also spared by Martinique's governor. Liberated from the prison, he found a job as a sideshow attraction at Barnum and Bailey's travelling circus where he appeared smudged with chimney soot which his promoters called volcanic ash. To heighten the effect, he was shackled to a ball and chain, much to the delight of visitors with a spare penny.

These volcanoes – Vesuvius, Tambora, Pelée – have some common qualities. They killed through huge pyroclastic flows of millions of tonnes of fiery ash. They expelled hot rocks like bombs from a rocket launcher. Their ultimate cause was heat and pressure acting upon magma under the Earth's crust, balloons squeezed until they burst. At the time, no one understood the source of all that energy. The Greek scientist Anaxagoras speculated that the Earth's interior held hot windy gas that simply needed occasional relief – a flatulent Earth, if you will. Seneca, a Roman philosopher writing just a few years before the Pompeii tragedy, figured there were great fires burning inside the Earth.

The idea of fires burning in the Earth's belly had occurred to others, of course, probably because of similar occasional eruptions and splashes of lava. Roughly 3,000 years ago, the Hindu Vedas started a tradition of belief still followed today regarding errant souls entering a scorching underworld. They called the place Naraka, and the souls of the dead were roped and dragged into Hindu Hell by devilish creatures called Yamadutas. Depending on the sinfulness of the human, hell could be eternal, or not, and might range from the soul being cooked in burning oil to being torn apart by red-hot tongs while standing on melting-hot copper. Hindu Hell is bordered by a river of nasty slime which you may be forced to swim if your earthly sins included destroying either a village or a beehive. Red-hot hell, prevalent in a number of religions as a post-life punishment, was almost certainly derived from real glimpses of red-hot lava.

Volcanoes are our window into the Earth – a dramatic way for us to see the melted rock that occurs everywhere beneath our feet. Volcanoes provide rare samples of the inaccessible inner Earth. They demonstrate the power and activity of the hot convecting rocks that build mountains, form islands, and release dust, sulphur, and carbon dioxide into the air. Billions of years of volcanic belches created the atmosphere we breathe and liberated the water we drink. Volcanoes continue to shape our weather and our continents. And, as we have seen, they sometimes kill thousands of us in a moment.

A Sunset of People

One of my geology professors at the University of Saskatchewan took great delight in describing what occasionally happens to volcano spectators. To be caught in a firewall from an active volcano is to be swept over by pelting fiery-hot ash. Your lungs suck in red hot dust – if the shards of volcanic glass carried by hurricane-speed winds have not already stripped your flesh to the bone, which would ignite and disintegrate, of course. When the professor gave that description, it seemed he had been close to a volcano or two in his research. But not as close as Katia and Maurice Krafft, a French husband and wife team who spent years filming unpredictable pyroclastic volcanoes from unsafe distances and repeatedly photographing lava flows within a whisper. In June, 1991, they were filming eruptions of Mount Unzen, on a southern island of Japan, when a pyroclastic flow unexpectedly swept out of a channel and onto the ridge where they and 41 other people were standing. They died instantly, along with journalists and several research observers – including Professor Harry Glicken. Glicken's death by volcano also seemed inevitable.

The American volcanologist was the scheduled observer at Mount St. Helens

The Mountain Mystery

ten years earlier, when that mountain exploded. Had he been there, it would have killed him. But Glicken had to attend business at a college in California and missed experiencing the 1980 St. Helens eruption. His replacement for the day, David Johnston, died instead of him, manning the doomed observation post that had been Glicken's. Johnston, a promising young scientist, reported the explosion in his final words over a two-way radio: "Vancouver, Vancouver. This is it."[80] And it was. Meanwhile, for a few years, Harry Glicken postponed his own predestined death by volcanic eruption.

Rocks, trees, and people were pulverized during that 1980 Mount St. Helens eruption. I was 1200 kilometres away from the explosion, living and working in a Saskatchewan village near the Montana border. Two days after the explosion, the daylight sky turned grey. Enough volcanic dust settled on the white cab of my truck that I could etch my name on the vehicle's hood. I did not realize that parts of 57 humans were in that thin grey grit. Volcanic ash and dust made the evening skies spectacular with vivid pink, orange, and red accompanied by unusual shades of green and purple. Particles of dust from the mountain, several million trees, and those few dozen people tinted our Saskatchewan sky that spring. The volcano, by the way, was the most deadly ever in the recorded history of the United States. But its devastation may pale in significance on the day Mount Baker, Mount Rainier, or Glacier Peak awaken. Of the 169 volcanoes in the 50 states, the most potentially devastating are in Washington, Oregon, and northern California. In

Mount Baker viewed from Vancouver Island – Beauty, Majesty, and Death
(photo by permission of Edward C Wiebe)

Living with Fire

North America, we await the explosive destruction of the Pacific Northwest.

The volcanoes of this region are strikingly beautiful. But it is a haunting beauty. Each time I have looked towards Washington's gorgeous Mount Baker from Victoria, in Canada, 100 kilometres across the Juan de Fuca Strait, I have had an unnerving sense of the terror that will one day come. On bright clear days, Baker fills the horizon, though often the foot of the mountain is shrouded in mist. On those days, the volcano seems to float eerily above the clouds. The haunting comes, in part, from the knowledge that this mountain will one day violently shake off the lovely glacial ice that coats its shoulders. When this happen, lahars, or volcanic mudflows, will churn hot gas, rocks, and millions of tonnes of slushed glacier into a slurry that will rush down the slope at a daunting velocity. Even at 100 kilometres per hour, a lahar has the consistency of concrete; when it solidifies, the material becomes cement, encrusting the mangled debris it has accumulated on its slide down the mountain.

Mount Baker is presently somewhat quiet, having expelled a few belches of gas in 1975, and has had little to say since. The United States Geological Survey is more concerned about nearby Mount Rainier, which has more ice than Baker, is a more active volcano, and sits atop Seattle. Six thousand years ago, Mount Rainier shot lahars down valleys where tens of thousands of homes now stand. Those flows were 150 metres deep and pushed all the way to Puget Sound, 50 kilometres away.

Touring, studying, and standing upon active volcano sites can be unsafe, but living on the lower levels of a volcano is almost always uneventful. Volcanoes build excellent soil for farmers while presenting exquisite backdrops for boardrooms and executive offices. A good view of Mount Fuji's perfect cone is such a premium that it adds millions to the value of Tokyo buildings. A writer for *The Economist* was recently so taken by Fuji the editors allowed a short piece commemorating an effort to save the very last tiny spot – a lane between two buildings – where people can still see Mount Fuji from street level in Tokyo. Soon a 45-storey block of flats will be completed. There will no longer be any natural ground level view of the iconic mountain from the city's downtown.[81]

The last eruption of this beautiful emblem was over 300 years ago, December 1707, when an explosion released billions of cubic metres of ash, some of which settled ankle deep in Tokyo, a hundred kilometres northeast. Cinders, ash, stones, and rocks were ejected from three new vents, but there was no lava flow. Mount Fuji has been rumbling lately and a fresh release is overdue. But our science of eruption prediction is too primitive to make even a broad guess: Will Fuji explode within the next decade, or the next century? We don't know.

As we have seen, Vesuvius, Tambora, Pelée, and Fuji are among those puffing smoke within the span of recorded history. All four are known as stratovolcanoes. This type is tall and cone-shaped with layer upon layer of hardened lava, pumice, tephra, volcanic ash, tuff, and shards of volcanic glass.[i] The layers build up when successive eruptions pile new material over old. This may take centuries to achieve, but people have watched the birth and growth of two stratovolcanoes in modern times. Both are in Latin America. Izalco was born on a coffee plantation in 1769, stayed active for two hundred years, and grew 2,000 metres. It has been dormant since 1966, quieting just as a nearby tourist hotel with an eruption observatory was scheduled to open. The other volcano, Parícutin, was discovered by Mexican farmers in 1943. Dionisio and Paula Pulido were clearing brush so they could plant corn behind their house. The ground cracked open. The couple watched stones fly up from their formerly flat field. In a week, they owned a fifty-storey mountain. The volcano grew another 300 metres over the next eleven years, finally gassing out in mid-1954. Theirs was an ant-mound of a volcano compared to most volcanic structures. For different reasons, Izalco and Parícutin destroyed the livelihoods, but not the lives, of nearby residents. Although any eruption can be deadly, it is usually not until a volcano achieves enormous height, after thousands of years of growth, that it brings true devastation. The weight of those layers, followed by erosion and in-filling of the vent hole, then a gradual build-up of trapped gas and lava may culminate in a blast of steam, sulphur, carbon dioxide, ash and rock – then the flatulent Earth rests again.

Volcanoes are Not the Answer

We will tarry amidst all these volcanoes just a bit more, for an important reason. Quite a few people – including some with a keen interest in the hills and dales of this planet – think all mountains are created by volcanoes. They are not, but it is easy to see why this notion persists. Humans are attracted to dramatic events and the volcano is not only a dramatic display of mountain building, but it occurs rapidly enough to be noticed. We don't live long enough to see the Earth's crust as anything but rigid and permanent, unless we witness either an earthquake or a rare, transformational volcano. However, more than 80 percent of the Earth's surface – above and below sea level – is volcanic material, the legacy of melted rock. Over a period of hundreds of millions of years, emissions from volcanic vents formed our oceans and atmosphere, supplying the gas, water, and elements that evolved into life and sustained it. Above the seas, over a thousand active volcanoes have been identified. They are mostly concentrated along the edges of continents and within island chains. More than half of the world's visible volcanoes form in a circle, a ring of fire, surrounding the Pacific Ocean.

---

[i] *Lava* is both the hot flowing liquid rock from the Earth's interior and also the same material later cooled and solidified; *pumice*, the rock that floats, forms when pressurized rock is ejected from a volcano then depressurizes, creating spongy pores; *tephra* are fragments of rock, from boulders to ash, spewed from a volcano; *volcanic ash* is dusty burnt and pulverized rock; *tuff* occurs when hot volcanic ash welds together after drifting, falling, or flowing from a volcano; and, *volcanic glass* is melted rock from volcanoes, cooled so rapidly in the air that crystals do not form.

Living with Fire

Although many of our most gorgeous landmarks have a core of fire, most mountains and mountain ranges were not built by fiery igneous rocks. Instead they are deformed layered sediments that once accumulated in shallow seas. Nearly all the great mountain ranges – the Rockies, Andes, Alps, Himalayas – are elevated and twisted layers of limestone (from coral reefs), shale (from coastal mud), and sandstone (from ancient beaches). They are not mounds of volcanic violence.

Yet, volcanoes intrigue us. Uncontrollable and destructive, never to be tamed nor trusted, they beg appeasing. Fortunately, sacrifices to volcano gods have been mercifully rare. Perhaps this is because volcanoes are unpredictable, making it difficult for mystics to claim credit for calming them through rituals or offerings. And, of course, human sacrifices at the mouth of a volcano are hard to accomplish logistically. The ascent is difficult and may include encounters with poisonous gases. It is not easy to drag a large colourfully bedecked entourage up to a crater's rim to watch a priest connect with higher powers. To solve this problem, pyramids in the shape of volcanoes were sometimes assembled, creating a controlled environment to celebrate heart-extraction rituals.

At any given moment, most of the Earth is safely volcano-free. Of the hundred billion people who have lived during the past ten thousand years, fewer than half a million died from volcanic activity. These unfortunates mostly evaporated in pyroclastic flows, as we've seen at Pompeii. But there are other things volcanoes may do to people. Erik Klemetti, at Denison University's Geosciences Department in Ohio, recently explained how one might expire upon falling into a pool of lava.

"Do you suppose throwing yourself into lava would have the same effect as falling into a lake? Probably not,"[82] he eulogizes. Klemetti explains that red-hot lava is usually melted basaltic rock, as dense as any other rock. One does not sink into rocks, liquid or not, but instead is buoyed upward – if not for the red-hot part. Lava has a temperature of 1200° Celsius so there is another reason you will not sink – the heat would cause your body to burst into flame and evaporate before you hit the surface of the glowing lava. Boris Behncke, a volcanologist specializing in Etna's affairs, recently said, "I once stood for hours on the rim of a small pit on Stromboli, which contained a vividly spattering lava lake, and thought, if I really wanted to commit suicide in a fast, nearly painless manner and without leaving a mess to be cleaned up, that would be the place to go. . . Jumping into that vent and encountering that gas stream, you'd produce a brief 'fizzzzz' and a tiny puff of vapor, and that would be it."[83]

Volcanoes are dangerous, but integral to the design of the planet. Although they don't build our mountain ranges, the pressure that squeezes mountain ranges into existence comes with great heat and frequently aligns with the pathways lava uses to reach the Earth's surface. We only recently figured out why volcanoes develop in rather well-defined locales. Three hundred years ago, when geology was a gentleman's hobby, everything was speculative. One of the greatest minds of the eighteenth century, probably the greatest scientist of his time, was a prolific French author, mathematician, and philosopher who speculated much on the planet's geology. Georges-Louis Leclerc,[i] who declared himself Count, or Comte, de Buffon, claimed it was volcanoes that forced primordial seas to recede, making way for the Earth's large animals, and in the process, splitting up the continents.

Buffon's descriptions were colourful: "a volcano is an immense cannon, from its wide mouth are vomited torrents of smoke and flames, sulphur, and melted metals, clouds of cinders and stones, the conflagration is so terrible, and the quantity of burnt and melted matters so great that they destroy cities and forests." Buffon continues his description, then attributes it all to an act of nature, even though the volcano's throat was mistaken by "ignorant people for the mouth of Hell. Astonishment produces fear, and fear is the mother of superstition. The natives of Iceland imagine the roarings of the volcano are the cries of the damned, and its eruptions the rage of devils and the despair of the wretched." Buffon would have none of this superstition: "all its effects, however, arise from fire and smoke."[84] Buffon would be numbered among the most brilliant of scientific thinkers if all his pronouncements were as accurate as this. But his blunders were as significant as his discoveries were enlightening. His research led him to conclude that the New World was dismally weak, disadvantaged, and prone to lapses of virility, caused by swamp gases and cold weather. As we shall see, the president of the United States set out to prove the world's greatest scientist was wrong.

---

[i] Georges de Buffon, known as Leclerc and Comte de Buffon (1707-1788), French science philosopher.

5

First Rocks, then Mountains

The brilliant Georges de Buffon was prone to surprising gaffes. Aside from his mathematics, which were dazzling, he is largely remembered in the Americas for his ill-considered ridicule of the western hemisphere, in which he described the continents as populated by weak creatures living amid impotent swamps. Perhaps this reflects an arrogance of opinion that may have arisen from his extremely privileged aristocratic life. As a child, he inherited a fortune from his mother's uncle. His great-uncle Georges Blaisot was a French tax-farmer. That term implies someone has been planting farmers into a field somewhere, then harvesting at pleasure. An apt description. The tax-farmers of France's *ancien régime* were granted the right to collect whatever taxes they could coerce from the people under their control – which often meant resorting to threats, cajoling, torture, imprisonment, and the use of private armies to collect as much money as possible from peasant farmers and merchants. From the collections, a fixed amount had to be given to the king, all the rest was the tax-farmer's share of the seasonal harvest. With the money (which was considerable), tax collectors built lavish estates and ennobled their names with grand titles ranging from Lord and Baron to Count. In de Buffon's case, his great-uncle (who was also his godfather) farmed all of Sicily, then died unmarried. He bequeathed his immense fortune to seven-year-old Georges Leclerc. The child's father promptly purchased the town of Buffon. In due time, young Leclerc renamed himself Comte de Buffon. Much later in his life, the king of France made the sobriquet official.

Buffon was a brilliant mathematician and a rather competent scientist. But that was later in his life. First, he exhibited a spoilt and brattish nature. University in Switzerland was interrupted when he met a young Englishman, the Duke of Kingston, who was engaged in the eighteenth-century equivalent of a modern youngster's backpack tour of Europe. In those days, however, it was a *Grand Tour*, expensive and indeed, grand. The escapade involved dozens of helpers, courtesans, and assorted groupies. Buffon dropped his studies and joined the tour. There were duels, abductions of young women, and general merry-making as the group pillaged and plundered its way through southern France and northern Italy for almost two years.

The Mountain Mystery

Buffon left the party when his father was set to remarry. A bit panicked, Buffon hurried home to ensure he received the 80,000 pounds (about 20 million dollars today) that was still left in his inheritance. After collecting his money and bestowing upon himself his new title, the Count of Buffon headed to Paris to study science, mechanics, and especially mathematics.

The early eighteenth century was an exciting time in the maths. Newton and Leibniz had recently introduced calculus to the world. With calculus it was possible to calculate volumes in uneven reservoirs, trajectories of motion, compound interest rates, and incremental changes wrought by really, really tiny things – something that Buffon's contemporary, the mathematician George Berkeley, called "the ghosts of departed quantities." Buffon used these ghosts to study all manner of curiosities, including a thought-puzzle he invented (and solved) that resulted in a whole new field of mathematics, geometric probability. His invention solved what is now called *Buffon's needle problem*: a needle falls on the floor. The floor is hardwood, made of strips. What are the chances that the needle will lie across the line between two strips? Intuitively, we know the answer has something to do with the length of the needle and the distance between floor splices. Buffon invented a whole system to tie together those various lengths and the infinite range of angles the needle might encounter. He solved this, and other problems, by merging calculus with probability theory. The esoteric question had applications all over the field of engineering mechanics, from an indirect way to approximate a circle's circumference to a cheaper method to build a bridge. All of it derived from an inspired study of the way a needle falls to the floor.

Besides solving math puzzles by integrating calculus with probability, Buffon applied his prodigious intellect to mechanics and engineering. For example, at the king's request, he directed an effort to find the best lumber for naval ships. He built a laboratory, worked with scaled-down models, analyzed tensile strengths and water-resistance. Ultimately, when small-scale didn't translate back to full-scale, he worked with real timbers. These were hands-on, versatile, data-rich investigations. His mind was open and curious. In geology, he determined the Earth is much older than 6,000 years, opting for 75,000 instead – a calculation that saw his books burned in Paris in the mid-eighteenth century by a briefly resurgent Catholic Church.

Buffon was a gifted and prolific writer. So much so, other scientists sometimes dismissed his work as cleverly presented trite ideas of little merit. This was not always fair, though sometimes true. Buffon was certainly given to well-worded speculation. Which leads us back to his major *faux pas*, something which certainly makes him notable to non-European non-mathematicians. He declared that the New World was so weak and infertile that it couldn't give rise to large strong animals the way virile Eurasia did. Buffon, author of a 44-volume, 6,000-page natural history encyclopedia, curator of the royal museum, the most famous scientist in the world, proposed that the Americas were weak and diminished,

First Rocks, then Mountains

because of "swamps, marsh gases, cold, and humidity."[85] This miffed the governor of Virginia, who was also a scientist, philosopher, and gifted writer. Thomas Jefferson was determined to prove Buffon wrong.

Jefferson, Franklin, and Fossils in the Pocket

About the time the Americans were creating a new nation, Jefferson sent a team of twenty soldiers into the woods of New Hampshire to hunt for a moose – an animal Buffon had heard of, but didn't believe existed. When the stuffed seven-foot animal arrived in Paris, Buffon gamely admitted the New World's environment wasn't necessarily completely lacking natural vigour. One wonders why Buffon tangled himself into this controversy when, in his inspired geology essays, he wrote about the way the New and Old Worlds could have once fit together as a supercontinent, with all parts equally hardy.

The debate, and Thomas Jefferson's determination to prove the elderly Buffon wrong, became part of the Americans' drive to discover what lay to the west of their original eastern colonies. Jefferson heard that gigantic elephant bones had been discovered in Kentucky. While president, he hired George Rogers Clark to go to Boone County to collect the largest of those bones for examination at the White House. Meanwhile, Jefferson sent Clark's brother and Meriwether Lewis to explore the rest of the continent and catalogue all the animals found along the way. Looking for a living specimen of a giant elephant was not Jefferson's main motivation in commissioning the Lewis and Clark Expedition, but Jefferson knew it would be grand to drag an elephant back east, and perhaps ship it off to Buffon in Paris. Fossilized elephant bones – but no live samples – were sent east from Big Bone Lick, Kentucky. Jefferson and the other scientists who examined the huge bones quickly realized these differed somewhat from modern elephant remains. They were more like mastodon fossils recently discovered in Siberia. Meanwhile, George Cuvier, in France, identified similar specimens in a private European fossil collection. Cuvier and the Americans recognized these animals as likely the same creature, now extinct. The theory of extinction led to speculation that dramatic changes had occurred over a rather long-lived planet. Although the concept of extinction was contrary to biblical scripture, the president of the United States held the proof of mastodon extinction in his own hands.

The science of geology had progressed, but not much, in the two thousand years since Palaios found his fish in a stone on the mountain. But neither Buffon nor Jefferson knew why fossils from the sea lurked among terraces high above oceans, even though they were both exceptional scientists. Nor could they explain why mountains existed. There was no visible proof of growth among the mountains, except for singular volcanoes which were as likely to self-destruct as expand. Erosion from wind and rain made disappearance – extinction not unlike the great elephants – a mountain's only likely fate. Mountains had always existed,

The Mountain Mystery

and now they were wasting away. That was the simple answer, confirmed by observation and common sense.

Mountains remained an enigma for another hundred years, and more. Genesis gave them short shrift until Noah's Ark lodged upon one, leading many to believe mountains were not part of the perfect creation. Mountains were the ugly residue of the Great Deluge. English travellers of the seventeenth century, venturing into the centre of Europe, sometimes described the Alps as "horrid, hideous, and ghastly."[86] This was likely a reflection of provincial preference. People from flat environments, trudging along mountain trails in foreign lands, have a natural repulsion to rocky spires. I know from my own experience of nearly twenty years on Saskatchewan's bald flat prairies, the rare trip west to the Rocky Mountains left me drained and anxious. I got over it – I now live in Calgary and enjoy a glimpse of a few peaks, including 3,000-metre Blue Mountain, through my living room window. I have become comfortable with spires looming in the background. However beautiful the peaks may be, people from the flats usually prefer their landscape flat, open, and arable.

Jefferson's friend and colleague, Benjamin Franklin,[i] had already formed his own opinions about the Boone County mammoth bones before Jefferson's celebrated excavation. He had obtained earlier samples and had even shipped some of his fossils to his old friend Buffon, in Paris. So it is time for Ben Franklin to make a brief appearance in this story about mountains and fossils – he had something to say about both.

You may be aware that Franklin invented bifocals and figured out lightning is electricity. He then invented the lightning rod to prevent the fires lightning occasionally sparked. He also created an energy-efficient furnace; and, as a joke, he once proposed daylight savings time. He printed newspapers, including the best-selling *Poor Richard's Almanac* which relayed his clever adages about saving pennies and rising early. Before becoming ambassador to France, Benjamin Franklin edited the Declaration of Independence after Thomas Jefferson sent it to him for a last review. One could write a book listing Franklin's nearly limitless accomplishments – indeed, many authors have. He did well for the fifteenth of seventeen children born to a poor merchant; and for someone whose education ended at age ten so he could help his father make and sell soap and candles.

Franklin had a keen interest in everything, so it is not surprising he had clever geological insights. He observed that the Icelandic volcano Laki, which erupted in 1783 while he was living in Europe, had caused the horrific winter weather felt on the continent in 1784. Franklin noted the season began with heavy smoke and fog, which he speculated were caused by volcanic ash.[87] Franklin was the first scientist to grasp a volcano-weather relationship. His interest in the Earth extended to the seas – after collecting data from merchant ships sailing the Atlantic, he discovered

---

i Benjamin Franklin (1706-1790), American author, statesman, scientist, inventor.

– and named – the Gulf Stream, publishing the first map of its route in 1768. A few years later, he made his most startling geological conjecture.

On the Theory of the Earth, a 1782 message to French geologist Abbé Giraud-Soulavie, Franklin described the way islands and continents might move about on the Earth's surface. In the letter, he construes how an island he observed might have obtained its twisted, convoluted layers of rock. "Some part of it having been depressed under the sea, and other parts, which had been under it, raised up. Such changes in the surface parts of the globe seem to me unlikely to happen if the Earth were solid to the centre. I therefore imagine that the internal parts might be a fluid more dense than any of the solids we are acquainted with, which therefore might swim in and upon that fluid. Thus the surface of the Earth would be a shell, capable of being broken and disordered by the violent movements of the fluid on which it rested."[88] Benjamin Franklin described an Earth with thick, dense fluid inside which causes the surface crust to be violently broken and moved – no one else would come closer to describing the modern theory of plate tectonics for almost two hundred years. But his comment, a mere speculation, was not noticed by geologists. And if it had been, they had no way to test its validity.

As the science of geology was being unearthed and understood more completely, it was becoming harder and harder to reconcile either a universal flood or a six-day creation to actual observations. Geologists were starting to dispense with the exercise entirely. Scientists began to accept biblical stories as allegories and parables, not actual scientific treatises. While president, Thomas Jefferson scrapped the Old Testament entirely, then he took scissors to the New Testament, snipping out all references to supernatural events and miracles. His edited Bible, *The Life and Morals of Jesus of Nazareth*, as he called it, excluded references to virgin births, healings, resurrections, and passages suggesting Jesus was divine. For Jefferson, the Earth was explained by physics and geology. It was ancient, as were the relics he carried in his pocket when he went to his inauguration – he had arranged to meet a geologist attending the ceremony who would help identify the future president's pocketful of fossils.[89]

Among those eighteenth-century scientists who rejected a global flood after years of studying geology was Georg Christian Füchsel[i]. To him, observations suggested most continental lands were once covered by seas – but never simultaneously as in a world-wide deluge. In 1762, Füchsel published *Historia Terrae et Maris (A History of Earth and Sea)*. In this, and other works, he showed a remarkable grasp of theoretical geology.[90] He realized the layering of fossils into dispersed groups meant many floods affected the Earth, but none inundated the entire planet at the same moment. Füchsel wrote, "Similar changes now take place: for the Earth has always presented phenomena similar to those of the present day."[91] This is an important concept in geology – in all of science, actually. The laws we observe today have always been. And like Jefferson, he

---

i   Georg Christian Füchsel (1722-1773), German geologist.

snipped miracles from the way the world works. The laws of physics and geology are immutable and logical. Füchsel said that the formation of rock deposits always followed existing laws. Rock layers form horizontally, under water, and inclined beds are "the consequence of earthquakes or oscillations of the ground."[92] This simple fact of geology became the foundation all subsequent Earth scientists built upon. The processes of mountain building, erosion, rifting of valleys, etching of caves, spewing of lava – all of Earth's transforming activities – have always worked the same way throughout the planet's long history.

Uniformity

Füchsel predated Scotland's most celebrated geologist, James Hutton[i], by twenty years. Their ideas of rock formation and the Earth's timelessness were similar. But it was Hutton, more than Füchsel, or Steno before them, who put these modern concepts squarely in the minds of geologists. Hutton was a physician and a farmer. These avocations helped him see the Earth as a living organism "not just a machine but also an organised body that has regenerative power."[93] A living organism with an indeterminate longevity. Hutton was among the first to seriously contemplate the true vastness of time. Deep time[ii], as it is sometimes called, is something he described as "a thing of indefinite duration."[94] Hutton imagined endless cycles of mountain building, destruction by erosion, draining of sediments to the sea, consolidation, and then another uplifting, forming fresh continents. Recycling forever. Continents destroyed; continents rebuilt. Nothing remains of the original Earth and there is no end to the cycles.

James Hutton championed a new theology. Everything was originally designed by God; God created the Earth to provide a place for man. But it had been created in a way that required no further divine maintenance. Even the process of recycling rock through erosion was designed to rejuvenate the soil for man's agriculture. Mankind used the Earth's resources, but everything would be replenished. Hutton rejected the theist-geologists whom we have just examined – his grasp of geology was far deeper and his role for a deity was much narrower. Burnet's *Sacred Earth* (1681), Woodward's *Essay of the Earth* (1695), and Whiston's *New Theory of the Earth* (1696) all have a similar thesis. Their subtext

---

i James Hutton (1726-1797), Scottish geologist.
ii Ever since James Hutton and Bishop Nicolas Steno, western culture has been adjusting to the concept of *deep time*, as opposed to the previously accepted, theologically endorsed, shallow time of 6,000 years of existence for the universe. Calculating the time required to cool a red-hot Earth, as physicists had, or to erode hillsides, as geologists had, resulted in ages varying from tens to hundreds of millions of years. To the limited experiences of human activity, millions and billions of years can be understood only as an abstraction, but modern Roman Catholic theologian Thomas Berry suggests a deep understanding of time and the evolution of the cosmos is a prerequisite to understanding man's spiritual place in the universe. Deep time, then, has taken on a philosophical connotation far beyond its original physics.

First Rocks, then Mountains

varied as each tried to explain the planet's evolution within a biblical context, but they all reflected a common understanding of the Earth's recent origin, brief existence, and rapidly approaching end. Hutton was different. Ironically, though Hutton's faith was deep and his geology included religious considerations, Hutton is often cited as the first modern geologist because he insisted that natural processes were the sole tool for geological transformation. His work reinforces deism (the idea that reason, observation, and science reveal God) and discounts theism (the belief that revelation of God is only possible through prayer or sacred texts). Further, deism's supreme being is an absentee landlord. The deist rejects the theist notion that gods intervene in the daily affairs of man and Earth. Hutton was central to the Age of Enlightenment in England, northern Europe, and America, and surging concepts of reason and individuality. The deist aspect of Hutton's explanations for Earth's evolution found a willing audience.

Hutton's work impacts us today. Exploration geology (discovery of coal beds, iron deposits, gold veins, oil fields) has its roots in his studies as does stewardship and care for soil, nature, and environment. Although educated in Paris as a physician, Hutton worked as a farmer. Improving the family's Berwickshire farms led to an interest in soils, which required a knowledge of minerals, which led to mastery of geology. While farming, at age 27, Hutton wrote to a friend that he had become "very fond of studying the surface of the Earth, and was looking with anxious curiosity into every pit or ditch or bed of a river."[95]

Hutton became a partner in a very profitable chemical factory which manufactured salts used for dyes and for metal smelting. With the money, he bought rental properties around Edinburgh, and continued farming. In his various business activities, Hutton was largely an adviser and investor, but with his farms, he was an active manager. He farmed for another ten years, taking time for occasional business adventures and geological field trips. He began asserting his belief that the various soils, stones, and rocks he observed on his farms were composed of the bodies of animals, vegetables, and various minerals "of a more ancient formation."[96] From this, he developed his theory of uniformitarianism, an unfortunately cumbersome term which we will now gently tackle.

Uniformitarianism is the idea that natural processes operate today as they have in the past. Supernatural intervention is rejected. Unlike ancient Greek playwrights who occasionally rolled a god or two across the stage to explain some twist in plot, uniformitarians believed deities weren't needed to explain geological transformations. Earth processes were gradual, predictable, natural. Uniformitarianism is still the accepted mode for gradual geological changes on the Earth, though geologists have incorporated a non-supernatural version of *deus ex machina*[i] by embracing the idea of catastrophic events. The hundred-year

---

[i] *Deus ex machina*, a phrase coined by Horace (65 BCE – 8 BCE) who cautioned writers not to imitate the authors of classical Greek tragedies who dropped gods from a crane or pulled them onstage by other machines to resolve some otherwise unexplainable entanglement of plot.

storm, they say, can change a coastline more significantly than a hundred years of gradual erosion. This interruption of incremental change is now generally accepted and is known as "punctuated uniformitarianism."

Another important idea credited to Hutton is *plutonism*, the idea that igneous rocks are the progeny of fire. Before Hutton, the theory of *neptunism*, proposed by Abraham Werner,[i] theorized that all rocks – even igneous rocks – were formed from the crystallization of minerals precipitated in primordial seas. Werner, the son of a mining engineer in the German town of Wehrau, now in southwest Poland, originally intended to follow his father's career, so he enrolled at the Freiberg School of Mines. But two years later, he switched schools and was reading law, then diverted again to study of the history of language. Finally, he quit school without any degree. However, drawing heavily on family experience, Werner soon wrote the first modern textbook on rocks, fossils, and mineralogy. That text, *External Characteristics of Fossils and Minerals*, was one of the few things he ever published. But he was a charismatic lecturer and drew enchanted students from across Europe. They venerated his pronouncements and eagerly accepted Werner's speculations. His theories carried unwarranted credibility in the geological community. He largely misled his disciples and imparted misconceived notions about geology, likely due to his shallow experience. Werner never travelled. His insight was largely limited to the coal and iron mines near his family home. For example, it was apparent to him that volcanoes were caused by underground coal fires, something he had seen in the mines. This corroborated his theory that the Earth's core was cold and all igneous rocks – granite, basalt, andesite – were sediments from an ancient ocean. Even as Werner's hypothesis was beginning to spread, there was already a great deal of evidence that igneous rocks were derived from melted materials. Hutton realized this, exposed it, but had a tough battle. Werner's Neptunists were everywhere.

The Neptunian account, named for the water god, claimed a great ocean once covered the entire Earth, then slowly dried up, while almost all the rocks of the Earth precipitated from minerals dissolved in those waters. Hutton, however, favoured Plutonism, named for the god of hot underworld things. Hutton realized surface igneous rocks had solidified from melted materials originating within the Earth. This is quite different from the Neptunists' position. Mostly, the water-god folks were wrong. The process involving evaporating seas only works in very rare situations where shallow seas or lakes desiccate, leaving behind salt, anhydrite, or potash layers – but not igneous rocks such as granite and basalt, which originate from heat, as Hutton described.

In 1768, at age 42, Hutton turned over his farms' daily operations to tenants and moved to Edinburgh, where his closest friends included the philosopher David Hume and economist Adam Smith. Hutton spoke regularly at the Royal Socicty of Ediuburgh, which became his stage for teaching what he discovered

---

[i] Abraham Gottlob Werner (1749-1817), German geologist.

First Rocks, then Mountains

about geology. His other venue for discourse was the Oyster Club, which he founded with his friends Joseph Black and Adam Smith. The club's weekly discussions were raucous affairs – chatting was rumoured to be only part of the fun. There was a wild mix of widely disparate intellectuals (most of Edinburgh's brightest, as well as regular visitors such as Ben Franklin) who gathered for heavy drinking, loud laughing, deep conversation, and oysters. Hutton, Black, and Smith formed such a deep friendship that when the famous father of modern capitalism died, Hutton and Black were executors of his estate. They divided up properties according to Adam Smith's instructions. Ever loyal, they also followed Adam Smith's last request – all his unfinished works were to be destroyed. Thereupon, Hutton and Black reluctantly burned 16 volumes of their brilliant friend's unpublished memoirs and manuscripts.

Hutton became widely known for his theories of the Earth – especially the cycles of regeneration and the indeterminately great age of the planet. Well-known, but not universally appreciated. One brutal battle was fought against Richard Kirwan[i], an Irish intellectual from the Galway district. Kirwan had studied to become a Jesuit priest (but quit within his first year), then was admitted to the bar (but quit within two years), and finally took up chemistry, meteorology, physics, and geology. Kirwan clung to two scientific beliefs which were rapidly losing favour. Like some of the ancient Greek philosophers, he believed that all matter consists of just a few basic elements. One of them, fire, is liberated when burnt. He spent years trying to prove the idea, but died as the last scientific phlogistonian on Earth. Ever clinging to fading traditions, he also fought vigorously to maintain Noah's Flood as the source of marine fossils found on mountaintops. Kirwan and Hutton sparred over this idea so bitterly that Kirwan finally declared Hutton to be an atheist. But it was an unfounded charge. Hutton was no atheist, he simply held different beliefs than Kirwan.

Hutton produced *Theory of the Earth* partly as a rebuttal against critics, especially Kirwan, who accused him of not only atheism, but poor logic as well. Hutton's 1,200-page two-volume set was a revision of his lectures and letters. Pieces of the work were first read aloud at the Royal Society of Edinburgh by Hutton's friend Joseph Black in March 1785. There was not enough time to read the first paper at that meeting, so Hutton himself finished reading it a month later. But a complete version of his *Theory* was not published until it appeared in the Royal Society's *Transactions* three years later.

Although James Hutton investigated geology with brilliant clarity, he had considerable trouble communicating his thoughts. Sentences filled paragraphs. Some of Hutton's vocabulary was already obscure by the eighteenth century. His huge and important book has been deemed unreadable by all subsequent generations of geology buffs. There are certainly some precious gems buried within his musings, but rambling was Hutton's literary forte. It took him several

---

i   Richard Kirwan (1733-1812), Irish scientist.

rewrites to arrive at his most famous quote, the Earth is "without vestige of a beginning, without prospect of an end."[97] Succinct, but ambiguous. Scholars have argued about whether he meant this literally (all the original rocks of the Earth have been recycled) or figuratively (the vastness of time – past and future – can not be contemplated), or both. He is generally considered such a poor writer, a true ambulator in thought and organization, that his ideas might have been lost if they hadn't been rewritten after his death by his loyal disciple, Charles Lyell.[i]

Legal Clarity

Like Hutton, Charles Lyell was born in Scotland. Eldest of ten children in a modestly well-off family, Lyell entered Oxford where he studied geology, but graduated as a Master of the Classics, as literature and dead languages were called. With such a degree, he was qualified to practice law. In his early 20s, he roamed around rural England defending peasants accused of stealing sheep or hunting in the king's forest. According to his biographers, he was more interested in the geological outcrops[ii] of the countryside he rode through than courtroom drama. Within ten years, he gave up law and began studying geology and writing about it full time. His own observations confirmed Hutton's theories. Lyell published revisions of Hutton's discoveries, writing with more clarity and charm than a typical lawyer with a classics education. His first book, *Principles of Geology*, was published in three volumes between 1830 and 1833. Within that collection the concept of uniformitarianism was presented so convincingly it became the fundamental concept of geology for a century. The theme of *Principles,* "the present is the key to the past,"[98] is essentially a restatement of Hutton's muddier "from what has actually been, we have data for concluding with regard to that which is to happen thereafter."[99]

Lyell was much more than the interpreter of Hutton. By studying faults and fissures, Lyell identified earthquakes as sudden shifts in underlying rock layers. At Vesuvius and Etna, he concluded volcanoes are mostly formed by gradual layering of materials from the vents, not from swelling of the ground as was commonly believed. His charting of stratigraphy resulted in splitting the recent Tertiary period into three parts which he called Pliocene, Miocene, and Eocene, epochs which saw the rise of mammals as well as the rise of the Himalayas. Lyell also named the more expansive Paleozoic, Mesozoic, and Cenozoic periods. All six terms are still used daily by geologists worldwide. Lyell's enthusiasm for national geological surveys helped establish stratigraphic databases and tied discoveries around the world together within a common nomenclature. To gather the information that went into his scientific papers, Lyell travelled widely –

---

i Charles Lyell (1797-1875), English lawyer and geologist.
ii An *outcrop* is the visible part of a layer of rocks that may extend great distances under soils and other rock formations, but becomes exposed when layers above have eroded.

First Rocks, then Mountains

making two voyages to North America in the 1840s. His journeys to Canada and the States resulted in *Travels in North America* and *A Second Visit to the United States*, published in 1845 and 1849.

At Niagara Falls, Lyell calculated the waterfalls would continue digging its gorge for another 10,000 years, then burst a huge chasm into Lake Erie, flooding a large tract of the American northeast. It will likely take much longer – he used an aggressive rate of erosion (about one foot, or one-third metre, yearly) to vividly illustrate the incessant activity of geological processes. The fact that he made two uncomfortable voyages across the Atlantic to meet geologists and visit the continent's interesting geological sites is evidence Lyell was enamoured by America. Late in life, Lyell donated a sizable collection of books to help rebuild the Chicago public library, replacing some of what was destroyed in 1871 by the infamous fire that razed the city's downtown.

Colouring the Rocks

Hutton and Lyell were keenly interested in the theory of the Earth's rocks and speculated about the planet's origins and history. Charles Lyell was amply aware that he was elevating the science of geology to a new respectability and a final divorce from clerical influences. Historian Roy Porter says that Lyell held a "vision of himself as the spiritual saviour of geology, freeing the science from the old dispensation of Moses."[100] But geology also has a practical, non-philosophical side to it. It took an Englishman with an unlikely background to fully enlist geology into the service of the Industrial Revolution. Mining, excavating, and constructing networks of canals became urgently important and William Smith[i] became the leading advocate of practical geology. With him, the science took its broadest step away from the armchair leisure of gentlemen and into the calloused hands of builders.

There was nothing easy in Smith's life or career. He was of extremely humble origins, without an education, money, or family connections. His brilliant work was ignored, then plagiarized. He was financially ruined and spent time in debtors' prison. His life was a slowly unfolding tragedy. A hardscrabble beginning was followed by desperate middle years. His father, a village blacksmith, died when Smith was eight. Raised by an uncle, he left grammar school at age eleven. His formal education ended, he was sent off to find a position as an apprentice. He worked with a surveyor, learned quickly, and became skilled at marking the boundaries of large estates in England's wealthy southern region. Eventually he joined the Somersetshire Coal Canal Company, working as chief surveyor for eight years. The company built a 30-kilometre canal to ferry coal, limestone, and passengers from the Somerset coal fields (which had 80 separate collieries) through a series of 22 locks into the major canals that fed London.

---
[i] William Smith (1769-1839), English geologist.

The Mountain Mystery

While surveying canals and coal mines, Smith observed that rock layers regularly occurred together, stacked as neatly as slices of buttered bread. Significantly, he noticed how individual layers could be distinguished by their fossils. He travelled throughout England for the canal company, meticulously collecting sacks of fossil-rich rocks. From his observations, he recognized that the same succession of fossils – simple to complex – was everywhere, establishing the principle of faunal succession and creating a method to identify layers. If, for example, numerous spines of Echini are found in a mass of rocks, one is undoubtedly holding a piece of what Smith called the Coral Rag formation.

Smith examined rock exposures in canals, railroad cutaways, quarries, cliffs, river banks – anywhere he could see slices of fossil-pocked earth. He took samples, mapped locations, rock types above and below, and even noticed how layers occasionally dipped at subtle angles into the ground. He taught himself to draw cross-sections, vertical columns, and then extensive maps – even predicting where subtly inclining rock formations found deep in mines should eventually break through to the surface miles away.

Although his work was exemplary, Smith was fired from his position at the coal-canal company. Little is known of the circumstances. Perhaps he spent too many hours collecting fossils instead of surveying, but his termination probably involved his purchase of an estate near a route he surveyed for the canal. He remained on good terms with his former director, who found a few consulting jobs for him. But mostly, William Smith was

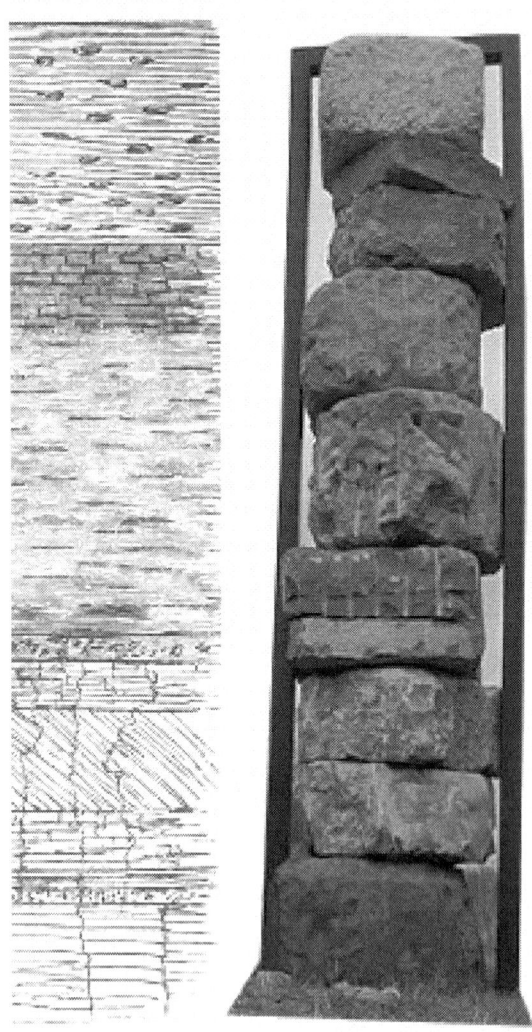

Left, Smith's hand-drawn stratigraphic column; Right, Monument to Smith along Colliers Way

First Rocks, then Mountains

unemployed, though he kept as busy as ever – he spent years travelling and cataloguing rocks, noting their positions, dips, and related fossils. He continually perfected mapping techniques, which resulted in both his greatest achievement and his most vexing defeat.

Smith released charts and drawings derived from fossil brachiopods and ammonites he had collected from the rocky clay and limestone underlying the coal fields. He made a small map, copying an idea from an agricultural plat popular at the time. It used colour coding for different soil types. In a similar way, he mapped the boundaries of outcropping rocks and filled the zones with colour. In 1801, Smith produced the type of map today's geologists would appreciate. It was the first of its kind. A true geological outcrop map, a valuable tool for both geological study as well as for surveyors and canal builders working in the field. It was simple and practical, yet detailed and thoughtful. For example, his initial map was at a scale of one inch to five miles and used local names for the twenty different rock layers he identified – such as the London Clay and the Kentish Rag.

The naming system Smith used for a rock formation usually combined rock type (clay, shale, lime) with the name of the place the layer outcropped. This is still done, but in some cases, a rock layer is first encountered inside a mine, or within an oil, gas, or water well. The formation may then adopt a nearby town's name, even if its rocks don't reach surface. For example, geologists prospecting the rich Keg River oil and gas formation in northwestern Canada are chasing a layer of carbonates deposited when tropical reefs thrived there millions of years ago. In 1955, an oil well was drilled near the community of Keg River and struck one of the long-buried reefs. The geologist on duty was obliged to give the new rock bed a name. The Keg River formation doesn't outcrop at its namesake, but the formation was so important to the oilmen they named it for the village where they hung their parkas at night. Eventually they pumped millions of barrels of oil from the Keg River's oil-saturated porous limestone and dolomite.

Smith named formations throughout England. Without a job or steady income, Smith nevertheless travelled and mapped, drew, sketched, and sampled Britain's rocks. He published a revised fossil study, *Strata Identified by Organized Fossils*, then created a brilliant geological cross-section, a layer-cake diagram of rocks below the surface from London to Snowdon, the highest mountain in Wales. Smith spent months drawing and colouring. In 1815, he published Britain's first extensive geological map. Continuing to use a scale of one inch to five miles, his finished map was over eight feet high and six wide. He printed and signed 400 copies, of which 100 are believed extant today, mostly owned by museums. Smith expected his map would be especially valuable to field geologists, so he printed and sold it in panels that could be mounted into a full wall map, or carried a piece at a time to the work site during geological projects. His map used standard symbols for rivers, roads and mine entrances, but he placed these over a background of various colourful blobs that indicated the exposed surface rocks.

The Mountain Mystery

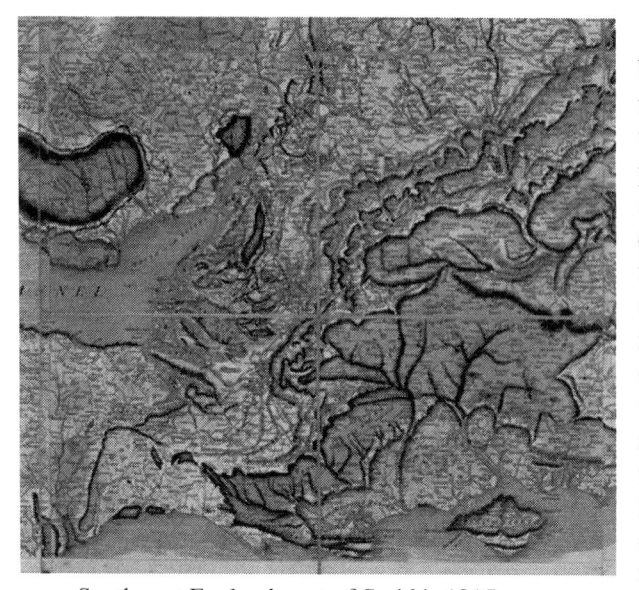

Southwest England, part of Smith's 1815 map

William Smith had become an obsessed, unemployed 46-year-old who borrowed money to publish maps and cross-sections. He was using expensive colour-printing techniques and could not sell enough maps to recover costs. Meanwhile, other printers enlisted well-connected geologists and simply copied Smith's techniques – or sometimes his complete maps. They discredited Smith as an unschooled amateur, but sold knock-offs of his work with pedigreed signatures affixed. The market wasn't large for geological maps and cross-sections. His masterpiece was priced at the equivalent of about $3,000 in today's currency. Smith himself had originally copied someone's geographic maps as his own base upon which to draw his geology, so it was standard practice to borrow and enhance. But the small niche was flooded by cheap imitations. Smith was bankrupted. Unable to pay the printers he had hired, he spent months in King's Bench Debtor's Prison. Released in August, 1819, he headed across London to the house he thought he owned. A court bailiff had seized it. The man who had created the world's first geology maps had lost everything.

    Ruined, broke, and fifty years old, William Smith took a menial surveying job. He carried his gear, chains and small telescope to farms and estates that were being sold and subdivided to accommodate England's growing prosperity and population. He did this for a decade – until one of his employers, Sir John Johnstone, recognized him as the man who had mapped England. Johnstone helped Smith, now sixty-one and doing the work of a thirty-year-old, reestablish his reputation. In February, 1831, the president of the Geological Society of London brought Smith before his assembly and awarded William Smith the society's first Wollaston Medal, identifying the aging surveyor as "The Father of English Geology." A few years later, Smith, accompanied by his nephew, was invited to a Dublin meeting. There he was startled to find himself recipient of an honorary Doctorate of Laws presented by the British Association. His last public act, in 1838, at nearly seventy years of age, was to serve on the commission that selected building stone for the new Palace of Westminster.

6

Everything Changes

One of the great observations from this awakening era of geology was that fossils found near the surface usually resemble modern creatures while more primitive fossils are encased in deeper older rock layers. For example, before an event geologists have come to call the Cambrian explosion, there were no fish fossils. Earlier creatures, discovered in older rocks, were blobby, usually lacking backbones, and likely not capable of profound thoughts. When fish appeared, they were at first small and jawless. They did not have much of a backbone – that appeared a few million years later when unmistakably fish-like animals show up in Ordovician rocks. But it took still more time for fish to flourish and diversify. Eventually, during the Devonian, or "Age of Fishes," as it was once called, some lineages grew to seven metres while others developed lungs and became comfortable sitting on the beach. And on it went.

Among the great fossil experts who followed William Smith was Mary Anning,[i] an English fossil collector, dealer, and palaeontologist. She is one of the few female geologists who make an appearance in this book – and we are three thousand years into trying to figure out how a fish fossil had arrived on a Greek mountainside. She, like her predecessor female geologists, was rather discouraged and mistreated by the men who dominated the science.

Anning was a child when she found her first huge fossil skeletons. They were on the treacherous cliffs that dropped to the sea at the English coast where she lived. Landslides exposed fresh fossils each winter. (One slide almost killed her when it swept her dog down the scarp and into the ocean.) Anning's parents were dreadfully poor – she and her brother were the only two of ten children to survive their crowded, rough quarters where smallpox, measles, and hunger took their toll. Mary Anning was sickly until she was fifteen months old.

But in her fifteenth month, lightning struck a tree near the spot where she was being held by a neighbour. That family friend and two other women were killed. Mary Anning was stunned, but was easily revived. It became part of local legend that her prodigious energy level, curiosity, and intelligence were due to that lightning bolt. However, her survival also made the village a bit suspicious of her.

---

i Mary Anning (1799-1847), English palaeontologist.

The Mountain Mystery

Her father was an unsuccessful carpenter who couldn't afford land for a house, so he built the family home on a wooden bridge in their coastal village. It occasionally flooded, covering the floor with all manner of filth. The family earned spare cash scavenging the nearby Dorset cliffs, recovering fossils otherwise destined for the sea below. Venturing onto the dangerous ledges to gather snake-stones and devil's toes[i] for curious seaside holidayers was always part of her chores. Anning was 12 when she unearthed her ichthyosaur. She also discovered the first two plesiosaurs ever found. These were big animals and they changed the way biologists viewed the creatures that preceded man. But Anning didn't restrict her work to oversized skeletons. She identified the ink sacs of ancient fossilized octopus-like varmints and she realized stones called bezoars in her day are what we now called coprolites, or ancient fossilized dung. She dissected modern fish and compared them with the bones of earlier types she uncovered on the cliffs. She was observant and knowledgeable, but grew bitter that her discoveries brought little recognition.[101]

Her insights might have advanced palaeontology further than they did, but women could not be members of the Royal Society, nor the Geological Society – in fact, women were not even permitted to attend meetings as guests. In addition, members were preferably Anglican but her family belonged to a splinter group of dissenting Congregationalists, spiritually related to the earlier Puritans. Religious and gender prejudice kept Mary Anning from being published, though she tried. Meanwhile, male colleagues discreetly sought her advice regarding fossils they couldn't identify. And when she became gravely ill from cancer, those geologists created a fund to help with her expenses. She died at 47. Members of the Geological Society erected a touching memorial in her honour. The inscription reads, in part, "in commemoration of her usefulness in furthering the science of geology, also of her benevolence of heart and integrity of life."[102] Her work helped create an appreciation for the vast evolution of species and the extinctions that had occurred among the planet's creatures.

Charles Darwin[ii] had been a renowned geologist before achieving his fame in evolutionary biology. He was awarded the Wollaston Medal, the highest honour of the Geological Society of London – the same award that had gone to William Smith thirty years earlier – for his contributions to geological sciences, particularly studies of fossils, coral reefs, volcanoes, and sedimentary processes. Before Darwin released *On the Origin of Species*, he had learned stratigraphic mapping at Cambridge from Adam Sedgwick, the best known geologist of the time. (Sedgwick was so important to geology at Cambridge that the university's earth sciences museum is named in his honour.) Darwin collected and analyzed fossils from his *Beagle* voyage and was Britain's foremost expert on fossil

---

i   Snakestones (ammonites) and devil's toes (belemnites) were not recognized as marine fossils by most of Anning's customers who instead saw medicinal and spiritual value in these curios.

ii   Charles Darwin (1809-1882), English geologist, botanist, and naturalist.

Everything Changes

barnacles.

On Darwin's voyage around the world, his geology textbook was Charles Lyell's *Principles of Geology*. Lyell advocated the gradualist, or uniformitarianist, approach to analyzing geological processes, which was at odds with what Darwin had been taught at Cambridge. It was a fortuitous choice of book. Not only did Lyell's book open Darwin's mind to the idea that incremental changes occur over vast spans of time – something he would later apply to explain the origin of all species – but Lyell's ideas had practical consequences during Darwin's voyage.

Darwin was the first person to correctly understand how coral reefs develop. Until Darwin, it was believed that reefs grew on underwater volcanoes. Darwin instead conjectured that the reefs slowly developed on the margins of islands and continued to grow as the sea floor subsided. It was Lyell's gradualism at work. The original island might completely erode, leaving only the reef to be observed. Not everyone agreed with Darwin's conjecture, but a hundred years later, the 1952 surveys of the US Atomic Energy Commission examined the same Pacific islands and proved Darwin correct. They discovered that subsidence had been occurring for millions of years, just as Darwin described, and resulted in stacks of coral thousands of metres high. Geology and gradualism loom large in *Origin of Species*. Darwin's study of geology, his enthusiasm for Lyell's gradualist concepts, and his detailed knowledge of fossils added fact and substance to his theory. Two full chapters of *The Origin* are devoted to geological evidence. Over the years, Charles Lyell and Darwin became close friends. Lyell's support and correspondence encouraged Darwin to complete his study and publish his book.

Time for Change

It takes a great deal of time to accomplish the gradual changes Lyell described. Darwin estimated it had taken 300 million years for erosion to create the deep wide valleys among the English Weald mountains. He also assumed it had taken millions of years to first build those disappearing mountains. Without invoking catastrophes or supernatural events, gradual transformations in geology – and, as Darwin famously pointed out, in the variety of species – require inconceivably vast amounts of time. It is almost impossible for the human mind – evolved to respond to immediate dangers and opportunities – to contemplate millions of

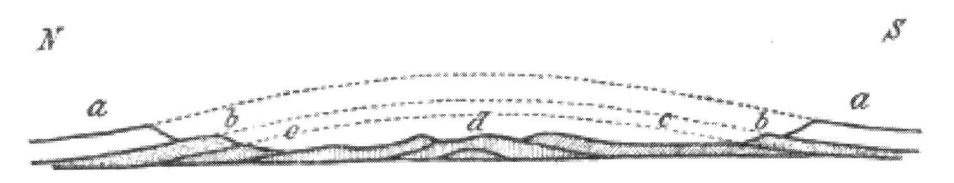

Erosion of Weald: Layer "b" once rose hundreds of metres above today's mountains.

years of gradual change. Except perhaps by analogy.

Consider that every few generations a small genetic mutation occurs. This may be due to environmental accidents such as radiation confusing a bit of DNA as the body prepares for the next generation. (This is one reason the dentist puts a lead blanket over your lap before zapping an X-ray of your tooth.) Genetic mutations may affect height, brain size, hairiness. Frequently, the allele results in weakness or susceptibility to disease, in which case the carrier might not survive to pass the change to the next generation. However, even undesirable mutations may be inherited, if those affected live long enough to reproduce and successfully compete for mates. Hence, colour-blindness abounds.[i]

Most alleles are indifferent errors in coding, but a rare few give some advantage that makes the mutated individual more successful in a particular environment. These changes are real: lighter skin helps absorption of ultra-violet light, enhancing Vitamin D synthesis where sunlight is weak;[103] lactase persistence is from a mutation that expands the food range available to a third of the world's population;[104] sickle-shaped red blood cells offer resistance to malaria.

Light skin, lactase persistence, and sickle-cells are recent evolutionary adaptations, occurring within the past few thousand years, and enhancing survival in each case. Mutations in viruses and bacteria may be detected within weeks. But for creatures with longer lifespans, changes due to the slower pace of generations usually require vast time to be seen. Darwin and Lyell realized the Earth is a patient old planet. Vast time is abundant.

To grasp the effect of incremental change across deep time, try this mental exercise: Imagine you can hold your mother's hand, arm's length away, and she her mother's, and the next your great-grandmother's, and so on, through a long line of your ancestors, stretching along a city block. The last person you would see on that block lived 3,000 years ago, and could be almost identical to your mother. There would have been some genetic changes between the person at the end of the block and the one holding your hand, but in 150 generations, you might need genetic analysis to find them. Your mother and that ancestor are so similar you would probably not notice a difference. You are even rather similar to that cave-lady waving at you from three blocks away.

But if you jump along the line far enough, you will see some not-quite-human-looking features. A shorter person, with an odd-shaped forehead, and much more body hair. These genetic evolutionary changes are a mix of gradual changes and sudden lurches. Evolution is imperceptibly slow on a generation-to-generation level, yet occasionally jerking along – short fingers to noticeably longer ones in a single generation, perhaps. Because longer fingers offer a leverage in grasping rocks and throwing spears, offspring will be fed better, protected better. Mutations

---

[i] An inability to discern red from green may be an evolutionary advantage. It likely contributed to an early division of labour. Colour-blindness is almost exclusively a male condition. Those afflicted are nearly useless as berry-pickers, encouraging them to hunt antelope instead.

Everything Changes

– evolutionary change – can offer great advantages. Of course not all mutations create healthier, stronger, better-adapted offspring. Most don't. Those unfortunates are eaten by lions – they are evolutionary dead-ends, extinct creatures who are not among your ancestors.

Let the line continue, say from Washington, D.C. to New York City. Or London to Manchester, or Paris to Brussels. If you are near the Capitol Dome in Washington, the distant ancestor standing in front of Trinity Church on Broadway is your great-great-great-great- (the next 400,000 'greats' have been omitted) -great-grandmother. This ancestor, living seven million years ago, doesn't look anything like a modern human.

The 400,000th person is a short slender hairy lady with gangly arms. We may call her LouAnne. LouAnne had twin daughters, one was your great-ish grandmother Lulu, the other Lulu's sister, your distant great-aunt, Luma. Aunt Luma also has a chain of descendants, each holding hands, stretching from New York City to Washington. Occasionally, the creatures along Luma's family line also experienced genetic mutations that helped the lineage survive in a changing environment. However, they occupied a different ecological niche and circumstance found them evolving differently. The cumulative effects of Luma's descendants' genetic alterations result in the creature at the end of Luma's line, standing next to you. She is your cousin Blinko, a chimpanzee. You, at the end of Lulu's line; the chimp at the end of Luma's line. You are not 'descended from an ape' – you are an ape. And the chimp is not an ancestor, but a cousin. This was the message Darwin conveyed in *Origin of Species*, and which he stated more emphatically in his later book, *Descent of Man*.

Charles Lyell inspired, prodded, and promoted Charles Darwin's pursuit of a gradualistic explanation for the diversity of species. Lyell was twelve years older than Darwin and had an established reputation. His help added credibility to Darwin's theory. However, it is clear Lyell didn't agree with all the consequences of evolution which Darwin recognized, even though

Ridiculing Darwin
The Hornet magazine, 1871

The Mountain Mystery

Lyell himself had long entertained the idea of evolved species. In 1827, thirty years before Darwin's book, Charles Lyell described a manuscript he had just received. It was written by the French biologist Lamarck.[i] Lyell wrote:

> "I devoured Lamark . . . his theories delighted me more than any novel I ever read, and much in the same way, for they address themselves to the imagination. I read him as I hear an advocate on the wrong side, to know what can be made of the case in good hands. [Lyell was trained as a lawyer – he is referring to any good courtroom opponent.] I am glad that he has been courageous enough and logical enough to admit that his argument, if pushed as far as it must go, if worth anything, would prove that men may have come from the Ourang-Outang.[sic] But after all, what changes species may really undergo! How impossible will it be to distinguish and lay down a line, beyond which some of the so-called extinct species have never passed into recent ones."[105]

Any book by Lamarck would have been an interesting read in the early 1800s. He was controversial and his interests were broad – zoology, biology, chemistry, geology – not only did he write extensively in these fields, he invented some of these words specifically for his books. Lamarck was an unlikely candidate for revolutionary science. He was the eleventh child of an impoverished aristocrat. The men of his family typically became soldiers, but Lamarck's father put him in a Jesuit school, believing he might become a great priest. Before that was accomplished, Lamarck's father died. Immediately, young Lamarck left his religious studies, bought a horse, and darted across France to join a battle against a German invasion. His company was slaughtered by artillery. Only 14 French soldiers were left, none of them officers. One of the older men suggested young Lamarck take command and order a retreat, knowing the new commander would face court martial. Lamarck took command, but ordered the survivors to hold their ground until reinforcements arrived. They came. Lamarck, age 17, became a national hero. In appreciation, a scholarship was awarded and Lamarck tried medicine – he studied for four years – but eventually ended up a science professor researching fossils. Lamarck was as puzzled by fish fossils on mountaintops as our ancient Palaios.

Lamarck suggested a way fossils of sea animals could end up on high dry land. He thought the fossils had been raised through a complicated scheme that involved continents drifting slowly around the Earth, under the influence of the moon's gravitational pull. He figured that the eastern coasts of continents eroded while new sediments were deposited along western shores as the landmasses slowly crept from east to west. This explained both the curvaceous fit of Africa to

---

[i] Jean-Baptiste Pierre Antoine de Monet, Chevalier de Lamarck, known simply as Lamarck (1744-1829), French biologist and natural philosopher.

Everything Changes

South America and the rather obvious occurrence of mountains along America's west coast while plains dominate the east. In 1802, publishers refused to print his theory. Lamarck self-published one thousand copies of *Hydrogéologie*, but sold only a few.[i] In *Hydrogéologie,* Lamarck expounded uniformitarianism and introduced his idea that within a changing environment, organisms need to adapt or become extinct. His book also delved into the idea of geologically vast periods of time. In short, it was a revolutionary book. But not at all popular. Except among progressives like Lyell who found much to admire in Lamarck's work.

Lyell maintained the same mix of excitement, curiosity, and skepticism when he read Darwin's drafts of *On the Origin of Species* thirty years later. Lyell struggled to reconcile his religious sentiments to what science was revealing, but he failed. Loyalty to Darwin limited Lyell's public criticism of the ultimate result of evolution – man's descent from earlier creatures. Besides, Lyell had earlier encouraged his friend's research. But Lyell disagreed with the mechanism of natural selection, partly because it didn't fit his idea of deductive scientific reasoning. Nor could anyone conduct an experiment to test natural selection during the nineteenth century. In his mind, if a theory was not testable, it was not really a scientific theory. In Lyell's last book, published after Darwin's *Origin of Species*, Lyell wrote that it remained a profound mystery how the huge gulf between man and beast could be bridged. Darwin thought he had built that bridge and was disappointed with Lyell's cautiousness.[106] But both agreed with the idea that natural processes had been slowly unfolding on Earth for millions of years.

Before Lyell, not many geologists looked at rocks and really appreciated the enormous passage of time between the moment a rock formed and the afternoon it was lifted into their hands. But within fifty years or so, most clever people accepted that the Earth was ancient and rocks represented millions of years of existence. As we have seen, Charles Darwin was perhaps the first to give a reasonable estimate of the age of the Earth by noting the rate erosion etched chalk valleys into the hills of southern England. From those observations, in 1859, Darwin calculated that the Earth had to be at least 300 million years old. This is significant, not because it is accurate (it is not), but because it finally put a number to an idea – rather than the vague "vast ages" that some geologists were recognizing. He offered something quantitative.

Using geology and the methods available at the time, naturalists determined the age of sediments, based on rate of deposition, fossils, and erosion. The calculations they reached for these shallow sedimentary rocks were sometimes rather good. But hidden below the layered rocks were the true Methuselahs, the longest-lived pieces of Earth's crust – bits of igneous and metamorphic material. There are no fossils in these ancient melted and twisted masses of Precambrian

---

i Because of Lamarck's significance to the fields of zoology and evolutionary biology, a copy of *Hydrogéologie,* which he couldn't give away while he was alive, was recently available through Nigel Phillips Booksellers in Chilbolton, England, for $5,000.

rock. For a long time, wild guesses were the estimates for the age of the most ancient of the crust, the granitic shields of the continents. To know the true age of the Earth – not just its upper sedimentary layers – geologists needed a new tool. Physics eventually provided a method based on radioactive decay, but that discovery was decades away. In the interim, physicists provided another way, based on measurements of heat, to estimate the planet's age.

Miner's Hell

As far back as Gilbert and Bruno, the Earth was described as a cooling globe of once glowing hot iron and rock. Assuming that the Earth's surface had once been scorching and had now cooled to something we find comfortable, scientists could calculate the rate of cooling and estimate the planet's age. The Earth's interior is still hot. Deep ore pits demonstrated this to miners hundreds of years ago. To quantify this obvious fact, scientists collected temperature data from mines around the world. Virginia coal mines, for example, become steadily warmer at a rate of one degree for each 20 metres of descent. To collect deeper readings, scientists developed self-recording thermometers which they lowered into wells. Near Columbus, Ohio, it was 31° Celsius at 846 metres below the surface while 38° Celsius was found at the bottom of an 813-metre well in central France. Dozens of readings from wells around the globe helped scientists determine an average global geothermal gradient. In general, temperature increases about 26° Celsius per kilometre of depth.

The increase with depth is fairly uniform around the world, except close to hotspots like geysers and volcanoes. In those places, the geothermal gradient can be 1200° Celsius in half a metre, if lava is flowing just under a thinly-crusted pahoehoe surface, for example. Just a bit removed from such glaringly active zones, rocks may be solid and safe enough to insert pipes and harness energy for heating greenhouses or for spinning electric power turbines. But everywhere else, the geothermal gradient is reasonably predictable, reaching the boiling point within about four kilometres.

One winter, when rivers were frozen and there was a tinge of arctic in the air, I descended into a deep potash mine. A kilometre below Saskatchewan's icy surface, I stood on dusty pink salt in a subterranean cavern from which potash was dug. The air was hot – and windy from huge fans bringing currents of fresh air through the underground labyrinth. I was surrounded by pink walls of potassium chloride while huge mechanical moles were grinding salts into powder as the machines crawled through caves of their own creation. Their grinding brought salty dust to my tongue, nose, and eyes. The grains of salt and chunks of rock loosened from the walls by the miners were translucent pink. From this mine, the potash was conveyed a thousand metres to the surface where the salts were processed. From there, most of the refined potash from Saskatchewan's

Everything Changes

depths traveled in hopper trains a hundred cars long, past the prairies, over the Rocky Mountains, down the Fraser River Valley, then by ship across the Pacific to China. Once there, potash becomes fertilizer on Asian soils. A third of all the world's potash is chiseled from the remains of a single ancient desiccated sea – mined, pulverized, and processed in Saskatchewan, Canada.

Potash is extremely useful – it is part of the reason the world has enough food for its billions of humans. Almost like a vitamin, potash strengthens plant tissue, builds resistance to disease, and enables photosynthesis. Sprinkled on fields of corn, soybeans, rice, wheat, fruits, and vegetables it increases yields, improves taste, colour, and texture, and keeps plants healthy. Each year, over 2 billion kilograms of raw potash are dug from deep mines in Canada and Russia, but potash is sometimes also spit up by volcanoes. This was one reason the farmland on Mount Vesuvius helped Pompeiians thrive.

Most potash is found over a kilometre below the surface. But it doesn't form at such a depth. Potash salt starts out the way most salt does, in a shallow evaporating sea. Millions of years ago, much of Saskatchewan had been the bay of an ocean. Mineral-saturated water washed the inlet regularly, then evaporated, leaving behind layers of salts that became metres thick, beginning with halite (table salt), then anhydrites (plaster material), topped by pink potash. In Saskatchewan, the mined fertilizer is found in four main layers, each about 7 metres thick, each separated by a metre of table salt. All of these evaporites were deposited layer upon layer, a few millimetres per season, year after year after year. Eons passed amassing these common salts. The Saskatchewan evaporation pan formed in a hot arid Devonian[i] climate, within a shallow sea that covered an area the size of North Dakota. A drive around its undulating 2,950-kilometre former perimeter takes days. The region of Earth that held that sea 400 million years ago slowly sank under its own weight and became buried by mud that produced shale, reefs that became limestone, and sands that cemented into sandstone. This is why the potash isn't scraped from the surface, but is excavated by monstrous digging machines assembled at the bottom of deep shafts where the temperature is always 27° Celsius. Comfortably warm for the crews in thin shirts and hard hats. But there are much deeper and hotter mines elsewhere in the world.

The deepest accessible holes in the ground are South African gold mines near Boksburg and Carleton. These extend nearly 3,600 metres below surface. The temperature of the rocks at that depth is 60° Celsius, the air temperature is 55, but air-conditioned to 30° by pumping an ice slurry into the zones being burrowed. Humidity, however, hovers close to 100 percent. If miners continue following the descending seams of gold, unmanageable temperatures will be reached. With the 26-degree-per-kilometre geothermal gradient, digging just a bit deeper will result in air temperatures that pass the boiling point. The miners will arrive in Hell.

---

i   The Devonian period was 360 to 420 million years ago. During this period, the seas were widely populated by fish and the first widespread colonization of plant life on the continents occurred.

The Mountain Mystery

Because the interior of the Earth contains tremendous heat, nineteenth-century geologists realized the Earth is cooling from a much hotter level. Gilbert discovered that the inside of the planet is iron. Gilbert's iron sank towards the Earth centre while lighter rocks buoyed upwards. Since then, it was believed that everything has been cooling, heat flowing upwards, eventually escaping into outer space. The mathematician Joseph Fourier[i] developed a mathematical technique to analyze heat flow in 1822.[107] Then he used his math to show how the Earth is cooling.[108] Fourier was a mathematical wizard. Among many other things, he created a branch of math used to separate repeating number patterns from each other and from noise – the Fourier Transform makes electronic communications possible. His equations were original and complicated, but when he described the way heat is emitted at the surface of the Earth, he was clear enough.

Fourier said that the interior of the Earth preserves part of the "primitive heat which it had at the time of the first formation of the planets."[109] However, he found that heat conducts to the surface extremely slowly. Without suggesting an actual temperature, he calculated the heat dissipation of a high-temperature object (he suggested a piece of the sun as an example) inserted 20 kilometres below the Earth's surface. It would take 200,000 years for that heat to seep to the top – and then it would only increase the local surface temperature one degree.[110] He did not suggest the Earth's age is 200,000 years, but his thought experiment implied something vastly longer than 6,000 years. Fourier also knew that if the temperature gradient were maintained all the way to the core, "Such a result produces a very high internal temperature."[111] Indeed it does. If sustained, the Earth's centre would be over 100,000 degrees – a value Fourier knew was unlikely. However, he knew the core of the Earth was fantastically hot.

By biblical tradition, the Earth was created flat and cold. Waters were separated into two parts, then, on the third day of Creation, water separated again to allow land to appear. There is no mention of the molten sphere which nineteenth-century geologists advocated. Even though some groups held fast to a belief in a cold flat slab for Earth's design, in general people seemed to have had little trouble accepting a red hot Earth as part of Creation. One of the most religious scientists to support an ancient, hot, and spherical Earth was a brilliant physicist who spent the first half of his life getting nothing wrong. It seemed every guess, every conjecture, and every pronouncement from Lord Kelvin[ii] was correct. Until his 50th birthday. Then, it seems, he spent the second half of his adult life getting almost everything wrong.[112]

---

i Joseph Fourier (1768-1830), French mathematician, climatologist, and physicist.
ii William Thomson, Lord Kelvin (1824-1907), British physicist and engineer.

7

The Apple Cools

William Thomson, later known as Lord Kelvin, was born in Belfast in 1824 and lived 83 years, most of those in Glasgow, Scotland. His father was offered a position as a mathematics professor at the University of Glasgow, so Thomson's family moved from Ireland to Scotland when William was a child. At age 9, the future Lord Kelvin was weak and sick from suspected heart problems. His father enrolled the frail youngster in a program that provided a curriculum for gifted children at the university. So, at age 10, William Thomson entered Glasgow University's grammar school. He would remain attached to the university for about 70 years. There is no question that Thomson was brilliant. At age 12, he won a prize for his translation of Lucian's *Dialogues of the Gods*; at 15 he won recognition in astronomy for an essay on geodesy, explaining the irregular shape of the Earth; at 16, Kelvin wrote about heat transfer, even clarifying some of Fourier's calculations on the subject. In 1844, at 20, Kelvin addressed the age of the Earth, showing that measurements of the rate of heat loss could be calculated backwards to place an upper limit of the planet's antiquity.

The young physicist's model described an Earth once extremely hot, but now cooling in a mathematically predictable manner. Kelvin imagined that the heat of the glowing hot globe was initially evenly distributed, centre to surface. The surface cooled and internal heat transferred towards it. He maintained that millions of years ago the surface cooled quickly and has held the same temperature ever since. Thus, the Earth began red-hot, its heat slowly leaking into the frigid cosmos. Eventually the poor planet will become a frozen rock, circling a sun which Kelvin theorized must also grow dim and expire. A bleak, but distant, future. There was enough time for the prophesies of Lord Kelvin's religion to unfold first.

Kelvin attended chapel service daily. His devout faith complemented his scientific work, as is evident from his address to the 1889 annual meeting of the Christian Evidence Society, a group founded to fight atheism in Victorian society. This was an organization which directed its activities to the "lower grades of society, to save them from infidelity"[113] partly through printed tracts, but also through private study and lectures.

The Mountain Mystery

Kelvin's 1889 address to the Christian Evidence Society includes cautious statements about the age of the Earth: "I may say, strenuous on this point, that the *age of the Earth is definite*. We do not say whether it is twenty million years or more, or less, but we say it is *not indefinite*. And we can say very definitely that it is not an inconceivably great number of millions of years."[114]

A few million years was Lord Kelvin's favourite estimate. This was contrary to the much longer time geologists and evolutionists needed to explain the changes they detected. But Kelvin was not a geologist, and certainly not an evolutionist. He was a physicist and his main argument – a good one – was that the sun couldn't possibly blaze for much more than a few million years. In time, Kelvin's estimate of the age of the Earth decreased – from several hundred million years (suggested by him as a young man during the first half of his life) to 20 million years (his guess in 1897, near the end of his career). This was before Einstein, before we knew that a tiny speck of mass could create an enormous amount of heat and light. Nuclear power in the sun was not even imagined by Lord Kelvin and his contemporaries. Instead, Kelvin speculated that the sun's energy could be due to some unexplained gravity-into-sunlight mechanism. He believed gravity supplied the sun's huge energy output, and finally decided it could last no more than 20 million years. But as we have seen, by measuring the rate of erosion of the Weald Mountains, Darwin had already deduced that the Earth had to be at least 300 million years old.

Kelvin's early career was stellar. As Glasgow's professor of Natural Philosophy, he invented math systems and investigated the nature of heat, creating discoveries still useful today. Among his early achievements were the correct calculation of the temperature we call absolute zero – the temperature where molecular motion stops. In his honour, other scientists named the Kelvin thermometer calibration, a system which begins at zero and forces all other temperatures to be positive. Rather than referring to absolute zero as minus 273.15° Celsius, the coldest possible temperature is simply 0° Kelvin.

Kelvin's greatest contributions were his discoveries of heat conduction and thermodynamics, but he also created the mathematics that helped analyze electricity's flow and dissipation. Because of that, he was offered a position as a director and engineer for the firm laying the first trans-Atlantic telegraph cable in 1865. The eventual success of that project made him quite wealthy. And famous. William Thomson became the first scientist in the United Kingdom to ascend to the level of nobility. He took the name Lord Kelvin. His ascension at age 68 was partly in recognition for his work in thermodynamics and electricity. But the title presented by Queen Victoria was also politically motivated. Thomson was loyally opposed to Irish Home Rule. Although born in Belfast, he didn't sympathize with any increase in self-government for any part of Ireland.

For all his brilliance as a physicist, Lord Kelvin was a stubborn and prissy fellow. Among his many gaffes, in the 1890s, Lord Kelvin declared that the newly

The Apple Cools

discovered X-rays were a scientific hoax. He is remembered for predicting, a few months before the Wright brothers' flights, "No balloon and no aeroplane will ever be practically successful."[115] Nor could he let go of a belief that the universe is permeated by a physical *aether*, an imaginary rigid framework for conveying radiant energy, even after Einstein disposed of the notion. And Kelvin fretted that the world's oxygen supply would soon run out. But Lord Kelvin is most famously quoted as stating, in 1900, "there is nothing new to be discovered in physics... all that remains is more and more precise measurement."[i] Shortly after declaring physics a dead science and physicists forever condemned to lives of monotonous tinkering with what was already known, atomic science was born. With the discovery of radioactivity, the nature of the atom, quantum physics, and new theories about photons, gravity, and the interchangeability of matter and energy, physics was very much alive. It was the science of the new century.

Many grappled with the problem posed by Kelvin's Laws of Thermodynamics which suggested the Earth was not nearly old enough to complete the processes geologists observed. But for many nineteenth-century scientists, a cooling planet provided the most logical way to explain mountain-building. Originally, they speculated, the smooth-surfaced hot Earth slowly formed a tough hide that thickened as it cooled. They compared the planet's surface with a blacksmith's hot metal plate developing tiny striations as it cools. Multiplied thousands of times, the resulting tiny grooves on a new metallic shield or on a cannon ball become Alps-sized compared to the adjacent smooth surface.

The Earth as a Dried Apple

In 1885, when Austrian geologist Eduard Seuss[ii] published *Das Antlitz der Erde (The Face of the Earth)*, the notion the Earth was a cooling wrinkling iron ball prevailed. It was the way mountain ranges were born. In his landmark three-volume study, the Vienna professor championed his Dried Apple Theory.

The earthly crust, once solid and continuous, ruptured as it cooled, shrinking and wrinkling much as an old apple does as it dries. Desiccated collapsed zones in the apple analogy became ocean basins; continents formed where scabs of apple skin remained

---

i These words have been frequently attributed to Lord Kelvin from an address to the British Association for the Advancement of Science in 1900. However, it is not known with certainty that he actually said physics is dead, though he certainly made pronouncements of similar sorts.

ii Eduard Suess (1831-1914), Austrian geologist.

elevated. It was a theory, one of the best at the time, to explain why the Earth had continents and oceans. Suess proposed that as the Earth cooled, some existing oceans would become continents, left high and dry, while other continental crust cooled, sank, and flooded. This resulted in cycles allowing ocean rocks – entombing fish fossils – to become stranded in mountains, solving the long-abiding mystery of fossils on dry land.

We have trouble, today, imagining a shrinking Earth but the idea is reasonable. The Earth is so large that just a tiny one-tenth of one percent decrease in radius is 6 kilometres. That could create some of our tallest continental mountains. You can see we are not talking about a dramatic reduction in size. A trivial contraction could wrinkle a bald surface into our present dramatic topography.

Between the years 1883 and 1909, Suess published refinements of his dried apple-earth theory. He described scabs of crust surrounded by younger basins, today filled with water. He even suggested that the deep-sea trenches found along the margins of the Pacific Ocean are zones where ocean floor has pushed under the continents during the contractions. But one problem with the desiccating apple-earth theory is that the results – from mountains to earthquakes and volcanoes – should be randomly scattered. They certainly are not. Instead, they appear in swarms or patches. The apple model doesn't fit all the data.

Eduard Suess had more than dried apples to contribute to geology. In 1858, he proposed that a southern supercontinent called Gondwana had once existed. Later he described a vast ancient ocean he called Tethys which separated Gondwana from another huge landmass to the north, Laurasia. Beginning 180 million years ago, wrote Suess, northern Laurasia separated into Eurasia and North America; southern Gondwana was the precursor to all the other continents. He did not envision mobile continents, but rather a series of upheavals as the Earth shrank and crust broke and churned. Parts of Laurasia and Gondwana sank, becoming oceans while his Tethys Sea disappeared. Suess needed a supercontinent to account for similarities he had found in widely dispersed fossils.

Fossils were the key that untangled twisted and contorted rock layers that had thrust atop other layers. For Suess, the field geologist and professor, unscrambling the events that created the Alps had been a life-long goal. There was no longer any doubt fossilized fish on mountains indicated the mountain rocks had formed in a sea. Suess presented the contracting Earth as the solution to the mountain riddle, the mechanism that elevated mountain ranges. Suess had created his theories after long hikes in the Alps; late in his life, those same mountains provided clues to other scientists that turned Suess's work on end.

Near the end of Suess's career, Albert Heim[i], a Swiss geologist with a similar aspiration to explain the rise of the Alps, saw evidence contradicting the entire notion of a cooling, wrinkling Earth.[116] Heim was a prodigious scientist. At age 16, he made a model of the Alps that brought attention from professional

---

i   Albert Heim (1849-1937), Swiss geologist.

The Apple Cools

geologists; in 1873, at age 24, he became a geology professor; in 1875, Heim married Switzerland's first female physician, Marie Heim-Vogtlin. Heim's work cast doubt on Suess's contracting, apple-earth theory by showing that rock layers within the Alps had slid hundreds of kilometres. He famously demonstrated that Switzerland, one of Europe's smallest countries, would become the largest if all the rocky layers, thrust sheet upon sheet, were unravelled. Using the trick of cross-section balancing, he graphically pulled apart sequential layers, placing them in the long line they had occupied before being squeezed into their new positions.[117] In places, lateral shortening, undone, stretched a thousand kilometres. To the dismay of contraction advocates, this meant the Earth would have been much, much larger than anyone could explain in order to contract enough to crumple up the long stacked layers found in the Alps. Heim's work, illustrated with remarkable cross-sections, seriously injured the idea of a cooling, contracting planet. The theory was wounded, but not dead. Other ideas would eventually replace it, but contraction's advocates kept the notion alive for decades.

Scottish geologist Henry Caddel invented this 1875 "squeeze box" to model mountain building. His vice pushed simulated layers of sediment until they deformed to exactly mimic what he and other geologists saw in mountain ranges.
(CP14/079 Reproduced by permission of the British Geological Survey © NERC. All rights reserved.)

The Mountain Mystery

America's Turn

The Americans produced one of the most clever geologists of the nineteenth century and he had an alternate suggestion to account for those pesky mountains. James Dwight Dana[i] had a very long career, working largely as a mineralogist. He entered Yale at 17, graduated at 20, then taught maths to midshipmen in the American Navy. His duties included a naval exercise in the Mediterranean. While there, he published his first paper, in 1834, reporting observations about Mount Vesuvius. During his off-duty time at sea, he worked out the basic system of mineralogy – the complicated mathematical relationships of crystal structures. Crystals form as melted rocks cool. Specific mineral crystals form with their sides, or faces, at fixed angles, leading to 32 classes of symmetry. Tackling the geometry that describes crystallography is not for the timid. Dana, at age 24, summarized it in his 1,430-page *System of Mineralogy,* establishing his deserved reputation at a decidedly young age. Dana's interests also included crustaceans, coral island formation, and Earth morphology. Just about everything geological caught his eye – he even wrote an influential volcanology text while in his 70s.

James Dana realized there were only two ways mountains formed – from volcanic actions or from tectonic pressures. The volcanic ones were easy to explain. Magma swelled up under the surface of the Earth like a bruise welting under the skin. It either created a dome or a layered volcano, similar to what Dana had seen as a lad in the Navy. Ejected rocks stacked atop each other and a new mountain was built. Thus individual volcanic mountains were explained. But mountain ranges made from twisted layers of stratigraphic rock, populated by marine fossils, were problematic. How did entire ranges of mountains lift from the seas?

James Dana, at age 70

Dana agreed with the common wisdom that mountains and continents were the products of a cooling and shrinking planet – but not quite like the dry old apple enthused by Suess and many others. Dana, the best mineralogist of his day, brought elements of crystallography into the discussion. He thought the continents had formed much earlier in geological time and were the result of

---

i   James Dwight Dana (1813-1895), American geologist.

The Apple Cools

solidification of quartz, feldspar, and other minerals. As the shrinking, cooling surface of the Earth chilled a bit more, minerals such as black orthopyroxene and yellow-green olivine formed and became the ocean basins. Hence, he concluded, minerals that were heavier and cooled later developed ocean basins while other crystallizing minerals formed earlier into the rocky granitic mountains and continental shields. In this, Dana recognized two modern ideas about the oceanic and continental crust: the two were made of fundamentally different materials and continents were older. This observation would later be extremely important to sleuthing the nature of the planet's evolution.

Dana's crystallography even clarified the frequent occurrence of mountains along continental edges, something Suess's apple model could not explain. Dana thought that as the Earth cooled, pressure intensified where different rock types were in contact – oceanic basalt juxtaposed against continental granite, for example. This, he said, built the mountains. His model used cooling, contraction, crystallization, and deformation. In Dana's description, eroding sediments wash from land and accumulate along the edges of continents, then, due to heat and pressure, rise as mountains.[118] We now know that Dana got much right, but his model still did not fit all the facts. And his theory that continents and oceans are horizontally unmoveable would eventually be pushed aside.

One clue that Dana might be wrong about the lithosphere being locked in place came from Dana himself. He noticed the youngest Hawaiian islands are farthest to the southeast with progressively older ones trailing behind like ducklings. Dana was the first scientist to recognize the same trend in the volcanoes cutting across the big island of Hawaii. Old dead eroding volcanoes are in the northwest highlands of the big island while youthful and active Kilauea and Mauna Loa spew lava in the southeast. It would later appear almost self-evident that the entire island chain and all its volcanoes rise from volcanic material cutting through ocean crust that slid past overhead. But Dana's world was one of permanence, the cause of the age-dependent string of islands wasn't even guessed by him. Landforms were allowed to move up and down, but not left and right. Neither continents nor islands moved in Dana's models.

Dana's theory of a crystallizing and contracting planet was reasonable for his time and widely accepted among geologists by 1900 – partly because Dana had had such a stellar career; partly because the idea incorporated another American invention – Geosyncline Theory. This is the idea that erosion washes grains of rock off continents, dumping them into nearby seas where the weight of the grit depresses the sea floor. The amount of accumulated sediment becomes so enormous that the material pushes down thousands of metres. In the case of the Appalachians, Dana and his colleague James Hall[i] figured the depth of accumulation was about 12,000 metres just before those mountains rose from the depths.[119] It is fairly easy to imagine huge depths of eroded materials. Finding a

---

i James Hall (1811-1889), American geologist and palaeontologist.

way to elevate these twisted hardened rocks back above sea level was problematic. The geologists assumed that the heat of the deep burial plus the pressure of the contracting Earth forced the mountains up – like toothpaste squeezed from a tube. It was a convincing theory, but unfortunately there were no existing active examples anywhere in the world.

James Dana was head of geology at Yale for 42 years. He married his Yale chemistry professor's daughter, Henrietta Silliman. Dana was a respected gentleman who played piano at church, led Bible studies, and prayed over meals with his family. In his early 40s, he wrote *Science and the Bible* which attempted to reconcile geology with religion. But he kept his religious sentiments peripheral to his science. Unlike most of the earlier geologists, he didn't try to distort his geological discoveries to match his spiritual beliefs. Dana worked in the opposite direction – he found biblical passages that confirmed what science was telling him. He fully accepted the Bible as God's revelation, "But there are also revelations below the surface, open to those who will earnestly look for them."[120]

There was an exception to Dana's practice of finding biblical scriptures to justify scientific discoveries. He refused to reconcile evolution with his faith, even though he maintained a cordial correspondence with Charles Darwin. And Darwin with Dana. However, it seems neither one found it necessary to read the other's books before criticizing them. You can see how this unfolded with the following exchange of letters in 1863. First, we find Dana writing to Darwin:

> "The arrival of your photograph has given me great pleasure, and I thank you warmly for it. I value it all the more that it was made by your son. He must be a proficient in the photographic art, for I have never seen a finer black tint on such a picture.
>
> "I hope that ere this you have the copy of *Geology* (and without any charge of expense, as this was my intention). I have still to report your book [*The Origin of Species*] unread; for my head has all it can now do in my college duties. I have thought that I ought to state to you the ground for my assertion that geology has not afforded facts that sustain the view that the system of life has evolved through a method of development from species to species..."[121]

In his letter, Dana then proceeds to list some basic errors in Darwin's logic. Dana finds there are "missing links" between many species (though he readily adds that he knows not all the world's fossils have been discovered). Some species, according to Dana's understanding of Darwin, developed from "higher groups of species instead of the lower," implying a reverse evolution that would suggest Darwin's basic theory was wrong. And some species seemed to go extinct in the rock record, but then somehow "started again as new species." All of these

The Apple Cools

criticisms from America's greatest geologist were valid at the time. They were all subsequently resolved when the fossil record became more complete.

Darwin wrote back to his colleague:

> "I received a few days ago your book and your kind letter. I am heartily sorry that your head is not yet strong, and whatever you do, do not again overwork yourself. Your book [*Manual of Geology*] is a monument of labour, though I have as yet only just turned over the pages."[122]

It is interesting that two of the greatest geologists of the century couldn't find time to read each other's most important works. Darwin goes on,

> "With respect to the change of species, I fully admit your objections are perfectly valid. I have noticed them. . . Nevertheless I grow yearly more convinced of the general (with much incidental error) truth of my views. . . As my book has been lately somewhat attended to, perhaps it would have been better if, when you condemned all such views [regarding evolution], you had stated that you had not been able yet to read it."[123]

Darwin's irritation with his friend at last surfaced. Dana had been publicly attacking Darwin for months without actually reading *On the Origin of Species*. It would take years, but incredibly, the book which sat unopened on Dana's shelf was eventually read, appreciated, and accepted as fact by North America's foremost geologist. In Dana's 1896 edition of *Manual of Geology*, Dana completed his long treatise on geology with an unequivocal acceptance of Charles Darwin's science – with one notable exception. In his final textbook, James Dana virtually gushed with admiration for the theory of natural selection and he admitted that in the thirty years between his first rejection and his whole-hearted acceptance, science had found the missing fossils that had caused him concern. He listed the evidence: progress from aquatic to terrestrial life; progress from simple to more specialized; modern embryos, with "part of the early life of the globe"[124] represented in their development; "unity in the system of life" regarding how creatures are organically related (all are carbon-based life-forms); and, the increasing levels of cephalization, or brain complexity, as a function of time.

Dana summarized, "According to the principle of natural selection, an animal or plant that varies in a manner profitable to itself will have, thereby, a better chance of surviving, and of contributing its qualities and progressive tendency to the race, while others, not so favoured, or varying disadvantageously, disappear." Dana conceded that the origin of the variations was unknown, but expected science to discover this, too. His book was published in 1896, the same year as the discovery of radiation, a key environmental cause now known to contribute to genetic variation. In his final book, Dana backed up his support for Darwin's discoveries with dozens of specific examples. Years ahead of evolutionary biologists, Dana even correctly speculated that dinosaurs had evolved into birds. But James Dana never accepted that humans had evolved from earlier creatures.

Dana concluded his magnificent *Manual of Geology* with a lengthy discussion of Man's pinnacle position in the biological order of life on Earth. "Man's origin has thus far no sufficient explanation from science. His close relations in structure to the Man-Apes are unquestionable. They have the same number of bones with two exceptions, and the bones are the same kind and structure. The muscles are mostly the same. Both carry their young in their arms."[125] And yet, James Dana, the piano-player at the church where he was a leader, cautioned against carrying the similarities too far, inviting the reader to crawl around on all fours like a great ape and see that humans don't have the massive neck muscles required to keep the head level. He ended with ". . .the intervention of a Power above Nature was at the basis of Man's development. Nature exists through the will and ever-lasting power of the Divine Being, and that all its great truths, its beauties, its harmonies, are manifestations of His wisdom and power."[126]

Over a lifetime of research, teaching, and writing, Dana published two million words in his scientific books and papers. His influence was phenomenal not just in the role evolution plays – or does not play – in man's ascent, but also in his approach to science. He was able to reverse his earlier instincts and accept most of the idea of evolution when the mass of evidence was finally clear. And, in his own mind at least, he was able to reconcile his dichotomous forms of revelation – stones from the Earth and messages from God.

Dana was a man of great integrity and intellect. When he said the Earth was shrinking due to some form of crystallization and that mountains formed partly from regurgitated geosyncline material, most geologists were convinced, even if there were no active analogs. But not everyone agreed the planet was shrinking. Other theories about the Earth's changing landscape were lurking in the shadows.

World's Largest Jigsaw-Puzzle

One relatively obscure geology professor who had originally trained as a physician had quite a different notion about mountain creation. Richard Owen[i] was a geology and chemistry lecturer at Nashville University when he published a book that included mobile continents bumping into each other, creating mountain ranges. He developed his own philosophy of science and Owen wrote a book about his notion that planet Earth is a living being. He was rejected as a crank and his book would be forgotten, but it did point towards the future of geology.

Richard Owen was the son of Robert Owen, a fabulously wealthy Scottish industrialist who emigrated to America to buy an experimental cooperative

---
i   Richard Owen, M.D. (1810-1890), American physician, geologist, educator.

The Apple Cools

(actually, the entire town of Harmony, Indiana) from the German-American Harmonist Society, a group that believed in communal property, the imminent return of Christ, and, unfortunately for the sect's future growth, celibacy. The Harmonists were moving east from their Indiana community to settle near Pittsburgh. The elder Owen bought Harmony in 1825, renamed it New Harmony, and used the community to design a perfect utopian society. It failed within two years. But the town's residents established a free library, a town drama club, and a liberal public school. It also served as a home for the Owen family – one son became a congressman and helped establish the Smithsonian; another was a noted federal geologist. Richard, the youngest member of the family, became the state's chief geologist, then a professor at Nashville University, Indiana University, and finally, the first president of Purdue.

Although from Indiana, Owen received his medical degree in Nashville, then taught geology there. He eventually resigned from Nashville University in 1858, citing his opposition to slavery as the main reason. He moved back to the family base in Indiana. He was in Nashville only a few years, but it was there he wrote the book he is best known for, *Key to the Geology of the Globe,* in which he likened the Earth to an organic being, noting that all the geological structures we observe on the Earth's crust are the result of the same sort of forces that govern the development of organic bodies. Owen presented an early Gaia Hypothesis – a philosophy honouring the Greek Goddess, Mother Earth. The concept of Gaia proposes the Earth is alive in a complex, self-balancing ecosystem. Owen's book combined biology and geology in a rudimentary way, but nevertheless was similar to ideas developed a hundred years later by James Lovelock and Lynn Margulis, who, with an almost spiritual passion, emphasized how ecology, geology, biochemistry, and climate affect the habitability of Earth and combine in a living connected system. Owen had the same idea, but was restricted by the quality of science available in the 1850s. Owen's book has sections titled "Showing that diseases as well as appropriate remedies have some connection with the geological position in which they originate" and "An attempt to demonstrate the analogy existing between organic structures and geological strata." Through examples, Owen illustrated an inter-connected, living Earth. His dynamic geology included continents that drift. That idea came to him in a flash, he said, while preparing a talk for Nashville University students.

He wrote that the university chancellor asked him to present some of the great principles of geology to a joint gathering of medical and English-literature students. This forced Owen to think of a simple way to present his science. As part of his preparation, he "placed on the floor of a vacant room all the geological maps which he possessed, in their correct relative position. There suddenly flashed the idea that the formations in the Western Continent corresponded in many respects to those in the Eastern; and fitted, adjusted, and moved apart and together, until it appeared they must have been detached at some period from each other. . . The next point was to find the law according to which they had separated."[127] This was in 1856. Owen struggled to quickly pull together a theorem, realizing, "in this age of railroad speed, a long delay is unpardonable."[128]

Regardless of his reckless train-like haste to get his book published, Owen nevertheless created a complicated, mathematically-based system to explain the present layout of the continents. He discovered that triangles transcribed on the globe are useful mapping tools for finding copper, gold, and coal. He said that solar energy is the ultimate source of all power – it lifted the Andes, elevated Scandinavia, expanded and contracted the Earth's crust, and separated the continents. He made a strong case for elasticity and glacial rebound, though he didn't realize he was describing the crustal adjustments that occur after kilometre-thick ice-age glaciers melt. He wrote that whenever we see volcanic lava we are reminded that we live on a solid crust but some of the materials under the crust are in a fluid condition. Unfortunately, *Key to the Geology of the Globe* received neither popular nor scientific attention – it was a key unturned.

Meanwhile, French geographer Antonio Snider-Pellegrini[i] drew "before" and "after" maps of a supercontinent. He inferred Europe and America had once touched because various coalbed seams – including the fossils found within them – were continuous. Snider's 1858 book, *La création et ses mystères dévoilé* (*The Creation and its Mysteries Unveiled*), was written to explain geology in biblical terms. Snider's reconstructed map of the supercontinent showed Brazil snugged against Africa, but only after Snider pinched and twisted the coastal outlines of the modern continents. What Snider could not know was the continents actually fit extremely well without all the pinching and twisting – if the parts of coastline now under the Atlantic Ocean are exposed and used as the jigsaw pieces. Snider seemed to believe the continents belong together, but he couldn't propose a convincing mechanism, apart from the biblical Flood, to explain how they split apart. But by 1858, the flood explanation was becoming increasingly tiresome.

At about this point, religious scientists began to drift further away from a literal interpretation of Noah's flood and a six-day Creation. Arnold Guyot[ii], for example, was a devout scientist who championed a way to understand science while maintaining a spiritual grounding. Guyot was Princeton's first geology professor. He was so respected that a future geologist named a swarm of submerged Pacific islands in his honour. Guyot, who once prepared to train as a minister, wrote *Biblical Cosmology* in 1883 to promote a reverential reconciliation of the Bible with geology.[129] It was not entirely successful, but Guyot's effort did much to popularize the study of geology in America.

---

i   Antonio Snider-Pellegrini (1802-1885), French geographer.
ii   Arnold Henry Guyot (1807-1884), Swiss-born American geologist and Princeton professor.

8

Time for a Shake

About Creation, Arnold Guyot said we must accept "on trust the truth of Creation as an ultimate fact, not to be reached by any reasoning process."[130] Even today, with the original spark of the Big Bang still unexplained, some of physics becomes metaphysics, or possibly theology. Mystery still abounds. But Guyot referred to a more recent Creation and more intimate matters when he asserted: "The Bible narrative is in perfect contrast with the fanciful, allegorical, intricate cosmogonies of all heathen religions, whether born in the highly civilized communities of Egypt, the Orient, Greece, or Rome, or among the savage tribes which still occupy a large portion of our planet."[131] Guyot continued that although all societies, civilized or savage, have tried to claim divine authority and offer stories of the Creation, only the Bible got it right. However, after establishing his credentials as a non-savage, Guyot explained the Bible is a book of spiritual laws, not a biology, geology, or astronomy textbook. The Princeton professor then proceeded to explain each day of biblical Creation in great scientific detail.

In his university lectures, Guyot taught that geologic epochs last millions of years. He rectified the apparent disagreement between his lessons and the Bible by pointing out that the Sun was created on the fourth day, therefore a day can not be a day in the literal sense. A Creation day – his *cosmogonic* day – could last a billion years, if necessary. The Great Flood that preoccupied so many earlier geologists was not mentioned in his best-seller, *Biblical Cosmology*. Evolution was demonstrated as possible within his system though Guyot suggested that God's hand led evolution through its stages and man is a separate creation of God. "Any length of time that Darwin might desire for his transformations, would never suffice to make of the monkey a civilizable man."[132] Guyot wrote that an animal, even of the most beautiful form, "is still an animal and nothing more. However, a poor misshapen Hottentot, endowed with spiritual faculties, rendering him capable of becoming a living member of the spiritual world, through faith in Christ, would still be a man."[133] With this sentiment, Guyot summarized the temporary resolution between faith and science at the end of the nineteenth century: science explains everything, the Bible is true (if taken as allegory), and there is something separate, special, and spiritual about the human animal.

The Mountain Mystery

Arnold Guyot earned his right to be a spokesperson for the scientists' views on Christian spirituality. He had planned a life as a preacher and enrolled in university in Berlin, intent on the ministry. But he became a geologist. At 18, he formed a close friendship with Louis Agassiz[i], who was the same age, also from Switzerland, and would also become one of the great geologists of his century. In fact, Agassiz was destined to head Harvard's geology department while Guyot would be Princeton's chief geologist. When Agassiz and Guyot were 41, and both were established European geologists, their continent suffered a failed revolution. The two men left for America.

The year 1848 has been variously called the *Spring of Nations*, the *Year of Revolution*, and the *Springtime of the Peoples*. But it was ultimately a dismal failure and resulted in a reactionary backlash against democratic reform. The failed revolution began in France, quickly spread to Germany, Poland, Italy, and the Austrian empire. Strikes, student protests, and open revolt resulted in tens of thousands killed in their effort to overthrow corrupt absolutist governments, negotiate independence for some nations (notably Hungary from Austria), and expand democracy and individual rights, especially for the growing working class. However, the only real progress to come from the year of revolution was the end of feudalism, freeing serfs throughout Austria's realm. Beyond that, the aristocrats destroyed the disorganized revolutionaries. The backlash was felt like a vice. Universities were suspended, including the school where Guyot taught. With the failed revolutions, and subsequent repression, tens of thousands of young, educated, liberal middle-class Europeans fled. The ultimate result of the 1848 European revolution was a huge boost to American development – for example, more than 30,000 *Forty-Eighters* settled in Cincinnati, Ohio, alone. Thousands of journalists, entrepreneurs, and scientists arrived in America, including Arnold Guyot, who was motivated by his job loss and by the persuasion of his friend Louis Agassiz, who arrived in North America some months earlier.

Guyot landed in America with no knowledge of English, no employment waiting, and half his life behind him. His friend Agassiz introduced him to the faculty at Princeton and Harvard, but it was not immediately helpful. Although Guyot was invited to occasionally lecture – which he did in French – for six years he struggled with English, teaching at various Massachusetts schools, working where he could.

Finally, in 1854, Guyot joined the Princeton faculty. He stayed until his death, 30 years later. His popular 1872 book, *The Earth and Man*, pioneered the social science of human geography and attempted to explain the customs, dispositions, skills, and intelligence of people around the world as the result of local geographic influences. The book was quite popular, partly because it affirmed the notion of the superiority of the white race and placed that belief in a scientific context. Climate and geology, from Guyot's perspective, kept Europeans from

---

i   Jean Louis Agassiz (1807-1873), Swiss-American palaeontologist, geologist, and glaciologist.

resembling "misshapen Hottentots." Guyot's next textbook, *Physical Geography*, published in 1873 and revised in 1885, was the most important geography text published at the time. It dealt with a wide range of material – from "the Races of Man" and the cause of monsoons, to the nature of volcanoes and the destructive power of earthquakes. Guyot's text is a good representation of university-level science in the late nineteenth century. It was accepted that the interior of the Earth held hellish heat, and Guyot proved this to his students through examples that included hot springs and volcanoes.

In trying to find a cause for volcanoes, Guyot began by separating volcanic action from more general forces which "uplifted the continents and depressed the basins of the oceans." He listed two areas of regular volcanic activity – along the edges of continents and among some chains of mountains, particularly near recently discovered deep ocean trenches. Mountain ranges were described by Guyot as due to the cooling and contracting Earth, "not to the heat of the interior mass." According to Guyot, volcanoes usually appeared on continents close to oceans because deep fissures from the Earth's contractions were found there. Sea water seeped into the fissures, became super-heated steam which built pressure until it blasted the overlying rock apart[134] And so again, we find a variation of the contracting-earth theory playing a role in forming our planet's mountains.

Terror and Force

Physical Geography was a popular textbook. Guyot held his students' attention, quickly moving from exciting images of geysers to vivid descriptions of earthquakes. Here is his example of an earthquake's terror and force:

> "The earthquake at Lisbon, Portugal, on the morning of November 1, 1755, one of the most appalling in its results, exhibits the nature of these commotions of the Earth's crust, and the phenomena attending them. The day was the festival of All-Saints, and the churches of the city were full to overflowing, when, at forty minutes past nine, a rumbling noise was heard like distant thunder, gradually increasing until it resembled the sound of heavy artillery. A faint shock was followed by a heavier one, and within six minutes 30,000 persons were buried under the ruins of the churches and other edifices; and 30,000 more perished before the end of the catastrophe.

> "The ground seemed to undulate like the waves of the sea, the surrounding mountains were seen rocking violently on their base, and broad chasms opened in the Earth and closed again. More than 3,000 people had taken refuge from falling edifices on a broad marble quay just built on the banks of the Tagus, when the sea, which had before receded, came back in a furious wave forty feet high and swallowed up the entire multitude; then rushing upon the city, it continued the work of devastation. Similar oscillations of the

The Mountain Mystery

sea were repeated several times; and when the commotion ceased, several hundred feet of water covered the spot which the quay had occupied.

"Fires, kindled in the fallen dwellings, spread over the scene of desolation, creating a vast conflagration which completed the work of destruction. The ground continued to be agitated for several weeks afterwards, and another shock occurred in December following."[135]

Guyot continued, describing the areal extent of the earthquake: "all western Europe was agitated; nearly all the cities of Morocco were destroyed" and, he added, the ensuing tsunami was even observed on the opposite side of the Atlantic, with the sea rising twenty feet in New York, Boston, and Caribbean harbours. It is interesting that Guyot chose to reference the Lisbon earthquake as his example of a shaking nightmare. For years, this geological disaster was an enigma to scientists. Guyot was long on description, rather short on explanation – the true cause of this killer earthquake is still not known, though we will see later that geophysicists recently proposed a surprising explanation.

Whatever the cause of the Lisbon earthquake, it showed what can happen when the planet is stirred and shaken. Professor Guyot and his nineteenth-century colleagues plotted three kinds of destructive motion. Wave-like rumbling was the most common and least destructive – Guyot surmised that the other earthquake movements might be "simply the result of various systems of waves intersecting

The Lisbon earthquake, from Guyot's 1885 *Physical Geography* textbook

one another." They are not, but science still had no understanding of the cause or propagation of earthquake waves, so it was a reasonable guess. Guyot noted that seismic waves were "like waves of the sea, spreading from a central point, like ripples produced by dropping a pebble into still water"[136] – an analogy still taught and an idea that suggested other types of earthquake motion were like interference patterns in a high school physics ripple tank.

Earthquake motion can include vertical waves which seem to act from directly below the surface. These can be so violent nothing resists their force. If the ground whips up, then spurs downward faster than the acceleration of gravity, casings from water wells can be left suspended for a moment while the Earth plunges. Another wave type Guyot described was a "rotary or whirling motion, the most dangerous, but happily the rarest of all." Here he used an example from a 1692 Jamaican event where "the surface of the ground was so disturbed that fields changed places, or were found twisted into each other."[137] It sounds like the work of a giant kitchen blender. This description must have been based on legends, almost two hundred years old when Guyot put it in his textbook. Such spinning defies physics and nothing matching a whirling of the surface is known to actually occur. Forces of nature were still mysterious in 1885.

To make sense of earthquakes, Alexis Perrey[i] compiled an earthquake catalogue. Perrey was a French seismologist, a pioneer in quantifying the power of earthquakes. He began by reviewing activity in French-occupied Algeria with his 1848 paper published in *Mémoires de l'Académie des Sciences de Dijon*. He documented his earthquake observations yearly until 1871, creating the world's first extensive earthquake database. He correlated events that he thought might trigger quakes – changes in air temperature and pressure, seasons, moon positions.[138] Perrey, searching for the spark that incited crustal upheavals, pored over his data, 10,000 earthquake and tremor observations,[139] and found a relationship between seismicity and moon phases. His statistics suggested tremors occur most frequently when the Earth aligns with Sun and Moon and when the Moon is closest in its orbit to the Earth. A full Moon was more apt to incite an earthquake, according to Perrey.

His ideas may seem quaint, but there is a possible link between the Moon and earthquakes. We know, of course, the Moon exerts a tremendous pull on the oceans, causing twice-daily tides as high as 16 metres (50 feet) at Nova Scotia's Bay of Fundy. The solid crust of the Earth also bulges regularly – about 25 centimetres, or one foot, as lunar gravity draws Earth's surface upwards. In addition to causing tides, the moon slowly drifts north, then south, of the equator on an 18.6 year cycle. Earthquakes are statistically more common in the hemisphere with the Moon overhead. Some correlation may exist between Moon location and earthquakes, but the numbers are not solid enough to convince modern scientists of a real cause-and-effect relationship. None of this is certain or

---

i Alexis Perrey (1807-1882), French seismologist.

regular enough to predict the next big earthquake. It is hardly advisable to pack the pets and kids in the van and head for safety on every suspicious lunar phase.

We know earthquakes relieve pressure built up on the crustal surface when masses of brittle rock crack under stress. One can bend a wooden ruler quite a lot, but eventually it will snap; similarly, the crust beneath our feet. Earthquakes only occur where the rock layers are solid, which is mostly within 100 kilometres of the surface. At greater depth, rocks usually deform without breaking. Instead, they are a rubbery pudding – they don't fracture, so at great depths, earthquakes are generally not possible.

It has taken a long while for us to piece together the cause of earthquakes, and we still can't predict the next one with accuracy greater than a broad guess. The Greeks assigned the job of making earthquakes to Poseidon, whom they also called the Earth-Shaker. But Poseidon, described as an older white male with curly hair and a beard, had chores that included serving as god of the oceans and as the gods' horse whisperer. Making earthquakes was apparently a mere past-time for him. Both Aristotle and later the Roman Lucretius had a scientifically speculative hunch that earthquakes were an aftereffect of winds that caused the collapse of mountains deep within subterranean caverns. By the nineteenth century, earthquakes were often linked to volcanoes, although there were many examples where the two were not closely tied. The Lisbon earthquake, so dramatically described by Guyot, is an example. Disastrous as Portugal's earthquake was, a similar, but little known earthquake, helped collapse a great European republic – and led to redrawn borders on the continent's political map.

Croatia, Interrupted

For over five hundred years, a tiny – but very wealthy – nation existed along the Mediterranean Sea, wedged between powerful and aggressive neighbours. Any of them could have easily swallowed the country, but it was an earthquake and tidal wave that destroyed the Croatian Republic of Ragusa, or Dubrovnik, as the Slavs called it. If you drive south along the Croatian coast, heading towards Dubrovnik, you are forced to spend fifteen minutes in Bosnia. There is a checkpoint on the border with Bosnia-Herzegovina. The afternoon I crossed in and out of Bosnia, a square-shaped border guard bent into the car. He had a gun strapped to his belt and he leaned far into the window. "You, no pictures here," he told me in a way that made me wonder how many days one might spend in a Bosnian prison for just one photograph of his bald round head. I agreed, dropping the camera to the floor of the car.

We entered the tiny coastal strip of Bosnia, drove past partially renovated blocks of bleak socialist-era hotels, a dozen restaurants, and in a few minutes, we passed another set of border guards and were back in Croatia again, nine kilometres closer to Dubrovnik. The story of how land-locked Bosnia, a

Time for a Shake

mountainous republic once ruled by a sultan, acquired this tiny piece of the Croatian coast includes the aftershocks of one of the world's great earthquakes.

Southern Europe is tectonically active. Here the continental plate of Africa is smashing into the European plate. The pressure and heat of two huge landmasses fighting for the same space has altered the history and geography of southern Europe. Hence the Alps, Mount Vesuvius, which smothered Pompeii, and the cataclysm which sank Atlantis (or at least, Minoan Santorini). Africa and its relentless pressure also helped cause the demise of the Republic of Ragusa. Today's Croatia includes vast inland plains smothered with maize and grains, separated from the ocean by the arid craggy Dinaric mountain range. But the country also has a thousand kilometres of coastline, peppered with a thousand islands, many mere specks, others supporting huge lavender and thyme farms linked together by the buoys of oyster farms. And it has Dubrovnik.

Long ago, the Greeks occupied villages along the southern part of the Croatian coast, an area sometimes called Argosy or Ragusa. To the Greeks, the region was mysterious and inspired some of the tales of Jason and his Argonauts. The Croatian island of Cres, known to them as Apsyrtides, is given as the place Jason and his girlfriend Medeas killed Apsyrtus, the girl's brother. Coastal Argosy has a weak link with the Argonauts, so the Renaissance Venetians called the area Ragusa, a name corrupted back into the word *argosy* once more and applied to a flotilla, or any impressive parade of boats. This fits neatly with the story of the sixteenth-century Croatian city Ragusa, a place renowned for its ships.

The Republic of Ragusa existed as an independent nation for almost 500 years, reaching its peak during the early Renaissance. The entire republic had fewer than 50,000 people, so maintaining independence from Venice, the Ottoman Empire, the Austrians, Hungarians, and other opportunistic giants involved extremely clever diplomacy, which the Croatians linked to trade. The tiny republic had the

Ragusa Republic's capital, Dubrovnik

second largest merchant, or trading, fleet in the world. Ragusian merchants bought from the enemies of the Ottoman Empire, passing goods to them which wouldn't otherwise be available. It was a lucrative but precarious position between foes. Ragusa exploited its fortuitous location for centuries. Potential competitors – Ottomans, Venetians, Spaniards, English – were not keen to disrupt the flow of goods the Croatians supplied.

Ragusian ships sailed and traded around the world under a white flag with the word *Libertas* prominently displayed. Without an army, they established settlements and embassies from India and Morocco to the Americas. Wealth from shipping, buying, and selling financed one of Europe's first governments ruled by a senate. It was a progressive nation. Europe's oldest pharmacy, started in a Dubrovnik Franciscan church, has been in business since 1311. The republic had free communal fountains that piped fresh water underground 20 kilometres from hillside springs. Establishment of a retirement home and orphanage, plus the abolition of slavery four hundred years before the English, and the welcoming of displaced Iberian Jews in 1493 were enlightened decisions. The Republic of Ragusa, in its 418th year, was the first country in the world to recognize the United States' independence from Britain. Dubrovnik was rich, tolerant, and diplomatically astute. It reached its peak of prosperity about the time an earthquake flattened the small city-state.

Destruction on April 6, 1667 was nearly total. Every stone building collapsed, except the Sponza Palace and the Rector's House. The city's rector, Simone Ghetaldi, was among the 5,000 dead. The earthquake began with a dull rumble shortly after sunrise. Survivors described an eerie crashing and banging that at first caused walls and buildings to wobble, then crash. From Mount Srd, looming on the city's eastern edge, boulders were dislodged and hurtled down hundreds of metres. Rocks and dust darkened the sky, adding to the confusion that saw people flee the port, then return. Tsunami waves forced the Adriatic to retreat then rage back several times, tossing ships into the partially crumbled city seawalls. Fires from hearths and bakeries spread more destruction, including the loss of centuries of art treasures. Witnesses claim that the confusion, darkness, and fire helped "swell the numbers of robbers, many of whom, incredible as it may seem, were from the rank of the nobility."[140]

Fires burned out of control for twenty days – the earthquake had crushed the underground system that brought water into the city. The disaster struck during the long Ottoman-Venetian Cretan War, and both sides made a move to grab the injured port city. But they were foiled by Dubrovnik's diplomats, who once again were able to play the sides against each other. However there were not enough survivors to rebuild Dubrovnik. The Republic struggled for twenty years, then finally ceded the area around Neum to the Ottomans as a way to keep their country free. The Ragusan Republic agreed to the Treaty of Karlowitz, giving up a small chunk of coast it had held for 300 years, allowing that little piece of coastline to enter the hands of the Ottoman Empire. This strategically placed Turkish troops between Dubrovnik and the rest of the Croatian coast, which was being slowly consumed by Venice. Ragusa felt more secure having the Muslim Ottoman Empire as their coastal neighbour to the north than having the Venetians adjoining their lands. In addition to security, the Ottomans also paid gold in exchange for that narrow strip of coast. Dubrovnik could finally rebuild after the horrific earthquake. The Ottomans gained a link to the ocean for their land-locked Bosnian sultanate. For thousands of visitors today, the exchange made three hundred years ago means two extra border stops and an extra country to pass through on the trip south to Dubrovnik.

Why the Shake-ups?

When Guyot published his textbook about earthquakes and volcanoes in 1873, he stated that no part of the globe is absolutely free from such disasters, but there are regions where convulsions are more numerous and violent. Those areas, he said, were within the Earth's two great volcanic zones – the coastal regions of the Pacific Ocean, and the transverse zone separating the northern and southern continents. For him, with roots in Europe, Italy's volcanoes and the various Greek, Turkish, and north African earthquakes were the "transverse zone" that vaguely separated north from south. He could not know the commonality between the Pacific Ring and Mediterranean is the restless wandering of the Earth's crust. Without understanding the hidden mechanism, Guyot correctly pointed out that earthquakes and volcanoes are most intense along "the great fractures of Earth's crust."[141] He also rightly avoided concluding that one is the cause of the other. Earthquakes and fractures certainly swarm together, but which is cause and which is effect? "The two sets of phenomena may have a common cause, but must not be confounded or considered as necessarily belonging to the same class," Guyot said. He also pointed out that volcanic eruptions often take place without earthquakes. Many severe earthquakes occur in regions far removed from any active volcanoes. Although reluctant to promote any single mechanism, the nineteenth-century geographer did have some interesting thoughts.

Guyot wrote, "Within the tropics, earthquakes are most frequent in that part of the year in which the greatest atmospheric disturbances take place. They are most dreaded at the beginning of the rainy season, when the monsoons are changing their direction. In the Molucca Islands, the inhabitants, at this period forsake their houses for greater safety and shelter themselves under tents or the lightest bamboo structures until the danger is past." Guyot repeated the ideas advanced by Perrey that the moon plays a role in earthquakes, and he also noted that another scientist – Rudolf Wolf – found a direct link between the frequency of sunspots and earthquakes. Today such links seem tenuous.

But it appears that Guyot was not suggesting a direct cause-effect relationship between sunspots or monsoons and earthquakes, but was simply pointing out some interesting apparent correlations. Regression analysis – a math tool with the power to separate dependent from independent causes within a jumble of data – had not yet achieved a level of sophistication that could allow a proper perspective for these observations. Guyot admitted that no satisfactory explanation for earthquakes had yet been proposed. His own best guess was "the cause may possibly be found in the constantly increasing tension produced in the Earth's strata, by the steady cooling and contraction of the heated mass enclosed by the hardened outer crust."[142] In other words, the Earth's turbulence was due to cooling and shrinking.

9

Clever New Ideas

We have already seen that Lord Kelvin insisted the Earth is a cooling, shrinking planet of modest old age. In his youth, Kelvin derived heat conduction equations to estimate the Earth is roughly one hundred million years old. As Kelvin himself aged, he reduced his estimate. Late in life, the physicist was championing a time 20 million years past as the planet's time of origin – yesterday, geologically speaking. However, nearly all geologists were convinced the Earth was much, much older. But Lord Kelvin's reputation as a physicist was formidable. Few dared publicly question his formulae or calculations. Unfortunately, the basic premise for all his computations was wrong.

In 1895, Kelvin's long-suffering lab assistant, John Perry[i], published a paper challenging his elderly master's estimate. Perry had created a revised model of the Earth, one that changed the way scientists were thinking about the hot mantle below the surface. His work indicated a very old Earth. Perry, like Kelvin, was an Irish-born engineer and mathematician at the University of Glasgow. Perry became a professor of mechanical engineering and studied the way heat travels out of materials. His field of study was similar to Kelvin's, but Perry had more practical training and had quite a different idea about the nature of the inner Earth.

Perry assumed part of the planet's inside was fluid rock. In such a case, hot mantle would circulate in currents which, in turn, would carry heat away from the Earth's centre. Kelvin, on the other hand, insisted the entire inner Earth was solid. Solids dissipate heat much differently than fluid materials. Perry's convection, compared to Kelvin's conduction, could deliver a lot more heat much more quickly. On the basis of Perry's calculations of the convection flow and the rate of the planet's cooling, Perry concluded the Earth was two billion years old. He took his numbers into Kelvin's office. The great man told Perry that the numbers were wrong, the work was faulty. But John Perry published his results in *Nature*, anyway. Perry's maths were right, his assessment of the inner Earth's construction was correct, and his conclusion solved a great enigma for geologists. But his paper was ignored by the scientists who should have embraced it, largely because it showed the great Lord Kelvin had erred.

---

i   John Perry (1850-1920), Irish engineer and mathematician.

The Mountain Mystery

Perry recalled that in his discussion with Kelvin, he told the man that the assumption of a fully solid Earth was wrong. But Kelvin continued to believe the planet had no fluid layers and insisted the Earth's rocks would never flow in a billion years. "And yet," Perry said, "we see the gradual closing up of passages in a mine, and we know that wrinkling and faults and other changes of shape are always going on in the Earth under the action of long-continued forces. I know that solid rock is not like cobbler's wax, but a billion years is a long time, and the forces are great."[143] Perry had hit upon the idea of convection currents rising from deep within the Earth, gradually moving viscous rocks, carrying heat energy as they flowed. Historians feel Perry has not received due credit for this discovery[144] – which is fundamental to the origin of mountains – nor for his awkward stand against Kelvin, a move that could have led to his scientific obscurity. And perhaps did. Lord Kelvin was wrong and would prove to be doubly wrong. The second bit of evidence that helped determine the Earth's age came from the newly discovered physics of the atom. It was just a year after Perry published his paper and the consequences of the new science would set the universe on end.

The Old Bird Sits Up

Late in 1895, Wilhelm Roentgen discovered X-rays, quite by accident, while passing electricity through various gas-filled tubes. He found what he first called "invisible light" given off from one of the tubes. The rays passed through heavy paper and flesh, but not metals or bones – for example, he made a delightful photographic plate of his wife's boney fingers and wedding ring, exposed to X-rays. Within months, hospitals were using his discovery to examine broken bones and to plan surgery following gunshot wounds. Also within months, Henri Becquerel, a French physicist, expanded Roentgen's method of creating invisible rays by examining fluorescent minerals. Such minerals absorb light, then glow with a different colour as new light is released at a different wavelength. In preparation for an experiment, Becquerel stored a few rocks in a closed desk drawer, atop a wrapped photographic plate. The fluorescent rocks contained uranium. Instead of exposing the mineral to sunlight (he said it was too cloudy that day in Paris) he impatiently skipped the loading stage and developed the plate. He found intense streaks on the plate in the same pattern as the overlying rocks. He realized the mineral was emitting "invisible light" on its own, in a manner similar to X-rays. Becquerel had discovered radiation.

The Curies were also experimenting with uranium, repeating Becquerel's procedures. Marie Curie named the phenomenon *radioactivity*. In 1903, her husband Pierre Curie discovered that radioactive decay releases heat. Lots of heat. Geologists realized Lord Kelvin's idea that the Earth was steadily cooling from a fixed initial temperature needed to be modified. They found the Earth's subsurface materials include much radioactive material – radium, thorium, uranium,

Clever New Ideas

potassium, among others – all generating heat. This meant the Earth was much older than Kelvin calculated. You can quickly grasp how this works if you imagine a pot of cooked oatmeal on a stove. If Kelvin were in your kitchen, he would say the heat is off and he would easily guess oatmeal had been cooked an hour earlier. But Pierre Curie would notice the flame was still on, so the oatmeal might have been cooked a week earlier and still be warm. If our math is up to it, we could calculate backwards and determine when the cereal had been started. So it is with the Earth. It is cooling from its original cooked state, at a predictable rate. But our stove is still on, a nuclear fire is burning, making Kelvin's age-of-earth dating entirely wrong. It took a brave future Nobel Laureate to publicly point this out to Lord Kelvin.

Ernest Rutherford[i] is usually described as the father of nuclear theory because of his radical idea that atoms have a nucleus surrounded by an indeterminate cloud of electrons. Rutherford's model explained much that was mysterious about the substance of existence. It earned him the respect of the science community, and a Nobel Prize. Rutherford, a farm boy from New Zealand, did his most important work at McGill University, in Montreal, Quebec. He worked there for nearly ten years, conducting research that described *"The Theory of Atomic Disintegration,"* the title of his seminal 1902 paper. While at McGill, Rutherford also discovered alpha particles and named the deadly gamma rays. It was within this context that Rutherford was invited to address the Royal Institution, in London, in 1904; and confront Lord Kelvin with new evidence of an old Earth.

In London, Rutherford lectured the assembly of distinguished scientists about his research in atomic structure and radiation – and also about the implications of radioactive heat and the age of the Earth. He was a rising star, age 33. He knew that his statements about the age of the Earth would offend the stately Lord Kelvin who, at age 80, had steadfastly held onto calculations that had become untenable. Rutherford's idea that radioactive heat continually warms the planet proved Kelvin's age estimates were seriously wrong. He and other scientists estimated the Earth was at least a billion years old. Rutherford anticipated Lord Kelvin would be in attendance at the presentation:

> "I came into the room, which was half dark, and presently spotted Lord Kelvin in the audience and realized that I was in for trouble at the last part of the speech dealing with the age of the Earth, where my views conflicted with his. To my relief he fell fast asleep but as I came to the important point, I saw the old bird sit up, open an eye and cock a baleful glance at me! Then sudden inspiration came, and I said Lord Kelvin had limited the age of the Earth, *provided no new source of heat was discovered.* That prophetic utterance refers to what we are now considering tonight, radium! Behold! The old boy beamed at me."[145]

---

i Ernest Rutherford (1871-1937), New Zealand-born Canadian-British physicist.

The Mountain Mystery

Rutherford's radioactive heat is constant and almost ubiquitous within the Earth's mantle and crust. Radiation fuels the stove that keeps our planet cooking. Earth, as Kelvin declared, was cooling. But Rutherford showed that it was cooling very slowly because of the constant additional heat of atomic decay. We are still not certain of the exact amount of extra heat from the breakdown of the elements below the surface, but it is a lot. In addition to the slow peaceful disintegration of potassium and uranium, some scientists have speculated there are presently violent nuclear chain reactions also taking place deep inside this planet.

Recently, physicists Rob de Meijer and Win van Westrenen have suggested that uranium, being a heavy element, may sink and concentrate at the base of the mantle and then engage in a self-sustained nuclear fission reaction.[146] Like a bomb. Various studies have estimated that the amount of heat from such a sustained fission would be between 3 and 5 terawatts.[147] Each terawatt is one trillion watts. That's the energy of ten billion old-fashioned 100-watt light bulbs. It may sound a bit disturbing that a nuclear reaction – or splitting of uranium – is going on under foot, but this shouldn't be any more disturbing than knowing that just a few kilometres below the soles of our shoes, temperatures are hot enough to melt steel. If the uranium fission idea should happen to be true – if there really is a monstrous reaction happening below the Earth's surface – it won't be the first time a nuclear chain reaction was sustained by nature on our planet.

In 1972, French physicist Francis Perrin[i] discovered a spent uranium nuclear-reaction site in Africa. The hypothesis that uranium could spontaneously initiate a self-sustaining nuclear reaction had been worked out on paper by physicist Paul Kazuo Kuroda[ii] in 1956.[148] Kuroda was largely dismissed as a crank. He recognized this, saying, "Scientists were saying if this idiot is an indication of the program at the University of Arkansas, there must be nothing there at all."[149] For decades, Kuroda lived with the rejection of his idea of spontaneous nuclear reactions – it was universally accepted among physicists that the only time sustained reactions had occurred on Earth were under artificial conditions – in Enrico Fermi's lab, or at Hiroshima, for example. But Kuroda described the geological conditions which might allow uranium to sustain a natural chain reaction; Perrin found those conditions in Oklo, Gabon. The reaction at Oklo lasted intermittently for hundreds of thousands of years. During its meltdown period, an average of roughly 100 kW of power was output. That's not much, but it was emitted steadily for a very long time.

Near the surface of a hillside in what we now call West Africa, wedged between thick layers of sandstone, was a concentrated ore vein of natural Uranium-235, an unstable isotope more abundant when the Earth was young. A combination of heat, pressure, and hot water triggered the reactions. These finally died out after the concentrated U-235 isotope was converted to nuclear energy and

---

i   Francis Perrin (1901-1992), French physicist.
ii   Paul Kazuo Kuroda (1917-2001), Japanese-American chemist.

Clever New Ideas

stable derivative elements. It all ended nearly two billion years ago. The theory among physicists such as de Meijer and van Westrenen is that something similar is currently brewing deep within the planet.

Whether or not nuclear fission chain reactions are heating up the inner Earth today, the continuous slow decay of dispersed radium, thorium, uranium, and potassium emits enormous energy. Even on the surface, we are surrounded by powerful radioactive elements. In a typical section[i] of farmland, the top 24 inches (about two-thirds of a metre) has roughly two tonnes of highly radioactive uranium salted through its dirt. That's just three parts per million, but there is a lot of soil in a farmer's section. The same farm would also have a nearly equal amount of radioactive Potassium-40, along with some thorium and radium. Altogether, there are 1,300 billion becquerels, or atomic disintegrations each second, just within those first two feet of soil on a single section of cropland. Small wonder genetic mutations take place. Some of these natural radioactive materials are even more common deep within the mantle, where they keep the planet's inner parts cooking.

Growth of the Expansionists

A combination of radiation and trapped primordial heat keeps a big part of the inside of our world melted. And perhaps all that heat expands the size of our planet. Thermal expansion was once a popular idea for explaining the growth of mountains – and indeed, the concept still has some enthusiasts. But even in the day of Charles Lyell, the Scottish lawyer-geologist, some scientists promoted the idea that heat forces rocks to expand and build mountain ranges. Lyell, in his influential *Principles of Geology*, calculated that 500 metres of a typical mountain's height is simply due to heated, expanded rock.[150] Not enough to make an Alp, but enough to cause geologists to consider the role played by thermal expansion.

Among the first to build a mountain range out of Lyell's molehills was the Italian amateur scientist Roberto Mantovani.[ii] In 1909, he published his theory that just as volcanic eruptions can rip apart islands (as he witnessed when he lived on Réunion, in the Indian Ocean), perhaps entire continents can also be split apart. Mantovani explained that "a slow expansion of the Earth" due to thermal distension broke up an ancient supercontinent, a single landmass that he said once covered the entire Earth. Expansion, he advocated, forced new continents to drift away from each other in rift-zones which became oceans.[151] He even constructed a map of the original ancient continent before its rending by heat, pressure, and expansion. But speculating about the original position of the Earth's continents

---

i A *section* of land is a legal description for a square-shaped property of 640 acres (259 hectares), with a length of one mile on each side.
ii Roberto Mantovani (1854-1933), Italian violinist, diplomat, and geologist.

was a hobby. Mantovani's real work was music.

The professional violinist was part of an orchestra sent to perform on the French Réunion Island in 1878. Mantovani tramped around the island's unsafe shore near the town of Saint-Denis, observing huge volcanic fractures along the Indian Ocean coastline. The island, he believed, was cracking, ripping apart. He extrapolated his observations to a global scale, arguing that the same volcanic process could have fractured continents. He came to Réunion for the music, but was fascinated by the warm smokey island, so he stayed. Mantovani became Italy's Consul. His family joined him. They remained for years, Mantovani working as a diplomat and continuing his study of the island's unusual geology. He noted some of his initial geologic inferences in Réunion's *Bulletin des Sciences et des Arts*. But after a few years, he abandoned the little island and moved to northern France, where he managed a music school.

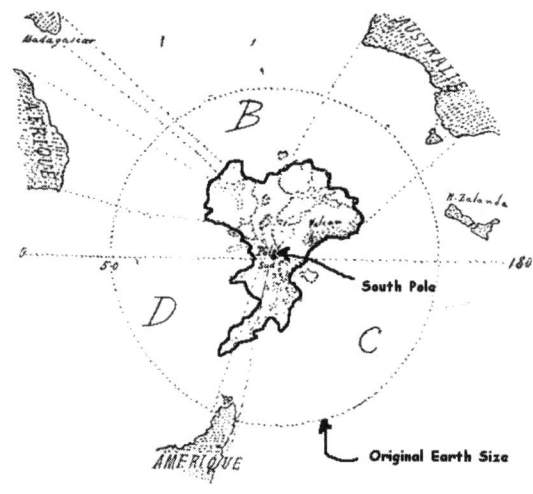

Mantovani's expanding Earth

It was in Normandy that Mantovani wrote his most famous paper, in 1909, and published it in a popular magazine, *Je m'instruis*. He presented his map of the Earth's supercontinent, noting that geological formations in the southern continents were too similar to have been coincidental. Mantovani was convinced of an Expanding Earth Theory, as was Charles Darwin, albeit briefly. During his voyage on *HMS Beagle* in 1835, Darwin made notes about an expanding-earth model as a mechanism for mountain-building. For him it was a short-lived theory. It arose from his observation of the way beaches in Patagonia were raised in steps. As Darwin made more observations, he changed his mind about that theory. Darwin, however, was in South America when a powerful earthquake lifted the coastline two metres.

He was told that such earthquakes occurred about once a century. From his observations, measurements, and the reports he heard, Darwin extrapolated that the Andes had formed in one million years. This supported Lyell's idea of gradual uniform change in contrast to the catastrophists who suggested the entire South American mountain range could have been put in place overnight, in one enormous event.

As Roberto Mantovani saw it, expansion theory satisfied some of the observations which had long held the attention of geologists. In his model, heat

Clever New Ideas

caused the Earth to swell considerably from a primordial small world into a less dense, larger globe. Heat could expand the planet. There have been, however, other scientists who have suggested the Earth is expanding due to quite a different cause. They contend the planet is gaining mysterious mass from the cosmos due to some undiscovered law of physics, suggesting the planet has been slowly growing ever larger.

One such idea was proposed in 1888 by Ivan Yarkovsky[i]. He explained how the invisible aether was morphing from energy into matter within the Earth. This rather clever idea came to the civil engineer while he was in the employment of a Russian railway company. Yarkovsky's theory included a mechanical explanation for gravity and the idea of transformation of energy into matter. It was a new physics, presented twenty years before Einstein derived a more elegant formula for mass and energy interchange. If a Swiss patent clerk could flip physics on its head, why not a Russian railway engineer?

But Yarkovsky's ideas were not given much attention until recently when his unrelated work on thermal radiation and asteroids was rediscovered by others and incorporated in the Yarkovsky effect, which describes how solar energy causes small asteroids to unpredictably shift in orbit. Also known as the YORP effect,[ii] the idea invited revisiting all of Yarkovsky's unorthodox ideas, including the absorption and transformation of aether inside the Earth into new chemical elements. Aether-energy transforming into mass seems strange, but was also independently considered by Nikola Tesla who studied the idea at great length, but was unable to prove it. Huge volumes of mass sucked out of an invisible – and likely non-existent – aether have never been observed; nor has any appreciable bumping up of the Earth's mass from forces acting somewhere within the core. Nevertheless, the Expanding Earth Theory has survived for a hundred years with some notable physicists advocating it as the cause of mountains on the planet's surface – some less notable geophysicists are still suggesting this idea. For that reason, we will return to the inflationary Earth concept a bit later. First, we need to look at the Earth's wrinkles and bumps from the perspective of the early twentieth century. The 1906 San Francisco Earthquake focused a lot of attention on the forces that shake and crumple the planet.

Shaking in California

Possibly because it lies so close to Hollywood, California's San Andreas Fault is the most famous earthquake zone in the world. Certainly it is the only fracture

---

[i] Ivan Osipovich Yarkovsky (1844-1902), Belarusian engineer.
[ii] YORP stands for Yarkovsky-O'Keefe-Radzievskii-Paddack. It was coined in 2000 by David Rubincam, an American geophysicist, to recognize some of the pioneers in the physics of the way the sun's solar energy radiated off asteroids, causing them to spin, twist, or otherwise change their orbits due to solar-induced changes in momentum.

zone most people can name. Despite its famous and fierce reputation, the San Andreas is almost always quiet. I hiked along the scarp of the fault in the desert just north of Palm Springs with some experienced geologists. There was no huge gaping hole in the ground, no sudden offset that the inexperienced, such as I, could point out. In fact, at that location, I did not notice the fault at all until a geologist revealed it as a bit of a ridge, which some of us proceeded to climb, thinking it was unlikely to toss anyone off its ledge that particular morning. It didn't. On one side, North America was occupying the ridge and the entire continent beyond; on the west side was a sliver of land that had been moving north for millions of years from somewhere beyond the equator.

The last truly devastating shaking along the San Andreas was the 1906 earthquake which destroyed San Francisco – 3000 dead; 225,000 homeless. In southern California, the last similarly powerful episode was in 1857. There the fault ruptured from central California to San Bernadino. Still farther south, near the Salton Sea, more than 320 years have passed since the last significant earthquake. As more time passes along the San Andreas – 110 years, 150 years, 320 years since the various violent shakes – it becomes more likely that a big one is imminent. A bit of stress has been released in the recent past: 1989; 1992; 2010; 2012. But the strongest of those was not even close to the energy involved in the 1906 San Francisco earthquake. The Southern California Earthquake Center predicts that the chance of having a 6.7 magnitude earthquake in California within the next 30 years in 99.7 percent. In other words, they are almost dead certain. Such an earthquake has enough power to level buildings and will happen within the lifetime of most Californians.[152] At present, even crude guesses of a future earthquake's location or date are impossible. Despite all that has been learned about the cause of earthquakes, prediction remains elusive. Even a few minutes' warning evades us. The Great San Francisco Earthquake could be repeated tomorrow with the same lack of prediction and with similar loss of property and lives as it had in 1906.

Science has progressed far from Aristotle's teaching that strong winds blew through caves inside the Earth, causing earthquakes. In his part of the world, earthquakes seemed centred around breezy sea caves, clefts, and caverns. By 1760, John Michell[i] was accurately teaching his Cambridge students that earthquakes were caused by shifting rock far below the surface. But it wasn't until 1891, when an earthquake ripped across the island of Honshu, Japan, and left vivid fractures with fault blocks lifted as high as houses, that anyone proved earthquakes and faults are related. The Japanese scientist inspecting the damage, Bunjiro Koto,[ii] was the first to realize fracturing of deeply buried rock materials caused earthquakes.[153] Twenty-five years later, San Francisco convinced American geologists that Koto was right.

---

i   John Michell (1724-1793), English geologist.
ii   Bunjiro Koto (1856-1935), Japanese geologist.

Clever New Ideas

The day after the 1906 California earthquake, Chicago geologist J. Paul Goode said the earthquake was due to the growth of the nearby Sierra Madre mountains, suggesting their uplift caused nearby blocks of rock to slip. Combined with studies by C.W. Hay of the Geological Survey, earthquake mechanics were becoming clear. Hay likewise blamed subterranean adjustments between the solid and molten parts of the interior. "The outer crust is solid, but after you get down sixty or seventy miles, the rocks are nearly in a fluid condition owing to great pressure. . . as the crust cools it condenses, hardens, and cracks, and occasionally the tremendous energy inside is manifested on the surface."[154]

Grand Canyon Poet

Just before the San Francisco earthquake, Major Clarence Dutton[i] wrote the best textbook about earthquakes up to his time. Dutton was a gifted writer and his book, *Earthquakes in the Light of the New Seismology*, does not disappoint. It reads like a novel and includes this passage about the earthquake experience:

> "When the great earthquake comes, it comes quickly and is quickly gone . . . a matter of seconds. . . The first sensation is a confused murmuring sound of a strange and even weird character. Almost simultaneously loose objects begin to tremble and chatter. Sometimes, almost in an instant, the sound becomes a roar, the chattering becomes a crashing. The rapid quiver grows into a rude, violent shaking of increasing amplitude. Everything beneath seems beaten with rapid blows. . . Loose objects begin to fly about . . . the shaking increases in violence. The floor begins to heave and rock like a boat on the waves. The plastering falls, the walls crack, the chimneys go crashing down, everything moves, heaves, tosses. Huge waves seem to rush under the foundation with the swiftness of a gale. . ."[155]

Dutton continues this description until the house is gone. "Or," says Dutton, "suppose we are out in the country and the earthquake comes suddenly upon us. The first sensation is sound, a strange murmur. Some liken it to the sighing of pine trees in the wind; others to the distant roar of the surf; others to the far-off rumble of the railway train; others to. . ." Well, you get the idea. Major Dutton, the colourful writer of geology, was later called "The Poet of the Grand Canyon" by novelist and Pulitzer winner Wallace Stegner, who said, "Dutton first taught the world to look at that country [the American west] and see it as it was."[156]

Dutton, a Yale graduate from New England, fought at Fredericksburg in the American Civil War, then spent much of the rest of his life in the American West as a geologist. He was head of the United States Geological Survey's Volcanic Geology Division – a role that placed him on volcanic mountains in California

---

i   Major Clarence Edward Dutton (1841-1912), American geologist and army officer.

and Hawaii. His exploits included sounding the enormous depth of Crater Lake in Oregon. That was not an easy task – it is the deepest lake in the States, but reaching its remote mountainous shore with exploration equipment was nearly impossible. His team – 35 soldiers with 65 horses and mules – carried a half-tonne boat up the lake's surrounding mountain rim, then carted it down a steep twenty-storey cliff to the water's edge. Finally they lowered their dinghy onto the cold water. Soldiers maneuvered the survey craft around the lake while Dutton plumbed the icy depths with heavy weights tied to piano wire. He needed those heavy weights to keep the wire straight as it unfurled to the bottom of the lake. He made almost 200 measurements. A plot of the depths revealed the throat of a caldera, created by a violent volcanic eruption which natives described through ancient legends.

Klamath and Modoc aboriginal groups had lived and traded around the Oregon lake for hundreds, possibly thousands, of years. The Klamaths had an oral history that described the birth of Crater Lake as a mystic event that included a war between sunny good forces above and evil dark forces within the Earth. A fight between Skell, the good, and Lao, the bad, included the legendary virtue of a human woman. The Klamath myth tells of a huge explosion which ended the good force's rule over the woman and her earthly lands. We now know that the explosion of a volcano created the lake's caldera in an event which would have released as much energy as one thousand atomic bombs. When the volcano of Mount Mazama erupted, just under eight thousand years ago, the mountainside was populated. The tale of the loss of those people, the continual tremors, and occasional earthquakes that followed, has remained in oral tradition for all the

Dutton's Geological Survey on Crater Lake, 1886

Clever New Ideas

intervening thousands of years.

Crater Lake was once known as Deep Blue Lake, an apt name as it is the deepest lake in the United States, and the perhaps America's bluest lake. When Dutton explored its depth with piano wire and weights, he expected the bottom to be a long way down. But it exceeded all expectations. His team threw their first cable from shore. It disappeared two hundred metres into the water. Then four hundred. It kept descending, so they retrieved the wire to check for mechanical failures. It seemed to be working correctly. During July and August of 1886, the crew took three soundings each day from their boat. They surveyed in concentric rings, narrowing inwards while surveyors stationed on shore used line-of-sight plane tables along the crater's rim to map the boat's positions. The greatest depth took a lot of piano wire – over 600 metres, or about 2,000 feet.

The depth sounding points, plotted on a map of the circular lake, revealed a flat-bottomed caldera with cinder cones, mostly submerged, though one cone rose to form Wizard Island. Major Dutton of the United States Geological Survey had discovered that Crater Lake is a topless, water-filled hollow volcano. For half a million years, the volcano had grown through steady stages of minor activity to become a mountain. Then Mount Mazama blasted apart – expelling 50 cubic kilometres of volcanic ash and rock. The pulverized pumice covered millions of square kilometres with gritty dust and ash. On the opposite side of the Earth, the sky darkened in Mesopotamia where people had settled small villages and were learning to cultivate crops and fashion copper into tools. The Mount Mazama eruption lasted just a few brief moments, then the remaining centre collapsed. It took a thousand years for the hollow mountain to fill with rain water and melted snow. The blast which created the Oregon caldera was the most violent eruption North America had ever experienced.

Dutton, after mapping the extinct volcano, returned to his study of earthquakes. He saw two possible causes: The world's various earthquakes were either created by volcanoes or by who-knows-what. The latter being some unknown tectonic force. Dutton had a feeling that the restlessness of the Earth was related to a "layer of plasticity" beneath the planet's thin crust. Although most geologists in the 1880s doubted the existence of such a fluid inner layer, Dutton was so certain it existed he wrote, "Reasoning or induction scarcely enters into it, it is substantially an observed fact."[157]

He suspected that this fluid layer was involved in the ever-shifting crust – building mountains, folding and shearing strata, and lifting or dropping masses of rock. Dutton noted that ever since Hooke, the man with the microscope, earthquakes were thought to be a cause of mountain growth and faults in rocks. Dutton said the cause and effect were backwards – breaking rocks caused earthquakes; earthquakes were not the cause of broken rocks. "Just as thunder is an effect of the electric discharge, and not the cause of it," he wrote[158]

At the time, however, it seemed that earthquakes, with their enormous

113

destructive force, caused rocks to break and faults to appear. Quite obviously, earthquakes shear buildings and split streets. But Dutton was seeking the deeper cause, the force that actually generates earthquakes. This drive to understand the phenomenon really came to the country's attention after the San Francisco earthquake. Geologists decided to map, measure, photograph, and trace all the faults they could in order to identify possible future dangers. Andrew Carnegie[i] stepped forward with cash for the Lawson Commission when government money didn't materialize. The Commission's goal was to try to understand and hopefully predict future events. From their studies, the geologists discovered that the San Francisco quake, and similar ones, occurred along pre-existing fault lines. Particularly, John Hopkins professor Harry Reid,[ii] who served on the commission, recognized the basis for how we view earthquakes today. Reid called it the Theory of Elastic Rebound. He showed that the Earth's crust slowly bends until it is suddenly returned to an undistorted position with the violent snap that creates an earthquake. This scientist and others of his generation didn't realize that the source of the bending was the relentless pressure of moving crustal plates. That was a discovery for another generation.

At the time, before the idea of plate tectonics and continental mobility, Swiss geologists created *Einsturztheorie*, or Downthrow Theory, to explain how rocks broke and shook the surface. Downthrow is the principle that earthquakes arise from the collapse of the overlying surface. This is not true for the deadliest earthquakes – Reid's bending, snapping rocks more ably describe the cause of large earthquakes. However, the Swiss theory explains minor modern examples: pumping oil and gas, mining coal, and extracting potash have all resulted in thousands of tremors as rocks subside into man-made cavities. In the Los Angeles Basin, the removal of 9 billion barrels of oil over the past century has caused the city to sink as much as 8 metres (25 feet) in some places. One collapse due to oil production cracked an earth dam which then flooded Baldwin Hills and killed five people.[159] Similarly, a significant New Year's Eve earthquake near Youngstown, Ohio, in 2011, was due to oil field activity. In the Ohio case, no one died, but by fracking (fracturing) shale, rupturing it with corrosive chemicals, then vigorously pumping oil, at least ten earthquakes were spawned near one troubling well.[160] More serious than minor earthquakes and a bit of surface shaking is the threat of injection fluid leaking from drilling and development activities. Unpredictable downthrow earthquakes can fracture rocks and shift their layers, polluting aquifers. Fracking-induced earthquakes – even minor ones – can open escape paths for the poisonous fracking chemicals to enter vulnerable water paths.

---

i   Andrew Carnegie (1835-1919), is remembered as a ruthless industrialist, a steel-maker, an impoverished Scottish immigrant to the United States who became the world's wealthiest man. He should also be remembered for his huge contributions to science. The Carnegie Institution and his direct gifts built hundreds of public libraries and museums, and sent scientific expeditions around the world.
ii Harry Fielding Reid (1859-1944), American geophysicist.

Clever New Ideas

Measuring the Shakes

To understand and monitor earthquakes, their speed and intensity are measured. For years, the seismometer was the tool. Seismometers really came into existence with the work of James Ewing,[i] a mechanical engineer who worked at the University of Tokyo. Ewing arrived in Japan as part of Britain's enlightened sense of responsibility. At least that's how the Victorians saw their duty as they sent representatives abroad, modernizing and anglicizing the world. But in Ewing's case, at age 28, he was invited by representatives of Meiji Emperor Mutsuhito, the 122nd emperor of Japan, to work at a western-style university the Japanese and British governments supported. Most of the staff were British; the students were Japanese. Ewing spent five years in Tokyo, teaching magnetic and electrical theory and mechanical engineering. A proper Scottish gentleman, he was rarely seen without his white butterfly collar and blue bow tie. He seems to have appreciated the country, writing it was "venerable in its traditions, its art, its manners, its high standard of patriotism and personal duty; but almost painfully lacking in the veneer of western culture."[161] But that's why Ewing and his ilk were there, to paint a thin veneer of the western world upon Japanese culture. James Ewing humbly appreciated teaching the willing students: "To an inexperienced teacher, there was stimulus and help in pupils whose polite acceptance of everything he put before them was no less remarkable than their quick intelligence and receptiveness."[162] Ewing was smitten.

And shaken. Ewing experienced over 300 earthquakes during five years in Japan. As a physicist and mechanical engineer, he could not ignore the chance to study the incessant tremors. Ewing's great contribution was that he suspended a wobbling pen above moving paper, combining time with seismic vibrations.[163] This created a seismic wiggle chart, the seismogram still in use today. Its most wonderful attribute was that it could be standardized. Similarly calibrated seismometers scattered around the surface of the planet triangulate earthquake epicentres. Triangulation was achieved earlier, but the new system was much more accurate. It allowed tremor intensity at the earthquake centre to be inferred. In the broader field of earth science, the seismometer, with its time scale set to synchronized clocks, could also measure the speed of the shock waves.

By 1900, seismometers were sensitive enough to record seismic events thousands of kilometres away. On the right is the seismogram generated by the great San Francisco earthquake, recorded in Göttingen, Germany.

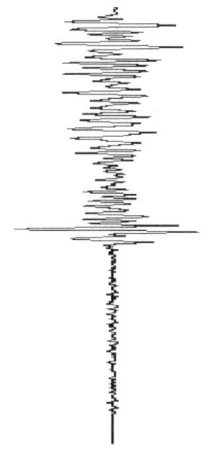

1906 San Francisco seismogram, recorded in Germany

---
i James Alfred Ewing (1855-1935), Scottish physicist and mechanical engineer.

The Mountain Mystery

Travelling at about eight kilometres per second, it took mere minutes for the San Francisco earthquake to show itself as a lively quiver on seismometers as remote as Wellington, New Zealand and Europe. When the shock waves arrived in Germany, the relatively calm background signal was interrupted by pressure waves of modest intensity, followed by more vigorous shear waves, seen on this actual seismogram, half-way up the chart. Shown here are the first 26 minutes of the San Francisco earthquake's activity. Not shown are the later ground-roll waves. Those surface waves were so strong, even 9,100 kilometres away in Germany, that they went off the chart's scale and couldn't be recorded.

The seismometer with its sliding paper and vibrating pen opened a window on the world beneath the crust. Soon geophysicists were probing lithospheric boundaries hidden hundreds of kilometres below the surface. Now the Earth had its own X-ray, as piercing as the type Roentgen used on human flesh. The first seismic ticker-tapes, sophisticated for their time, are crude by today's standards. But the basic principle remains the same.

There were other ways the ancients monitored the energy rumbling through the ground. Chinese seismoscopes were sometimes as simple as a water bowl, though often elaborately cast of copper or gold and mounted in a palace observatory. The vessel sloshed its water away from the earthquake's origin, leading to the general, but usually erroneous, impression that an earthquake is one large shock coming from a particular point, rather than a rip along a fracture zone. Such seismoscopes also gave a rough indication of the relative power of the tremor – an intense earthquake would leave more water on the floor than a softer tremor.

Sophisticated earthquake monitors, built by astronomer Zhang Heng[i] around 132 AD, dropped metal balls from dragons' mouths when the ground shook. Zhang's bronze seismoscope stood two metres high. Eight dragons slithered down the outside of the "samovar," each holding a metal ball, which could drop into the gaping mouth of an overweight bronze toad when the earth shook. It was apparently a hit with the court at the Han Dynasty palace and the kingdoms which followed – they kept the huge artifact for two thousand years. We are not completely sure how the royal court used their seismoscope. We can assume an earthquake's first shock wave would unbalance a

Zhang's seismoscope

---

i   Zhang Heng (AD 78-139), Chinese astronomer, mathematician, inventor, and statesman.

Clever New Ideas

central rod hidden inside, knocking it towards one of the dragons. Zhang's device was an artistic step forward, but lost the intensity information splashed from the sloshing water bowl. Instead, intensity information was gathered in the field by noting the destruction of villages, a method that persisted long after the Asian dynasties. The Chinese have stupendous experience in assessing such damage – no people have suffered more from earthquakes.

Shaanxi, the most deadly earthquake in history, killed over 800,000 people in 1556. Nearly all homes within one thousand kilometres of the epicentre collapsed. In some counties, over half the residents were dead within hours of the initial shock. Survivors reported that rivers changed course, the ground swelled up, forming new hills, or sank into new valleys, while streams burst out of the ground. The huge loss of life was mostly due to unstable housing. The homes, carved from soft silty soil, offered no resistance to the earthquakes that caused landslides and aftershocks. Those aftershocks rumbled for weeks.

But perhaps no series of severe earthquake aftershocks rivals those that lasted over a year in Venezuela. In 1766, the coastal city of Cumaná, Venezuela, was destroyed in minutes; the ground continued to shake destructively almost every hour for the next 14 months. With less vigour, but even more persistence, were aftershocks in southern Italy. Incredibly, after the destruction of Messina on the island of Sicily in 1783, the ground shook daily for ten years.

Some geologists during the nineteenth century believed earthquakes were caused by an accumulation of eroded sediment, piled so deeply near the mouths of rivers that the weight eventually snapped the rocks below, triggering tremors. But the greatest river deltas, the Mississippi, Amazon, and Nile, rarely have earthquakes while some of the worst quakes are far from any rivers. The idea was suspect, as was the idea that all earthquakes are due to the Earth's cooling and shrinking. Clarence Dutton knew it. He disagreed that earthquakes can be "attributed to the cooling of the Earth's interior and the slow, continuous readjustment of the cold outer crust to the shrinking nucleus. This is an old view and it received a remarkable development at the hands of Doctor Robert Mallet[i] in 1871. It was wrought out with such ability and attractiveness that for some years it seemed to have gained the adhesion of most geologists and not a few physicists."[164]

Mallet, whom Dutton disparages here, was a fabulously wealthy industrialist, owning the iron foundry that built Ireland's railways and bridges. He was also an engineer and is regarded as the father of seismology because of a fascinating experiment he and his young son conducted on a Dublin beach in 1849.[165] They dug a hole, buried a keg of gunpowder in it, then detonated the bomb. About a kilometre away, the father and son measured the energy wave that rippled through the sand, arriving at their primitive seismoscope. They were the first to show energy can travel in waves through the ground. It was the initial step in

---

i   Robert Mallet (1810-1881), Irish geophysicist.

understanding the sinusoidal wiggle of earthquake motion. Their experiment – with technical refinements – became the basis for modern seismic exploration.

Although Mallet performed the pioneering seismic experiment, Dutton felt Mallet's more general theories about the cooling planet were off the mark, "if the Earth ever was a uniformly heated globe, left in space to cool, the amount of cooling up to the present time has been inconsiderable. . . The Earth, excepting a thin outer shell, is as hot as ever, and the interior has not yet contracted at all."[166] Dutton also correctly pointed out that the contours and folds of the Earth are not geometrically aligned in any way that would be predicted by the cooling, shrinking-earth model. Dutton dismissed the notion that this could be the cause of either earthquakes or mountains. Instead, he proposed a way mountains might actually float high above the surrounding plains, achieving their airy heights in a way that resembled icebergs on an ocean. His idea, which he named isostasy, helped resolve a problem we will explore next – the hollow structure of the Himalayan mountains.

10

Hollow Mountains

By the start of the twentieth century, geologists knew what was under the Earth's skin – at least in a broad and simple way. An iron central core, a thin crust on top, slowly cooling rock between. A few geologists surmised that the sandwiched layer suffered enough heat and pressure to flow – not like water but rather like very, very stiff tar. They were right. We now call that zone of plasticity the asthenosphere. It starts just below our crunchy surface, the lithosphere, and extends several hundred kilometres below it. The unseen asthenosphere's effect on our daily lives is enormous. In addition to allowing the continents to move and thus promoting volcanoes and earthquakes, the asthenosphere keeps mountains balanced, their peaks pointing up. Mountains don't topple, splattering lakes and squishing villages. Nor do they sink under their weight. Intuitively, we might expect them to – most structures become unwieldy with height, or may disappear into the soil. The mechanism keeping mountains balanced and standing upright upon the asthenosphere is isostasy. Like an enormous iceberg looming over the sea, the largest part of a towering mountain is hidden below the surface, rooted in the asthenosphere. Isostasy keeps icebergs and mountains stable and upright.

The term isostasy was invented in 1889 by Major Clarence Dutton. He combined the Greek words *iso* (equal) and *stasia* (standing) to describe this "equal-standing" system of equilibrium. In Dutton's own words, "Isostasy is the tendency to maintain profiles in equilibrium, not to raise or lower them."[167] Isostasy is somewhat challenging to explain, but easily pictured by an example. Imagine a toy wooden boat floating in a tub of honey. If you press the boat down, not only do you drown the imaginary crew, but you displace some of the viscous honey and you feel pressure against your hand as the system tries to restore equilibrium. Remove your hand and the boat slowly rises. Stack some blocks on the boat and its top is higher, but its keel moves deeper. You have just demonstrated isostasy. Your boat floats and maintains equilibrium by displacing some of the thick honey. Geologists in the late 1800s realized isostasy explains the rise of mountains above the surrounding landscape. Heavy parts of the crust sink, lighter parts are held high. Mountain ranges have roots like the keel of our toy boat. And they are less dense than the asthenosphere, so they float upon it.

The Mountain Mystery

If we consider the iceberg again, its volume displaces some of the ocean's water. Under the iceberg, downward pressure is the same as the downward pressure below any water at the same depth. The iceberg towers high and its root is deep because the weight in a column of ice equals the weight of any water covering the same area, at the same depth. If ice were heavier than sea water, it would sink like a stone. Since it is lighter, isostasy forces part of it out of the water. If the sun melts a metre of ice from the top of the iceberg, it becomes shorter, but not by a metre because pressure from below buoys the floe up a bit. So it is with mountains, too. Even while suffering erosion, mountains maintain some of their former glory. This was confirmed from observations compiled by George Everest[i] who was asked to survey land around the highest mountains in the world.

Measuring India

After Britain had seized the Indian subcontinent, the British decided to measure their conquest. They began a survey to determine India's size, account for its resources, and map the entire area. This, as the new rulers explained, was for revenue and administrative purposes. The survey of India began in 1802 when William Lambton, a British soldier-surveyor, was placed in charge of the task. Lambton was chief surveyor for 28 years and died in India, still on the project. His assistant, George Everest, replaced him as Surveyor-General in 1830, a post Everest held for 13 years.

Everest's techniques were conscientious and thorough. Expensively so. He imported a famous telescope maker, Henry Barrow, to work with the surveyors' optical tools. In places, Everest built stone towers 15 metres high to give his men clear views above the trees and rolling terrain. Don't imagine that Colonel Everest lifted the stones himself – his field crew had 700 labourers, 42 camels, 30 horses, and 4 elephants. Even with an army of conscripted labourers, it was a frustrating task. The instruments (cables, chains, theodolite, telescopes) were temperature sensitive – on hot days, they would expand, stretching the measuring equipment, so the crew built shaded tables and moved them along the survey route. (For each degree the temperature rose, the chain lengthened six ten-thousandths of one percent – 12 metres in a 2,000-kilometre survey.) The survey lagged behind schedule because of monsoons, dust storms, searing deserts, jungles, and rugged terrain. Managers of the East India Company, a private monopoly licensed by the British government, thought the survey would take five years. The Great Trigonometric Survey of India, as the project was called, lumbered on for over 60 years before it was finally shut down. It drained so much investors' cash that the whole scheme was finally handed over to the deep pockets of the British government.

---

i Colonel Sir George Everest (1790-1866), Welsh surveyor.

Hollow Mountains

Ultimately, Colonel Everest measured the huge swath of real estate – from the southern tip of India to the Himalayas along the Nepal border in the north. He used two different survey methods. Everest favoured triangulation, reserving astronomical readings as a quality check. Astronomical measurements depended on alignment of telescopes to star positions. It would seem the easier task, but it took months, sometimes years, to find a location this way and the accuracy of the astronomical system depended on perfect clocks, cloudless nights, and infinite patience. Even then, kilometre-sized errors crept into the calculations, seemingly at random. Triangulation, the more accurate, but physically more gruelling, of the two systems, was used for most of the survey.

The Great Trigonometric Survey of India required perfectly straight lines, sometimes hundreds of kilometres in length. At each end of the line, angles were measured using a cumbersome one-tonne theodolite which focused on a third, rather remote point. Using trigonometry, the new third point's distance was calculated from the two known angles and one very accurately measured length. Hence, the distance to the next point was calculated, not measured. And the new line became the base to grow yet another triangle, measure new angles, and fathom a new point. In this way a series of triangles were constructed across India, pieces of a geometric puzzle. On a flat Earth, this is rather straight-forward,

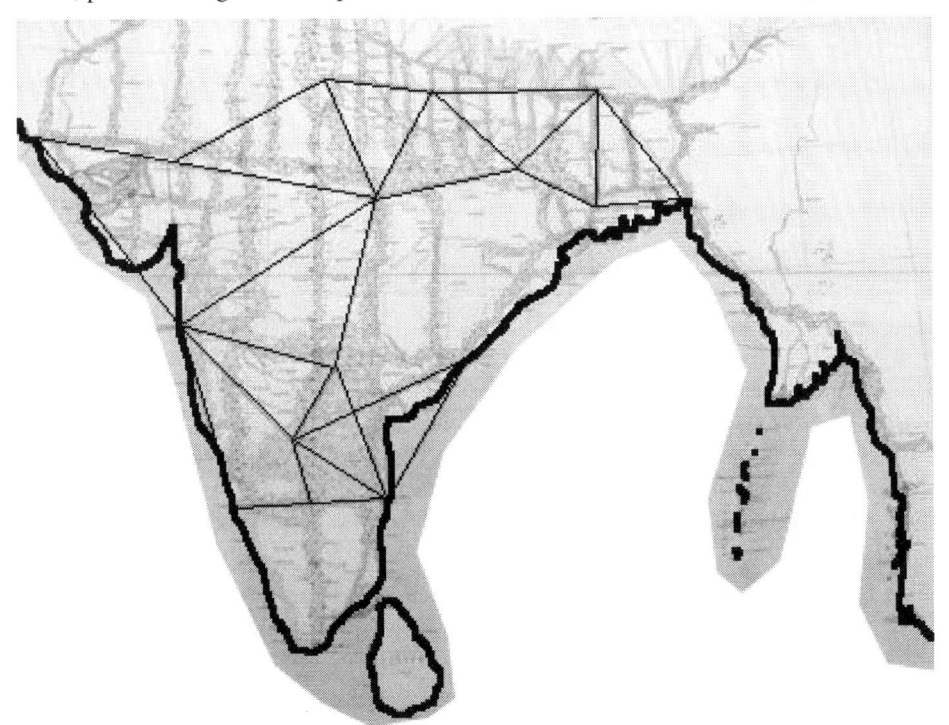

Triangles of The Great Trigonometric Survey of India, 1870

The Mountain Mystery

but tedious. On our spherical Earth, triangles have curved lines and their angles don't sum to 180 degrees. In addition, the planet bulges at the equator so there is not a perfectly spherical path to follow as one approaches Nepal from Madras. This trigonometry is challenging.

Corrections were made for the curve of the Earth, the bending of light through the scopes, the altitude gained and lost along the way, and daily changes in temperature. When the air was too hazy during the heat of day, the surveyors worked at night, using powerful lanterns, visible from 50 kilometres. Even gravitational pull had to be considered as the team neared the Himalayas. It was the sideways tug of gravity on the surveyors' plumb lines due to the massive mountains that yielded the most startling discovery during the survey. Gravity introduced an error into the entire project which Everest felt was unacceptable. Everest had done his best, but after tying a loop in his measurements, he was off 168 metres in his 2,400,000-metre study. Minuscule, but Everest had expected better. After decades in Asia, Everest retired to England. Eventually knighted, Sir Everest's successor named the world's tallest mountain, standing 8,848 metres above sea level, in his honour. But it was the 168-metre discrepancy that dogged Everest. The best scientists of the day decided the error was introduced from gravity anomalies while levelling the lines near India's northern border. The easiest way to eliminate the mistake was for geographers to assume that the great Himalayan mountain chain was hollow.

Hollow Himalayas

Plumb lines have freely hanging weights that help control alignment. In the case of surveying, they ensure straight lines are indeed straight. Near the northern mountains, the team realized the massive Himalayas should pull those plumb bobs towards the range. The geologists on the team took rock samples, and calculated the weight of the mountains. From those, they computed displacement at 15 seconds of one degree of arc. However when all their measurements were added, the error – Everest's missing 168 metres – was not fixed, it was actually worse. The mountains were not tugging on the plumb bobs as expected. The gravitational pull was just one third of what it should have been. It seemed as if the insides of the Himalayas were hollow. Not just a few caves; according to their calculations, two-thirds of the mountains' mass was missing.

Sir George Airy[i], the British Royal Astronomer, checked the calculations and presented an explanation to the Royal Society. Some months later, Archdeacon John Pratt[ii] at the Church of England's Calcutta office, an amateur geologist, used a different approach to also solve the baffling mystery. Although each used different methods, both Pratt and Airy dismissed the hollow Himalaya notion as

---

i   George Biddell Airy (1801-1892), British astronomer and mathematician.
ii   John Henry Pratt (1809-1871), British clergyman and mathematician.

Hollow Mountains

the least likely solution for the missing material. They had a hunch it was something else. They concluded that the reason the plumb bobs were not drawn towards the mountains as energetically as expected was based on the distribution of the mountain ranges' mass. This turned out to be more complicated than mere hollowness. Geologists were just beginning to realize that mountains have deep roots extending far below the Earth's surface. Continental rocks, like those that build mountains, are lighter than the dense rock below. The plumb bobs were being pulled towards the mountains by the towering masses, but the bobs were also being attracted in the opposite direction by heavier subterranean material opposite the mountain range. Lighter crust material in mountains floats above denser heavier rock. This is Clarence Dutton's isostasy again, though it was not at first recognized as such.

Sir George Airy realized that mountains include very deep and comparatively light-weight roots of varying depths, while John Henry Pratt thought the crust was of uniform thickness everywhere (he was wrong), but either way the missing gravity was found, saving the embarrassment of drawings of empty mountains in college textbooks. Pratt and Airy worked independently, their basic assumptions about the structure of mountains differed, and that, in turn, inspired quarrelling disciples (Prattites against Airyists) who would bicker over those differences for a generation. But together their work confirmed that all mountain ranges have deep roots that are lighter than underlying materials and their high peaks are balanced aloft, like icebergs, by isostasy.

The yielding asthenosphere below our crusty surface enables isostasy and also contributes to the Earth's distorted shape – flatter at the poles, wider at the equator. The Earth is no perfect ball, but as a first estimate of its shape, *spherical* sufficed for generations. The idea that the Earth is spherical, rather than flat as a pancake, was accepted by most scientists before the first century – Pythagoras had figured this out 600 years earlier and Eratosthenes measured the circumference, rather accurately, in the third century BCE. But on a practical basis, we live on a flat Earth. Our vision confirms a simple disc-like flat view. Things fall down; we stand up. For a hundred thousand years, this was enough to know.

Two thousand years ago, it was thought the Earth was a perfect globe – created *perfectly* spherical, the design of an expert craftsman. But Isaac Newton had the idea that the planet might bulge at the centre and be a bit flattened at the poles, due to its rotation. Therefore, he suggested, the Earth is not a perfect sphere. He called the squat shape an oblate spheroid and he proved its existence in 1687 within his wide-ranging book of physics, *Principia*. With this theory, the science of geodesy – the study of Earth's shape – was invented.

Later scientists refined Newton's model of our planet's figure. Geodesy has taken us from the Greek's perfect sphere to Newton's perfect oblate spheroid to today's deformed orb with highs and lows unevenly distributed around the surface. But you would hardly notice a difference on a drive around the planet. A

The Mountain Mystery

trip looping the equator would entail 40,075 kilometres. A north-south trip from pole to pole and back again is just 66 kilometres shorter. The perturbations that distort the oblate are even less obvious. In our ordinary experiences, it makes no difference that the Earth is not a perfect sphere, or that it is a pimply oblate spheroid. But whenever satellites are launched, our imperfectly-shaped planet forces complex calculations upon their orbits. An accurate measurement of the planet's shape defines satellite orbits more perfectly. This makes global positioning from satellites possible, improves street maps, makes communications more stable, and aviation safer.

To understand why the Earth is flat on top and bottom, but bulgy in the middle, imagine a whirling dervish, if you will. As the Sufi spins faster, his white gown expands away from the waist. The whirling Earth is similar, its waist puffs out at the equator. By the way, the mass of the Earth is so enormous and angular momentum is so great that if the Earth's energy of rotation could be harnessed, the power captured would light all the world's cities. Engineering students sometimes calculate the Earth's power of rotation during their first physics classes. Then they make proposals on how to convert that angular momentum into useful energy. Just applying a small brake – slowing the Earth down so each day stretches one second longer – would yield enough energy to power all of the globe's human activities. And we'd probably appreciate the extra second every day. All this power is simply from the energy of rotation, which is busy flattening the poles, bulging the Earth's belly, and (due to rotation's effect on tidal bulges) edging the moon deeper into space.

The Earth contains an enormous reservoir of untapped power. Some of it is related to isostasy, some to the planet's spin, and some to the thermal energy which explodes volcanoes, propels our continents, and builds mountain ranges. These forces are intimately tied to each other, but geologists at the time were only slowly recognizing the relationships. Solving geology's puzzles required the synthesis of a range of ideas and data – much of which was still undeveloped.

As we have seen, Lord Kelvin's brilliant but unnoticed assistant John Perry suggested that convection currents move the viscous mantle, distributing heat. In 1906, Austrian geologist Otto Ampferer[i] speculated that the Alps were formed by folding of the upper crust which occurred when magma sank into the mantle, pulling pieces of the crust downwards. This is an idea very similar to the modern description of subduction, although Ampferer's *Subfluenztheorie* didn't receive as wide an appreciation as it deserved. Among other scientists, English geologist Arthur Holmes was just beginning to develop a similar model of the Earth's inner workings, but it would take years before he had the data he needed to publish his idea. However, by the beginning of the twentieth century, a slowly growing number of scientists explained some of Earth's greatest mysteries in terms of a moving mantle and mobile continents.

---
i   Otto Ampferer (1875-1947), Austrian geologist.

Hollow Mountains

Fissiparturating with the Moon

Two American geologists, Frank Taylor[i] and Howard Baker[ii], independently and simultaneously suggested that the continents had moved in the past. Taylor believed the mountain ranges of Europe and Asia formed because the continents drifted toward the equator. Taylor was an amateur self-educated American geologist who noticed

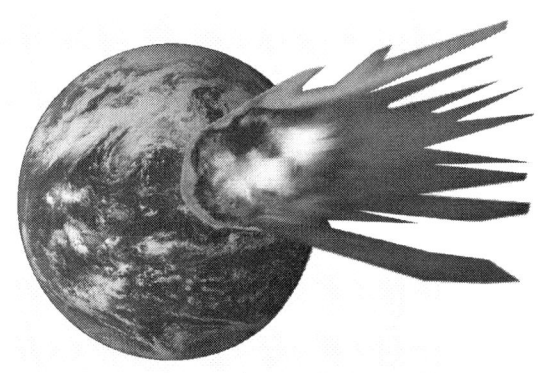

The Earth experiences fissiparturition.

– as had many before him – that the coasts of South America and Africa made a reasonably pleasant fit. He was one of the first to seriously insist the two had drifted apart. Taylor proposed that the crust of the Earth was influenced by lunar tidal forces. This, he surmised, pulled apart the continents in some areas, pushed them together in others. Experienced geologists pointed out that tidal forces were too weak to crumple mountains or collide continents. They also didn't like that Taylor, the son of a wealthy Chicago lawyer and a rather restless Harvard drop-out, was making proposals he couldn't prove. Such as his idea that the moon was a recently captured wayward space rock trapped by Earth's gravity. Although Taylor receives credit for proposing mobile continents in 1910, his idea that continents are drifting away from the poles and his sparse details on rock types and fossils across the disconnected continents resulted in weakening the idea of continental drift rather than strengthening it. This reinforced the notion that continent mobility was an odd speculation promoted by amateurs. Taylor's ideas were no more convincing than the proposals produced fifty years earlier by Robert Owen, the Indiana geology professor.

Baker had a more intricately-developed hypothesis. The continents, he presumed, slipped from their original locations because a gigantic hole suddenly appeared where the Pacific Ocean is today. Baker's 1911 proposal used tidal forces to explain continent movement, but his twist involved the planet Venus, which Baker imagined spun towards Earth from somewhere beyond Neptune. As it swung close to our planet, its gravity gouged a huge hole into the Pacific Ocean.[168] Ever since, the Earth has been adjusting to the gap with the continents sliding to their present positions. Meanwhile, Venus took up her spot as the Morning Star and the rocks that were sucked off the Earth during the Venus fly-by became the Moon. Something similar to this idea had originated a few years before Baker's hypothesis was published.

---

i Frank Bursley Taylor (1860-1938), American amateur geologist.
ii Howard B. Baker (1872-1957), American geologist.

The Mountain Mystery

Charles Darwin's son, astronomer George Darwin[i], also speculated the Moon originated from the Pacific Ocean – his collaborator, Osmond Fisher, pointed out the moon would fit neatly back inside the Pacific. They called this event *fissiparturition,* a cumbersome term for the idea that today's continents came about when the moon was ripped from the Earth, splitting the once-continuous granite crust and filling the resulting gaps with basaltic ocean crust.

Fisher and Darwin speculated that a passing star's gravitational attraction was midwife to the Earth's birth of the Moon. Fisher, in 1882, elaborated that the continents broke apart and drifted to fill the gap.[169] Neither Fisher, Taylor, nor Baker could support their theories with enough evidence and few gave their ideas much attention. We will later revisit the role the Moon's creation played in forming the Earth's ocean basins and mountains. The theory would reappear a generation later, but within the rigorous work of a more serious scientist, the head of geology at Harvard.

Riding the Moho

Although the ideas of George Darwin, Howard Baker, and Frank Taylor later proved only partially correct, each credited continental movement for our current landscape. In order to move our big clunky continents, there needs to be a slippery base at the underside of the crust. The abrupt discontinuous slippery zone was discovered by a Croatian geophysicist with an almost unpronounceable name, Andrija Mohorovičić.[ii] He was born in the village of Volosko, near the Adriatic Sea. The village is noted as Mohorovičić's birthplace (a metal plaque tacked on an old building there says as much), but it is usually crowded with windsurfers, drawn by the area's incessant winds. Mohorovičić said it was those constant breezes that encouraged his study of meteorology, his primary interest – but perhaps they also drove him away from the place. He was awarded scholarships abroad, which carried the gifted scientist far from the village where his father worked as a blacksmith making ship anchors. At 15, Mohorovičić was studying climate, experimenting in physics, and could read Croatian, Italian, English, and French – to which he added German, Latin, and Greek in university.

Mohorovičić learned German so he could study physics in Prague, where he was schooled from 1875 to 1878. His school was outstanding. Instructors included Hornstein, a sunspot and magnetism theorist who taught astronomy and analytic mechanics; Lippich, a theoretical physicist who taught the physics of energy (and was replaced by Albert Einstein on retirement); and Mach who concentrated on diffraction, refraction, and propagation of sound waves –

---

i George Darwin (1845-1912), English astronomer and mathematician.
ii Andrija Mohorovičić (1857-1936), Croatian geophysicist. If you take some liberties and pronounce it slowly, his name becomes An-DREE-yah Mow-Hoe-Rah-VICH-Itch. Or you can mimic Earth Science grad students and refer to him as Moho (Mow-Hoe).

essential elements of seismology. He also explored the shock waves produced when a projectile exceeds *Mach* velocity.

Mohorovičić's doctorate was granted in 1893, at age 35, for his studies of cloud formations which he tracked using a nephoscope he perfected – a device that tracked clouds and determined their heights and speeds. Before submitting his thesis, he was already head of a key meteorological observatory and he had created the national weather service for his entire country. But Mohorovičić had been a physicist before he was a meteorologist. His work at the weather observatory began to include the geophysics of geomagnetism and seismology.

The switch from meteorology to geophysics was somewhat abrupt and risky for the respected scientist in his early forties, particularly within the hierarchical system of the old Austro-Hungarian Empire. His meteorology papers had been as well received as they were diverse – Mohorovičić even recorded "atmospheric gravity waves" caused by the 1908 Siberian meteor which had destroyed a huge swath of forest 5,000 kilometres away, along the Tunguska River.

Mohorovičić abandoned meteorology because climate study "should have at disposal about a thousand years of observations, not a hundred" before there was enough data to allow meaningful conclusions. He figured climate studies should be deferred "to our far descendants"[170] who might then have the necessary information to understand it.

He also voiced some

Mohorovičić

frustration with the fact that his meticulously derived short-term weather forecasts were only 77 percent accurate. Mohorovičić completely abandoned meteorology and focused on seismology. His shift to geophysics somewhat reflected the changing focus of earth science around the world. His own institute, Zagreb's Meteorological Observatory was renamed the Institute of Meteorology and Geodynamics, then rebranded again as the Geophysical Institute in the years after Mohorovičić, indicating the growing importance of geophysics. Not long after

The Mountain Mystery

Mohorovičić made his change, a fortuitous earthquake struck near his home in Zagreb. From that earthquake, at the age of 53, Mohorovičić made the most important discovery of his life.

The 1909 Pokuplje Event toppled chimneys and a few stone buildings, but was a relatively benign earthquake with no recorded fatalities. However, the earthquake supplied a wealth of data from newly installed seismographs across central Europe. Mohorovičić asked colleagues from Munich, Strasbourg, Rijeka, Vienna, Sarajevo, and 24 other stations to send their records to him. He was able to derive startling conclusions from the obscure squiggles on those seismic sheets. Mohorovičić compared arrival times of the shock waves at the various stations and plotted what he called *krivulje vremena,* or time curves, for the earthquake's phases.[171] Interestingly, though his landmark graph has fewer than 30 words on it, titles and notes appear in three languages – Croatian, German, and Latin. Mohorovičić learned physics in German, was trained in the Classics, but was most comfortable in his first language. It was a time of transition in classical education – in another generation, Latin would no longer be inscribed on final drafts of scientific papers.

As he analyzed the earthquake data, a serious point of confusion arose. It should have been a fairly simple step to triangulate the location of the epicentre, but Mohorovičić failed again and again with his calculations until he gave up the notion that, with depth, the velocity of seismic waves travelling inside the Earth increase in a linear fashion. After plotting the Pokuplje earthquake data from those 29 different sets of seismograms, Mohorovičić noticed that the pressure wave arrivals required two separate curves on his time-travel graphs. Plotting time in seconds on the Y-axis and distance in kilometres on the X-axis for each station's records, his travel-time curves "revealed two individual primary waves at different velocities," he wrote.[172]

It was a bold step, a leap into scientific uncertainty, when he decided to ignore conventional wisdom and assume that the earthquake shock waves had entered a strange realm, about 50 kilometres below the surface, where they abruptly travelled much faster than expected. When he derived the correct expression through geometric drawings based on the raw data, he discovered the physics function scientists would later call the Mohorovičić Law.

He noticed that the same sort of seismic shock wave was travelling at two different speeds through the Earth. It was as if the voice of a friend, calling your name from some distance away, reached you at two different times. This might be possible if in addition to travelling through the air, the sound wave also vibrated along a stone wall and approached your ears by that faster route as well. Mohorovičić realized the seismic P-waves he was observing had found two different routes to the seismometers – a slow shallow route and a fast deep route. The energy from the earthquake's epicentre had split up, some of the energy waves finding a faster path along an interface between layers deep within the

Hollow Mountains

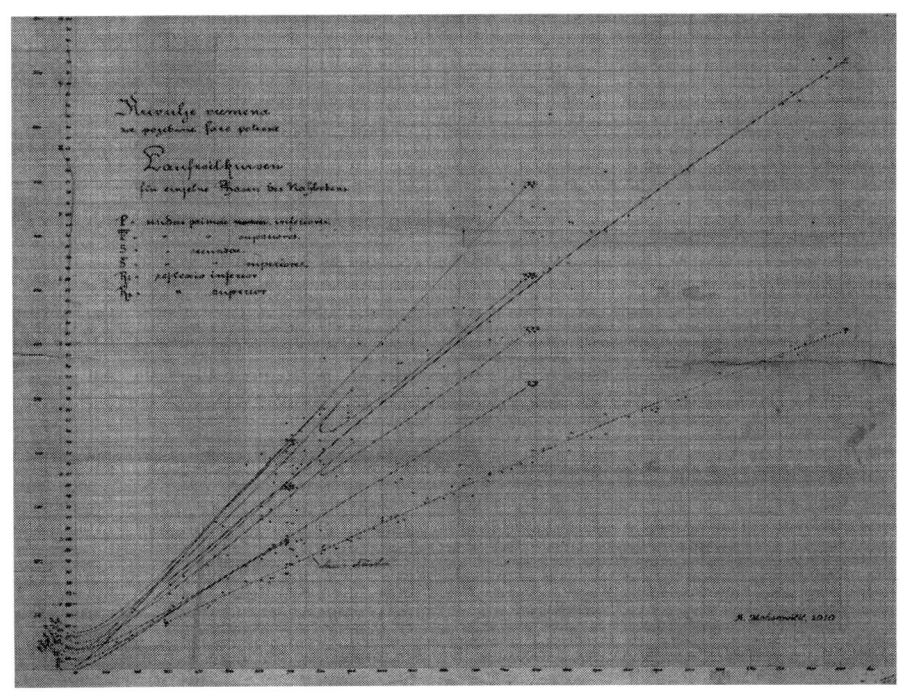

Seismic waves from the Moho – a graph labelled in Latin, German, and Croatian

Earth. Mohorovičić worked out the refraction equations, the ray-paths of the P-waves, and their reflection points as the sounds knocked around inside the Earth's crust, and he concluded there had to be an abrupt change in the material that comprises the Earth interior.

From his seismic records, Mohorovičić was able to show that the inner Earth has distinct layers, and the crust has a definite thickness, ending at the slippery zone which scientists now call the Moho Discontinuity. The change of composition at the mantle and crust boundary creates the seismic velocity change that puzzled him. Mohorovičić estimated the crustal layer of the continent in his part of Europe was roughly 54 kilometres thick. He wrote, "I decided on a rounded-down depth of 50 kilometres, since this was the approximate depth of the P-waves, thus the lower boundary of the topmost layer of Earth's crust is located at this depth. At this boundary, the material of which the Earth's deeper interior consists must change abruptly, because the velocity of the earthquake waves' propagation changes abruptly."[173]

Of his amazing feat of calculation, Mohorovičić's biographer Dragutin Skoko wrote, ". . . in the Earth's interior, seismic waves travel invisibly and inaudibly. They can be followed only by mathematical equations."[174] Mohorovičić was inferring details about a dark inaccessible place. No one will ever visit the boundary between crust and asthenosphere, but Mohorovičić described it. Using his techniques, refinements and more data now show us that the crust of oceans is comparatively thin – just 5 to 9 kilometres – but the continental crust is exactly as Mohorovičić had determined. This finding also supports Dutton's idea of isostasy – the thin ocean crust is heavy while the thick continental crust is light, but at depth they exert the same pressure on the asthenosphere. And very significantly, Mohorovičić showed that the planet has a discontinuity, or detachment forming a slippery zone, allowing continents to slide. What the continents needed next was a daring scientist who could demonstrate they actually do move.

11

The Ice Man's Solution

At the beginning of the twentieth century, the Earth still held four great unsolved mysteries. None was as significant as the origin of mountains. But there were others: Why were rocks from the seafloor different from continental rocks? How did ancient creatures leave identical fossils on distant continents? Why are glacial deposits scattered across the world, including the tropics and the north African deserts? Geologists and geophysicists were poking at various theories, looking at their implications and contradictions. Answers were elusive. In fact, as new evidence arose, it became increasingly difficult to fit a story together.

As we have seen, among nineteenth-century geologists, the most popular grand theory of the Earth was the idea that the hot planet had cooled, leaving bumps and hollows in the process – the bumps are our continents; the hollows, our seafloors. This theory had the logic of physics and the support of prominent geologists – Eduard Suess and James Dwight Dana among many others accepted various versions of this idea. It partly explained pieces of the four mysteries – particularly mountain growth and the differentiated oceanic and continental crust.

However, contraction theory didn't answer all the enigmas geologists unearthed. And confusingly, contraction as explained by Suess was different from Dana's model. The American geologist James Dana believed continents are permanent and ancient, formed early in Earth history, and have remained much the same ever since. Eduard Suess, the Austrian professor and expert on the Alps, disagreed. He felt Dana's description didn't properly account for fish fossils at high elevations remote from seas. His contraction model included cycles of sinking continents and rising oceans, a process that repeated again and again over millions of years, elevating new marine fossils each time. Suess used his apple analogy, suggesting mountains formed like ridges on desiccated fruit. For him, the continents sunk to become oceans while old oceans rose to become continents, as if riding a teeter-totter. Pressure from continents rubbing shoulders with oceans during these transitions squeezed mountain ranges up along coastlines. But the Suess model couldn't account for mounting evidence that continents were very, very old – and made of fundamentally different material than seafloors.

The Mountain Mystery

There were other problems for the supporters of contraction theory. A shrinking Earth could never shrivel enough to form the massive folds and the stacks of thrust sheets observed in today's mountains. Within Europe, the planet's surface would have had to shrink to just one-quarter of its original size to grow the Alps. No known cooling of iron or rock shows such shrinkage – something akin to taking the molten steel of a semi-truck and shrinking it to sedan-size as it cools and hardens. To circumvent that nasty detail, Dana's theory depended on uplifts to cause the highest mountains rather than lateral buckling and stacking. But field geologists found that the drawings they sketched from the flanks of mountains did not show the straight vertical rises Dana's planet required, except in a few isolated places. And if the Earth were cooling and shrinking, experimental models showed that the pattern of mountain ranges should be regularly spaced around the globe, not scattered haphazardly as we actually find them. These were fundamental problems with the once appealing theory.

There was also a problem with fossils. By 1900, most geologists and biologists accepted Darwin's description of species evolution. Darwin noted that the offspring of various creatures, isolated from each other and exposed to different environments, evolve into quite different beings with the passage of time. For example, bison arose on the American plains while the wildebeest fills a similar ecological niche in Africa. Both form huge herds, mostly survive by grazing (eating grass and seeds), but also by a little browsing (munching on the odd shrub). Both animals have manes, wild beards, and look like trouble. But you would instantly distinguish one from the other. Significant changes have taken place in the millions of years since the two animals shared a common ancestor. In addition, with very little practice, you would quickly discern the fossilized bones of each. No creatures evolved in Africa that exactly resemble the American bison and no wildebeest herd ever roamed the Kansas grasslands.

African Wildebeest

American Bison

So it was hard to explain, a century ago, how fossils of identical creatures, such *Lystrosaurus*, *Mesosaurus*, *Cynognathus*, and a myriad of lesser fossilized snails and lizards could be found on separate continents. Geologists also observed fossils of the tropical fern *Glossopteris* in Antarctica's coal beds as well as parts of India, Africa, and South America. None of these lifeforms could swim an

The Ice Man's Solution

ocean. Without admitting a supercontinent's former existence, its rupture and its subsequent drifting, it was difficult to explain how those plants and animals could have populated several remote continents simultaneously – unless the former continents were once merged in a unified land upon which they traipsed.

Most geologists believed that some sort of land bridges, now sunken below the waves, had helped creatures (including plants and snails) roam freely between the continents. There are modern land bridge examples – Central America links the American continents, allowing migrations; Alaska and Russia have periodically shared a land bridge that animals such as horses, camels, and humans have crossed. This, with a stretch of imagination, could solve the problem of the distribution of fossils. However, by 1900, no signs of any long-lost sunken bridge between South America and Africa had been discovered. Nor between Newfoundland and Norway, nor Australia, Asia, and Antarctica, nor any of the other dozen places they were needed to account for fossil similarities. An alternative solution was to assume that the problematic fossils had arisen upon a single supercontinent which split apart, pieces drifting off to become autonomous landmasses, thus disconnecting those ancient plant and animal fossils in the process. But to most geologists at the time, that seemed highly improbable.

The Unlikely Geologist

The person who best explained how fossils might have become scattered without the help of land bridges was not trained as a palaeontologist, botanist, nor even as a geologist. He was, instead, a polar explorer, a university professor, and a meteorologist. Although Alfred Wegener[i] is at the heart of the theory that would eventually explain how fish fossils appeared on mountain slopes, he was not an Earth scientist. He was more comfortable on icy plains than mountain ridges. But as an outsider, he offered a fresh perspective on the Earth's changing landscape. Wegener weighed the evidence unencumbered by colleagues or mentors attached to established geological concepts. Such detachment also made him a target for rejection, a rebel dismissed as an uninformed outsider producing misguided speculations. Ultimately, his contribution to geology led to the best model for explaining the Earth's dramatic scenery. With Wegener everything – earthquakes, volcanoes, distribution of rocks, glacial debris, and fossils – falls neatly into place. He advanced the crucial theory that answered the most enigmatic puzzles about the Earth. But his big idea was rejected for fifty years. His contributions were only appreciated decades after his death in an arctic blizzard.

The youngest of five children in a middle-class family, Wegener was born in Berlin, November 1, 1880. His mother, Anna Schwarz, and his father, Richard Wegener, lived in Berlin Mitte, the city's main centre and cultural and intellectual hub. The elder Wegener taught classical languages and lectured as a theological

---

i   Alfred Wegener (1880-1930), German arctic explorer, meteorologist, and professor.

philosopher.[ii] Young Wegener studied at nearby Köllnische Gymnasium, where he ranked first in his school, particularly excelling in physics. Afterwards, he studied in Berlin, Heidelberg, and Innsbruck. He returned to Berlin's prestigious Humboldt University where philosopher Walter Benjamin and physicist Max Planck, as well as Albert Einstein, Karl Marx, and Friedrich Engels had been either instructors or students. During graduate studies, Alfred Wegener assisted at Berlin's Urania Astronomical Observatory. Although armed with a PhD in astronomy, Wegener acquired a keen interest in climatology. He took a position at the Lindenberg Observatory where his older brother was already a meteorologist. Among other duties, Kurt and Alfred Wegener pioneered weather balloon tracking of storms. They also tested a new celestial navigation quadrant by climbing aboard an experimental balloon. One of their flights, in April 1906, set a record for the longest time anyone had ever been airborne – they spent 52 uninterrupted hours floating above Europe.

Later in 1906, Wegener was invited to join a Danish expedition to Greenland. It was the first of his four field studies on the frozen island. He later regarded the expedition as a turning point in his scientific career – exploring the arctic became his great passion. Greenland offered opportunities for discovery. By the early part of the twentieth century, much of coastal Greenland had been charted, but there were remote unknown stretches along the northeast coast which he explored. The arctic experience tested Wegener's resolve. His mentor and team leader died during the expedition while exploring Greenland's interior by dog sled in conditions similar to those that would later take Wegener's own life.

Wegener's curiosity, and some of his Greenland observations, led him to stray into the unresolved issues of geology. He devoured journals and corresponded widely with geologists. He admits he was seduced by the potential snug fit of the African and South American coastlines. In December 1910, Wegener wrote to his future wife, "Doesn't the east coast of South America fit exactly against the west coast of Africa, as if they had once been joined?"[175]

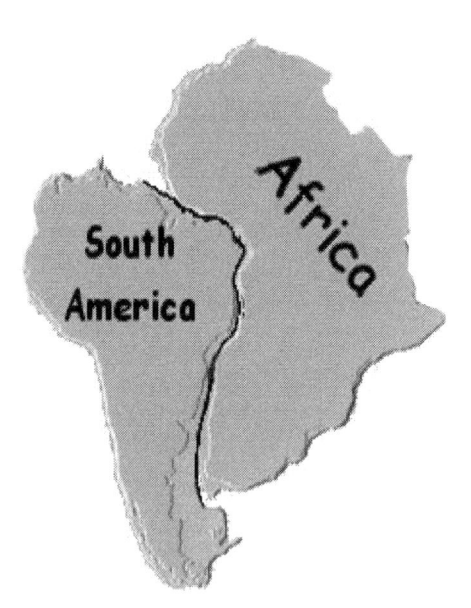

As snug as spoons

---

ii Most English biographies of Alfred Wegener describe his father as an "evangelical preacher" but this is a misrepresentation of *evangelical* which, for the era and location, simply meant non-Catholic. Further, Richard Wegener was an educator, not a clergyman.

The Ice Man's Solution

After returning from the first of his four Greenland expeditions, Wegener spent the next six years lecturing cosmic physics, applied astronomy, and climatology at the University of Marburg.

He was an unusually spirited professor – his students appreciated the small, energetic teacher who presented clear, organized lectures. His course notes were so well-formed that he was able to gather them into one of the first text-

Alfred Wegener in Greenland, 1912

books explaining meteorology. His 1911 text became the standard for several decades and it was the first to include detailed observations and weather data from the Arctic.

At first, Wegener claimed, he was unconvinced of his own theory. But in the fall of 1911 he stumbled upon a paper that listed fossil similarities between Africa and South America. Wegener pursued fossil distributions and palaeoclimate evidence. He finally concluded that the continents had split apart, plowed through oceans, and drifted to their present temporary positions. He saw the Earth's surface as dynamic and alive with motion. He even thought he saw signs of continental drift when he compared his observations in Greenland with maps produced a few generations earlier. On January 6, 1912, Wegener took the risky step of publicizing his unorthodox thoughts about mobile continents.

The German meteorologist presented a public lecture proposing his hypothesis of a supercontinent as the source from which modern continents split. Three years later, in 1915, he expounded his theory of a unified continent, or *Urkontinent*, when he published *Die Entstehung der Kontinente und Ozeane* (*The Origin of Continents and Oceans*), a book in which he collected hundreds of examples of evidence related to continental movement:

1) The outlines of continents fit together like a jigsaw puzzle;
2) There are geological similarities including mountain belts, river trends, ore deposits, and rock types along the Europe-North America and Africa-South America coasts;

3) Fossils of land vertebrates and plants extend across those same continents, though now separated by oceans; and,
4) Tropical plants once thrived in Antarctica while glaciers scratched striations, or grooves, into rocks in North Africa – occurrences best explained by continents moving across climate zones.

Wegener introduced his revolutionary idea at a session of the German Geological Society at the Senckenberg Museum in Frankfurt. In order to explain the unresolved geological confusion confronting scientists, he presented Pangaea[i], a conglomeration of all the continents clustered into an ancient supercontinent where the fossils' previous lifeforms freely roamed and where mountain ranges were continuous. According to Wegener, some unknown force caused Pangaea to break up. The old supercontinent's pieces slid about on the Earth, arriving at the positions we are familiar with on today's maps.

Although the circumstantial evidence was significant, without a massive power source to displace the continents, it was difficult for established scientists to seriously consider his idea. At the Frankfurt meeting, Wegener said, "the forces that displace continents are the same as those that produce great fold-mountain ranges. Continental displacement, faults, and compressions, earthquakes, volcanoes, transgression cycles, and polar wandering are undoubtedly connected on a grand scale."[176] They undoubtedly are.

Wegener described continents splitting, gliding, then colliding above the slippery interface Mohorovičić had just discovered. But he could not propose any engine strong enough to propel them. Nor could he explain why there were apparently no trails

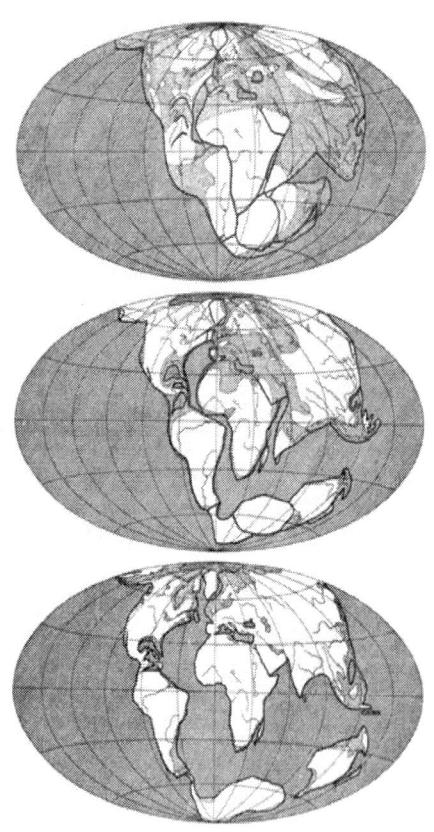

Wegener's break-up of Pangaea with continental displacement, from *The Origin of Continents*, 1929

---

i   Wegener used the term *Urkontinent* to mean *Supercontinent* in his 1915 papers. At a geology symposium in 1927, the term *Pangaea*, from Greek *Entire Earth*, was adopted to describe the hypothetical supercontinent.

gouged into the seafloor behind the continents, scratched into the substrate as they plowed along. The idea that rigid heavy continents could wander the Earth's surface was unacceptable to geologists. But Wegener continued sifting through the evidence that supported mobility. Over the next few years he followed his Frankfurt lecture with three papers about his theory. He found answers to detractors and built his defense from his observations and his immense library of correspondence with geologists.

Wegener developed his theory independently, but he credited earlier work, including W. L. Green's 1875 *Vestiges of a Molten Globe,* which observed that segments of the Earth's crust float on a liquid core.[177] Wegener also noted W. F. Coxworthy, who in 1890 wrote that today's continents are the disrupted parts of a formerly cohesive landmass. Wegener then gave a nod to Mantovani, the violinist who had constructed a map of the original supercontinent. Although there were major differences, Wegener used Mantovani's 1909 map to support his own theory. However, it seems Wegener was unaware of the work of another contemporary, the amateurish American geologist Frank Taylor, who had a rather poorly developed idea of mobile continents. Wegener eventually mentioned the American – in the final edition of his book, which was released after Wegener's death in 1930. But Professor Wegener offered substance and specialization the others lacked – particularly in his study of ancient climates.

Just before the Permian Age, 300 million years ago, there had been a hot swampy period with leafy foliage dominating the landscape and creating the billions of cubic metres of carbonaceous rocks – coal – that would eventually add smoke and energy to our industrial age. Today, those coal layers stretch around the world, even to Canada's most northerly island, Ellesmere, just a few hundred

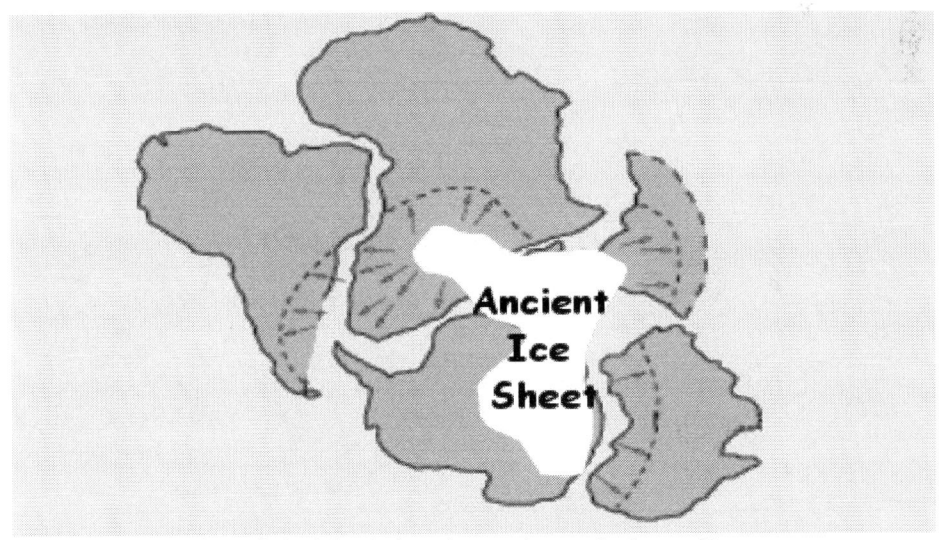

Patches of glacial striations (arrows) scratched upon Pangaea.

kilometres from the North Pole. While others assumed the entire Earth was once tropical, Wegener's explanation for coal in the arctic was continental movement. He showed that all the land once clumped together into one large continent centred on the steamy equator where the coal developed. Then the landmass slid to extreme southern latitudes where it was covered by glaciers. Finally, Wegener broke his supercontinent into drifting pieces which distributed coal seams, fossilized coral reefs, glacial striations, and salt beds to places that are impossible in today's climate. Continent mobility, as improbable as it seemed to geologists a hundred years ago, best explained such strange climate relics.

For meteorologist Wegener, the world-wide distribution of ancient glacial remains such as till, drummonds, eskers, and especially striations, were compelling evidence the continents had moved. Something very odd had happened to the Earth. The planet's surface had been hot, then experienced a deep freeze at the start of the Permian. Wegener noted glacial striations are found in North and South America, India, Asia, Europe, and even the Sahara. Such grooves are uniquely etched by the weight of glaciers as they drag stones across underlying rocks, leaving telltale scratches, or striation signatures. Nothing else causes them. This was a major problem for geologists. How could so many locales – including places on the equator – have been massively iced?

And, of course, there were those troublesome fossils. Palaeobiologists felt the probability that identical creatures could evolve on separated continents

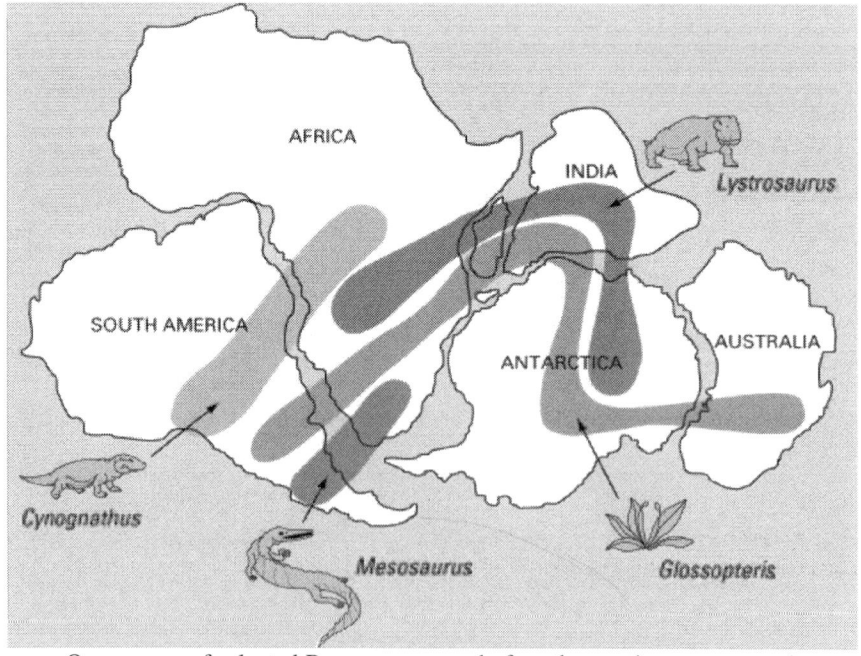

Occurrence of selected Pangaea genera, before the continents scattered.

approached zero. Environmental differences (in precipitation, elevation, vegetation, predators, diets, or genetic-altering radiation) always make cousin creatures diverge. But the *Mesosaurus*, a giant extinct reptile, had its fossils in both South America and Africa – and nowhere else in the world. There were indistinguishable snail fossils in Vermont and Norway. Elsewhere, coal beds that had formed from identical semi-tropical vegetation stopped at water's edge, then continued on the next continent. Clearly things were not where they belonged.

Wegener solved this problem with his idea that these examples were members of the same species, which became isolated when the supercontinent split. But others stubbornly challenged this distribution idea. The established geological community preferred to imagine that sunken land bridges, isthmus links, and island stepping-stones helped plants and animals stray across the seas.

Bridging the Gap

Geologist Joseph Barrell argued that basaltic ocean-spanning land bridges were animal roadways which eventually sank from view. However, others showed such bridges could not sink without contravening the way isostasy works. Nevertheless, in 1933, influential geologists Charles Schuchert[i] and Bailey Willis suggested that the land bridges could have been thick accumulations of basalt, similar to modern ocean islands that rise above the surface, then erode out of sight. Schuchert described the trans-Atlantic land bridge as large and wide, but couldn't explain how it had disappeared entirely without a trace. Although problematic, most geologists still preferred the land bridge theory to account for fossil homologies.

Around 1915, Canadian-American biogeographer William Matthew[ii] also argued that the problematic species distributions were adequately explained by Panamanian-style isthmus connections. Matthew combined land bridges and cycles of climate change, suggesting these regularly caused some animals to move away from northern regions towards the south, populating South America and Australia, for example.[178] For more problematic areas, he invoked the idea of "rafting" which is what you might guess – creatures clinging to floating logs, drifting across oceans, colonizing new areas. On the basis of his field work in Asia, Matthew also proposed that humans evolved in Mongolia, then spread to the rest of the world. He was remarkably influential in broadcasting his speculations, many of which later proved erroneous. Matthew was a gifted teacher and became a curator at the American Museum of Natural History, a job he held for over thirty years. From his museum post, he taught huge undergraduate palaeontology classes with as many as 900 students in attendance. He was elected to the National Academy of Sciences, but was ultimately disqualified when it was

---

i   Charles Schuchert (1858-1942), American palaeontologist.
ii   William Diller Matthew (1871-1930), Canadian-American palaeontologist.

discovered that even after living in the United States for forty years, he was not a citizen, having maintained his Canadian nationality.[179] Because of the enormous respect held for Matthew by North American scientists, his rejection of mobile continents[180] and his alternative solutions for fossil distribution helped persuade a generation of geologists to simply chortle at the idea of continents adrift.

Alfred Wegener persisted in his drift theory, mainly relying on palaeoclimate clues and displaced fossils as support. Even with all his evidence, most geologists dismissed it as circumstantial, reiterating the possibility of land bridges and claiming that palaeoclimate anomalies could be explained by a movement of the poles rather than continents – in other words, some were willing to accept the notion that the entire planet had drooped on its axis rather than agree continents could move. Attacks on Wegener's theory began to include assaults on his strong-willed personality as well. Geologists argued he was selective in his choice of examples and he had no real scientific background in geology, nor appreciation for their work or the traditions of their craft. And almost universally, all the Earth scientists pointed out the most serious flaw in Wegener's unifying theory.

Scientists repeatedly noted that continents are huge stable slabs of Earth and could never be moved by any force weaker than God. Wegener once suggested a type of tidal action due to gravitational forces from the Sun and Moon, but physicists quickly proved those forces were too weak. Others advanced notions of either a shrinking or expanding Earth transforming the crust, although these accounted for neither fossil distributions nor ancient climates. Wegener believed the Earth was neither expanding nor contracting. But he still had no convincing mechanism for pushing continents. Regardless of the power source, in the journal *Petermanns Geographische Mitteilungen,* Wegener attributed much of the folding and quaking on the Earth's surface to pressures caused by colliding and moving continents. He realized this properly explained folds and thrust sheets in mountain ranges like the Alps. Clashing continents were too buoyant to sink, so upon collision, they piled atop each other.

Some of Wegener's ideas came from his observations of icebergs during Arctic explorations. For him, large ice floes were analogues for continents. Some of the dynamics of fluid motion which he taught in his university lectures were derived from the arctic. Contrary to the staid establishment geologists, Wegener saw the Earth's lively crustal surface as responsive to internal fluid motions. Coastal mountains, suggested Wegener, formed because the continents were trying to plow through ocean crust – the resistance led to mountain building. He was almost right. Coastal mountains often grow where heavier ocean crust is colliding and submerging under the continent, pushing continental crust upwards. But this concept was not quite what Wegener had in mind. Crustal subduction would be discovered later, when others built upon Wegener's ideas.

A key difference between science and philosophy is the relentless quest to test, correct, and amend scientific theories. The idea of mobile continents passed from

The Ice Man's Solution

a peculiar notion to a real scientific theory when Alfred Wegener suggested that scientists might be able to measure the pace of continental motion. This was a testable question for his hypothesis. Measurement might authenticate his theory, or prove it wrong. Using old, but unfortunately unreliable maps, Wegener estimated North America and Europe are separating at a rate of 250 centimetres each year. Wegener was wrong, the actual velocity is a hundred times slower. Wegener thought astronomical measurements would eventually become precise enough to measure the growing gap between Ireland and Newfoundland. He optimistically expected new techniques "will soon remove the last doubts about the reality of this movement."[181] But when newer data arrived a few years later, it neither confirmed nor denied continental drift.

Wegener would not live long enough to learn he was right. The measurements that finally confirmed his theory required laser beams bounced between widely-separated pylons. Recently, Earth-orbiting satellites have marked movement with a precision deemed impossible to people in 1912. Much would be learned about the Earth during those intervening decades. Meanwhile, Wegener's grand theory was not given significant notice. Except as an object of derision.

Crap, Crap, Crap

Scientific disagreements are not polite. They are often crude, messy and undisciplined. Scientists, even today, cannot easily separate their personalities from their convictions. For example, in the 1990s, when I was considering a PhD in geophysics, I delivered a student paper to the Canadian Geophysical Union at the group's annual Banff, Alberta, conference. We were in the Rocky Mountains, near a chunk of limestone called Mount Rundle, a peculiar outlier tilted far to its side. A creation of one of the last thrusts of the growing mountain range, its edge forms a 12-kilometre blade overlooking the national park's eastern approach. A late May snowstorm left a white glow on the edges of the leaning ridge, but inside the conference centre below, the atmosphere was hot and muggy with the sweat and opinions of hundreds of scientists. I presented my paper, which won polite applause and a small award among the student presentations. It even attracted a few polite questions. But my paper was uncontroversial, and frankly, not very important. Later in the day, a subject of far greater significance was presented by an innovative geophysicist. Not far into his delivery, one member among the keen observers thought he should make his own thoughts known. "Crap," he said. And then, rather loudly, "Crap, crap, crap!" The moderator reminded the heckler that all questions, comments, noises, and crap should be reserved for later, after the speaker had finished. No, it was too important to wait, so there were more "craps" and even a few "total craps." Amid serene mountains, whose form and substance were hotly debated by scientists inside a stuffy hall, the speakers' chairman suggested the disgruntled dissenter take his noisy dissent outside. The Earth herself would find such a mix of poor manners and passion for debate trivial. Differences of scientific opinion can grow nasty. We cannot presume greater civility from our ancestors when they heard incredulous conjectures about the evolution of the planet.

Some of the scientists who glanced at Wegener's thesis felt he was arguing that the continents are sailing around the oceans. The respected geologist Franz Kossmat[i] was particularly unconvinced. He had published over twenty geology and mineralogy books and lectured for thirty years in Graz and Leipzig. But he is now remembered as among the first to dismiss Wegener's continental drift, insisting the oceanic crust was too firm for continents to simply plow through. In 1925, French geologist Pierre Termier[ii] said continental mobility was "a beautiful dream, the dream of a great poet. One tries to embrace it, and finds that he has in his arms a little vapour and smoke."[182] To which Alexander du Toit, a rare Wegener supporter, responded that such a statement only "reveals the gloomy spirit of its author."[183] A likely observation. Within a short period, Termier's son had been killed in an accident, his wife had died of Parkinson's, and his son-in-law, a geologist, had died in combat during the First World War.[184] Termier was renowned for inventing a novel approach to graphically decompress the Alps' stratigraphy and measure the amount of folding and thrusting within the mountains. He was not entirely in opposition to Wegener's drift theory. However, he did not accept it as a global phenomenon, believing instead the crust might shift a bit here or there, but nothing more.

In Britain, a year after the first English translation of Wegener's *Origin of Continents and Oceans*, Philip Lake said this about Wegener at a morning meeting of the Royal Geographical Society: "He is not seeking the truth; he is advocating a cause, and is blind to every fact and argument that tells against it. It is easy to fit the pieces of a puzzle together if you distort their shape, but when you have done so, your success is no proof that you have placed them in their original positions. It is not even a proof that the pieces belong to the same puzzle or that all the pieces are present."[185] The notion of drifting continents was almost universally rejected. Geophysicist Sir Harold Jeffreys[iii] was especially unyielding, certain that the crust's rigidity made continental drift impossible. His influence was enormous. His disciples would denounce Wegener's hypothesis for decades. We will meet some of those followers later, those who would lug Sir Jeffreys' animosity into fifty years of future debates about the Earth's mountains. But most European geologists were reserved in their criticism. It was in America that Wegener was most severely berated.

"Utter damned rot," William Scott, geology professor at Princeton (and President of the American Philosophical Society) said in 1923, describing the

---

i Franz Kossmat (1871-1938), Austrian-German geologist.
ii Pierre Termier (1859-1930), French geologist.
iii Sir Harold Jeffreys (1891-1989), British mathematician and geophysicist.

The Ice Man's Solution

theory of continental drift.[186] Edward Berry, an American palaeobotanist, called Wegener's theory "a selective search through the literature for corroborative evidence, ignoring most of the facts that are opposed to the idea, and ending in a state of auto-intoxication."[187] Bailey Willis, a renowned earthquake seismologist and geologist for the US Geological Survey, reportedly said "further discussion of it merely encumbers the literature and befogs the minds of fellow students. [It is] as antiquated as pre-Curie physics. It is a fairy tale." Willis also claimed Wegener was more "an advocate rather than an impartial investigator."[188]

Ralph Chaney, an American expert on plant fossils and ancient climates, wrote "It is amusing to note that in taking care of their Tertiary forests, certain Europeans [Wegener] have condemned our forests to freezing."[189] Chaney dismissed Wegener's palaeoclimatology as amateurish, apparently unaware that Wegener, with his father-in-law Wladimir Köppen, wrote the world's primary textbook on the subject. Others, such as Chester Longwell of Yale, rejected the concept of mobile continents in the 1920s and stayed opposed into the 1960s, even as the evidence became overwhelming. In 1968, Longwell sniffed, "Although partisans favoring drift may have been right, they based most of their case on the wrong reasons and were unable to visualize a mechanism."[190] His statement rings of revisionist history with a touch of sour grapes.

As recently as 1943, the most influential of all evolutionary biologists in America, George Gaylord Simpson,[i] vigorously opposed the idea of continental drift. He stated that palaeontological data unequivocally opposed the various theories of continental drift and he sneered that Wegener's "looseness of thought or method amounts to egregious misrepresentation." To remove any doubt that Simpson stood on the wrong side of history on this issue, he concludes his landmark paper, *Mammals and the Nature of Continents* with "The known past and present distribution of land mammals cannot be explained by the hypothesis of drifting continents . . . and the supposed evidence of this sort is demonstrably false or misinterpreted. The distribution of mammals definitely supports the hypothesis that continents were essentially stable throughout the whole time involved in mammalian history."[191]

Criticism was sometimes personal. Many geologists dismissed Alfred Wegener as a meteorologist invading a field he was not equipped to discuss. Although he'd trained in physics, his critics didn't realize that physics and fluid motion were sometimes more relevant to this issue than geology. Wegener was criticized for not doing his own field work, but instead, relying on other scientists' work and apparently gleaning only supporting documentation – yet he directly applied what he learned in the Arctic to his theories. In 1927, Wegener himself said, "I believe that the final resolution of the problem can come only from geophysics, since only that branch of science provides sufficiently precise methods."[192]

America was the hotbed of drift hostility. Many there were opposed to the

---

i George Gaylord Simpson (1902-1984), American evolutionary biologist.

climate scientist's theory, but the most powerful opponent to Wegener's idea was Rollin T. Chamberlin[i], a geologist at the University of Chicago. Chamberlin wrote about drift, "Can geology still be considered a science if it is possible for such a theory as this to run wild?"[193]

Later, in 1928, Chamberlin quoted an unnamed colleague, "If we are to believe Wegener's hypothesis we must forget everything which has been learned in the last 70 years and start all over again."[194] He was right, but not in the manner he expected. But his words carried enormous weight and his opposition to continental drift impeded geology for decades. So, who was Chamberlin?

There were two of them, actually. R.T. and T.C. There was something inherently odd in the father and son Chamberlin team that kept them on the conservative – and decidedly wrong – side of much scientific development for 60 years. The elder Chamberlin, Thomas Chrowder[ii], was the ambitious son of a Methodist circuit preacher and part-time Wisconsin farmer. This Chamberlin led the choir in his father's church, collected rocks and fossils, and eventually earned a degree in geology. He mapped the glacial geography of southeast Wisconsin – cataloguing the kettles and moraines left by the receding Ice Age. He proved himself thorough and diligent and became head of Wisconsin's geological survey. A bit later, in 1892, T.C. Chamberlin was asked to organize a geology department at the new University of Chicago. He remained there as geology professor until 1918. During most of his time in the city, he was also president of the Chicago Academy of Sciences. Shortly after starting at the university, T.C. Chamberlin founded the *Journal of Geology*. It became the most influential geology

The anti-drift movement didn't really take to the streets. But this fanciful image correctly reflects the sentiment.

---

i Rollin Thomas Chamberlin (1881-1948), American geologist.
ii Thomas Chrowder Chamberlin (1843-1928), American geologist and educator.

The Ice Man's Solution

journal in America.

T.C. Chamberlin remained editor until his death, passing the job to his son, R.T. Chamberlin. Together, they made the journal a powerful voice for mainstream geological dogma. One did not dare argue with a Chamberlin and later hope to be published in America's leading geology journal.

As founder and editor of the *Journal of Geology*, as founder and chair of his university's geology department, and as president of the local science academy, Chamberlin was politically the most powerful geologist in the United States. Chamberlin also taught a generation of new scientists his philosophy of science, an approach which, he said, was the only valid way to solve scientific puzzles – the system of multiple working hypothesis. It is a valid and useful approach to unravelling science mysteries, but it seems its chief promoter was among the least likely to take advantage of the tool.

Chamberlin promoted some strange ideas. He believed humans originated in Europe (". . .and gave rise to the most virile and progressive branches of the human family, the fair-white and the dark-white races");[195] he taught that the Earth was formed cold, "not white hot" as nearly all other scientists accepted; and he strongly supported Lord Kelvin's extremely low estimate of the Earth's age, even after it was commonly recognized as wrong by a factor of 200. Chamberlin did not agree with Kelvin's final calculation (in 1899, Kelvin asserted a 20 million year age for the planet), but he nevertheless dotingly endorsed almost everything the old man asserted. Chamberlin alleged, "individual geologists, reacting impatiently against the restraints of stinted time-limits imposed on traditional grounds have inconsiderately cast aside all time limits." Being inconsiderate of Chamberlin's failing and aging ideas about geology could ruin a young geologist's career. He further complained that some geologists assumed the Earth was ageless, allowing them to explain any process by "reckless drafts on the bank of time."[196] Chamberlin is especially remembered for his uncompromising opposition to Wegener's theory. If there had ever been an example of a scientist in an extremely high position influencing the course of inquiry in a determinedly wrong direction, it was Thomas Chamberlin. Or perhaps his son, Rollin.

T.C. Chamberlin was elderly when the first American edition of Wegener's *Continental Displacement* appeared, so it was mostly left to his son Rollin to dismiss Wegener as a crank. Rollin's friend F.J. Pettijohn,[i] writing for the National Academy of Sciences in 1970, asserts, "though he was not an innovator, not one to depart much from traditional thought patterns, Rollin Chamberlin established a place for himself as an individual and as a scientist of some stature . . . He had, indeed, severe handicaps to overcome – having so preeminent a father and remaining in the institution in which he was educated."[197] As we have seen, Rollin Chamberlin's father founded the University of Chicago's Geology Department. That's where Rollin was an undergraduate, was awarded his PhD in 1907 – and

---

i   Francis John Pettijohn (1904-1999), American geologist.

spent all but three years of his entire career as a member of that same school's geology faculty. One struggles to find a more intellectually reserved scientist. Immediately upon his father's death in 1923, R.T. Chamberlin became editor of his father's *Journal of Geology*, a position he clung to for 24 years. Chamberlin, described as a conservative in every way, opposed Alfred Wegener's theory so staunchly that very few geologists rose in support of continental mobility. Pettijohn, who also saw no merit in moving the continents, described Chamberlin's contributions to tectonic theory as based on field studies and on "Philosophical considerations . . . he held rigidly to the notion of the permanence of the continents and ocean basins and he gave short treatment to the concept of drifting continents."[198] At a 1926 conference, Chamberlin denounced every aspect of Wegener's drift hypothesis, adding it "does not fit the generally accepted record of geological time. The framework of the present continents was developed in pre-Cambrian time. Geological evidence does not show that a great continental mass split apart."[199] To be sure he was not misunderstood, Chamberlin added Wegener's hypothesis "takes considerable liberty with our globe, and is less bound by restrictions or tied down by awkward, ugly facts than most of its rival theories. Its appeal seems to lie in the fact that it plays a game in which there are few restrictive rules and no sharply drawn code of conduct."[200]

Advocating a Cause

Wegener remained convinced of the reality of mobility, which he had referred to since 1912 as "*Die Verschiebung der Kontinente*," literally, "The Displacement of the Continents." By 1928, Americans had renamed the displacement process *continental drift*, conveying a more whimsical sense. They used the phrase as the title of a collection of papers presented at a symposium sponsored by an oil explorers' organization. As *continental drift*, the derision was clear. The American Association of Petroleum Geologists (AAPG) symposium seems to have been organized primarily to discredit Wegener and the few other proponents of his theory of displaced continents. The AAPG published the collected papers in a volume which has become a permanent record of the many New World scientists who rejected displacement and even engaged in personal attacks against Wegener.

In a recap of the conference publication, reviewer F.A. Melton wrote that he was approached by a New England professor who described the book as an embarrassment. But Melton countered, "If a sense of shame is to be identified in any way with this published collection of papers, it will probably be experienced at some time in the future by a few of the well-known contributors."[201] Melton was optimistic in his expectation that the anti-drifters would one day recant. Decades later, when geologists finally began accepting Wegener's idea, most of the 1928 anti-drift group were dead. By the time drift theory was recognized as correct beyond all but a smidgeon of doubt, most of the fortunate few who lived

The Ice Man's Solution

long enough to be embarrassed simply clung to their long-standing skepticism.

The outfit that had organized the symposium – the wildcat, or exploration, geologists – were among the most doubtful of the lot. But, to their credit, the oil men organized the continental drift conference and even invited Professor Wegener. The New York meeting was managed by the improbably named Willem Anton Joseph Maria van Watershoot van der Gracht[i], who published the proceedings in the 1928 compendium, *The Theory of Continental Drift*[ii]. Van der Gracht, from Holland, was sympathetic to the idea of mobile continents. He and G.A.F. Molengraaf, also a Dutch geologist, were among the few presenters who could see value in Wegener's theory. An astute synthesizer might have united the various ideas presented at the conference and created a more solid drift theory. For example, Molengraaf's paper used seafloor spreading to explain the opening of the Atlantic Ocean and the East Africa Rift. It was a novel, and correct, suggestion, and it helped explain the force that moved continents. But like Wegener's own theory, it was overwhelmingly rejected.

Within the symposium collection, American palaeobotanist Edward Berry wrote, "It is inconceivable that masses of continental size should move over such large arcs and preserve their outlines of either coast or continental margin intact."[202] Berry had a point. The concept of drift as presented by Alfred Wegener invoked an image of continents plowing through oceans. In such a situation, they would be expected to become distorted – if they survived at all. It was difficult to imagine their shapes would still match as neatly as South America and Africa. For geologists in the 1920s, it was also appropriate to wonder about the enormous forces propelling slabs of crustal surface as huge as continents through thousands of kilometres of ocean crust. Continents are incredibly massive structures. Some geologists correctly noted that continents have deep roots, extending downwards tens of kilometres. Edward Berry was right, it is inconceivable that such a mass could move by plowing its way through oceanic crust. Even a small continent – Australia, for example – weighs a quintillion tonnes.[iii] The amount of energy moving it is unfathomable. Unknown to these geologists was the fact that continents don't plow though ocean seafloor like arctic ice-breakers in frozen seas; they slide over the Moho, like a stack of furniture being pulled on a rug across a tile floor.

---

[i] Willem Anton Joseph Marie van der Watershoot van der Gracht (1873-1943), Dutch-American lawyer, mining engineer, and geologist. It might be said that he had one of the biggest names in geology at the time.

[ii] The full title was *Theory of Continental Drift; A Symposium on the Origin and Movement of Land Masses Both Inter-continental and Intra-continental, as Proposed by Alfred Wegener.*

[iii] Australia covers 7,741,220 square kilometres. Continents are about 45 kilometres thick, so the volume of rock is 360,000 trillion cubic metres, weighing over a quintillion tonnes. At the time, geologists didn't know the precise continental thickness, but they knew how heavy a single cubic metre of rock can be. It was inconceivable that a continent – even the smallest one – could be pushed through oceanic crust.

The Mountain Mystery

One of the most aggressive anti-drift scientists at the AAPG gathering was Charles Schuchert. His presentation included a clay model of the continents. He slid it around in front of the audience, showing that the continents could never fit together. The illustrations, British geophysicist Edward Bullard[i] said much later, "were so bad that it is difficult to trace the reason for this extraordinary and quite false result."[203] Schuchert's solution lay in the narrow land bridges connecting continent to continent. He said an unnamed friend thought the fit of Africa and South America was "made by Satan" to confuse geologists.[204] Schuchert also had a troubling habit of referring to continental drift as a "German theory."[205]

Opposition to Wegener's continental drift occasionally took a poetic tone. In 1938, Australian geologist E.C. Andrews presented a structural geology talk and described Wegener's hypotheses as "a high-level flight in matters geological, and one of great stimulation in many ways, nevertheless it smacks suspiciously of the waxen wings of Icarus and courts a similar fate when leaving the Earth too far below."[206] A 1930s equivalent of a sound-bite. For decades, it was difficult for established geologists to flirt with unorthodox positions, especially with a pair of Chamberlins lurking behind the curtains. No one wanted to offend senior colleagues who reviewed doctoral theses, signed grant applications, distributed research funding, and assigned teaching positions.

Allan Krill of the Department of Geology and Mineral Resources in Norway, recently wrote, "I have found none who published first as fixist and then as mobilist. Apparently scientists who changed opinions felt they had to save face for themselves and their institutions in their publication records. Better to perish than publish."[207] As Professor Bullard also pointed out,

> "There is always a strong inclination for a body of professionals to oppose an unorthodox view. Such a group has a considerable investment in orthodoxy: they have learned to interpret a large body of data in terms of the old view, and they have prepared lectures and perhaps written books with the old background. To think the whole subject through again when one is no longer young is not easy and involves admitting a partially misspent youth. Further, if one endeavours to change one's views in midcareer, one may be wrong and be shown to have adopted a specious novelty and tried to overthrow a well-founded view that one has oneself helped to build up. Clearly it is more prudent to keep quiet, to be a moderate defender of orthodoxy, or to maintain that all is doubtful, sit on the fence, and wait in statesmanlike ambiguity for more data (my own line till 1959)."[208]

Twenty years after the gathering in New York, continental drift was still generally rejected without discussion, said film producer David Attenborough. He attended a British university in the late 1940s. In a 2012 interview, Attenborough

---

i   Sir Edward "Teddy" Crisp Bullard (1907-1980), British geophysicist.

The Ice Man's Solution

recalled asking one of his lecturers why continental drift wasn't being discussed. He says, "I was told, sneeringly, that if I could prove there was a force that could move continents, then he might think about it. The idea was moonshine, I was informed."[209] Thirty years later, Attenborough presented some of that moonshine in his 1979 series, *Life on Earth*, demonstrating the evolution of life and planet. But much had been discovered between Attenborough's college days and 1979.

Adapting to Wegener's ideas took decades. Geophysicist Bullard observed, "It is interesting to consider why Wegener's arguments did not carry conviction, since it is now clear that many of them are, in principle, sound. The reasons were, in part, associated with the nature of Wegener's presentation. He argues too hard and was often accused of advocating a cause rather than seeking truth."[210]

As we have already seen, the elder professor T.C. Chamberlin was opposed to Wegener's thesis for geological considerations. But he also advocated the idea that scientific advances must progress in a rigid style which Wegener had violated. Chamberlin's formula: Begin with objective data; introduce multiple working hypotheses; postpone any interpretations and judgements until everything has been thoroughly tested. Chamberlin's influential 1890 paper, "The Method of Multiple Working Hypotheses" begins with the remark, "With this method the dangers of parental affection for a favorite theory can be circumvented." He adds that good science requires multiple hypotheses relevant "to investigation, instruction, and citizenship."[211]

Working as a good citizen scientist was important to Chamberlin. Historian Naomi Oreskes suggests that a lot of the American animosity to Wegener and his drift theory stems from cultural differences in the way science was pursued. She points out that Chamberlin's method "reflected American ideals expressed since the 18th century linking good science to good government. Good science was anti-authoritarian, like democracy; good science was pluralistic, like a free society. . . And if good science was a model for a free society, then bad science implicitly threatened it."[212] This is quite true and explains much of the Wegenerian rejection, but Chamberlin's personal conduct was itself authoritarian. At any rate, it seemed to Chamberlin and many others that Wegener had quickly adopted a faulty conclusion, then worked backwards to find supporting data. This may have been true, but it didn't stop Wegener from being correct. And as we have already seen, Chamberlin's dedication to the multiple working hypotheses method didn't save him from an error-prone academic career.

Dismissing continental drift theory as simplistic, Chester Longwell[i] suggested that for advocates of the idea, "a definite choice of creed brings some peace of soul that is denied to the scientific skeptic."[213] It was not the last time Wegener's theory would be described as a pseudo-scientific cult. But there may have been more to the disdain for Wegener's theory. Rarely mentioned, but sometimes alluded to, is the fact that Wegener had been a German soldier. Americans such as

---

i   Chester Ray Longwell (1887-1975), American geologist.

his fierce critic Longwell had once had Wegener in his cross-hairs. Captain Longwell was a graduate student at Yale when the First World War broke out. He spent part of his two-year army service overseas where "his composure under unusual circumstances made all officers of the regiment admire and respect him."[214] The ugliness of the Great War cannot be overstated.

The Americans reluctantly joined the conflict nearly three years after it had begun, but once involved, their commitment was unreserved. They drafted 2.8 million soldiers, landing 10,000 troops every day in France. Anti-German propaganda was relentless. I had an elderly acquaintance who was a German-speaking American-born child when the United States entered World War I. "Immediately," he said, "German was no longer spoken, not even in our California house. Our name was changed in April, 1917, from von Remden to Remdy." In Wisconsin, towns with Germanic roots were rebranded, with many villages adopting anglicized names. And another friend told me that his uncles were raided while travelling across the country. They had stopped at a cheap roadside inn, were playing cards, laughing and talking loudly in a strange-sounding language. Police burst into the room – the old men had been reported as German spies. Apparently the motel owner who alerted the police had never heard spoken Yiddish.

Congress passed the Espionage and Sedition Acts in 1918 to punish citizens who spoke against the World War. It was illegal to interfere with conscription or publish opposition to the war. Parts of the law were eventually ruled unconstitutional, but they stifled dissent. Pacifist religious groups such as Hutterites, living on secluded community farms in the Dakotas, were harassed for speaking their Swiss-German dialect and for resisting the draft. Two brothers, Joseph and Michael Hofer, were taken from their farm in South Dakota and assigned to a base in California. They steadfastly refused to wear army uniforms, which they felt were contrary to their understanding of the Bible. They were sentenced to 37 years at Alcatraz, but sent to Fort Leavenworth, starved, tied to a ceiling, beaten with clubs, and tortured by other American soldiers. Both of the young Hutterites died. The boys' bodies were sent back to their family's farm – dressed in the uniforms they wouldn't wear while alive.[215]

Meanwhile, in the European trenches, soldiers on both sides slept in slimy mud-holes, then occasionally blazed across a few metres of farmland, landing in their enemy's trench where they fought with bayonets. Antibiotics were still undiscovered – filthy conditions resulted in uncontrollable infections. Shrapnel from shells was accompanied by chlorine and mustard gas, mostly deployed by the German Army. Alfred Wegener, already a renowned climatologist with two expeditions to Greenland behind him, was drafted in 1914 as a reserve lieutenant of the Queen Elizabeth Third Regiment, serving the German side.[i] His unit was

---

i The Queen Elizabeth Third Grenadiers (Königin Elisabeth Garde-Grenadier-Regiment Nr. 3) was raised in 1860. It was garrisoned in Berlin where Wegener lived.

The Ice Man's Solution

soon assigned to the front. During their advance into Belgium, Wegener was shot through the arm. After two weeks in hospital, he was sent back to the line and hit again – this time a bullet lodged in his neck. Finally, the directors of the German Army sent their great scientist off to work in the meteorological service.[216] After the war, both sides sought cooperation in the sciences, but it did not help Wegener that his methodology displayed "cultural differences." His theory languished on the edge of acceptability for years. Geologists saw no evidence Greenland was trudging away from Europe at a rate of two or three metres per year. They argued continents are not strong enough to plow through ocean crust yet weak enough to buckle into mountains. And there was the lack of a force able move continents.

However, geologists were ignoring an extremely powerful source of energy. Arthur Holmes,[i] a British geophysicist, demonstrated that convection currents – flowing streams of mantle slowly circulating below the planet's surface – could power continental displacement. It took decades before other geologists agreed. By then, Wegener, the meteorologist, geophysicist, and arctic explorer, would be dead, one day past his 50th birthday, lost in a Greenland blizzard. He and fellow-explorer Rasmus Villumsen were attempting to deliver supplies to stranded colleagues. Wegener's body was found months later, lying upon a reindeer hide, placed there by Villumsen, who was never found.[ii]

With the death of Wegener, the continents stopped drifting. Few geologists would take on the lonely challenge of adopting the orphaned and despised theory. That task finally landed in the hands of Arthur Holmes, a brilliant young English geophysicist who discovered the true age of the Earth, explored southern Africa and southeast Asia, and identified the powerful force that moved continents. But in the process, he became so desperately poor that he sold vacuum cleaners to village housewives to support his family.

---

i   Arthur Holmes (1890-1965), English geologist.

ii   Upon Wegener's death, leadership of the Greenland expedition passed to his friend Fritz Loewe. Loewe had trained as a lawyer in Berlin, but developed a passion for science and exploration, earning a PhD in physics. He became a meteorologist and understudy to Alfred Wegener. Before the expedition, Loewe had earned the Iron Cross as a young soldier in the German Army and had already spent time in the arctic. During Wegener's fatal 1930 expedition, Loewe's feet froze. A colleague at their Greenland camp clipped off nine of Loewe's toes with tin-snips and a pocket knife to avoid gangrene. Returning to Germany, Loewe, a Jew, was soon dismissed from his post with the Meteorological Service. He was able to relocate with his wife and two young daughters to England until he found permanent work, in 1937, as a lecturer in Australia. In Melbourne, Loewe co-discovered the southern jet stream. Few students knew the remarkable background of their professor with the awkward gait who clomped the university corridors for 25 years.

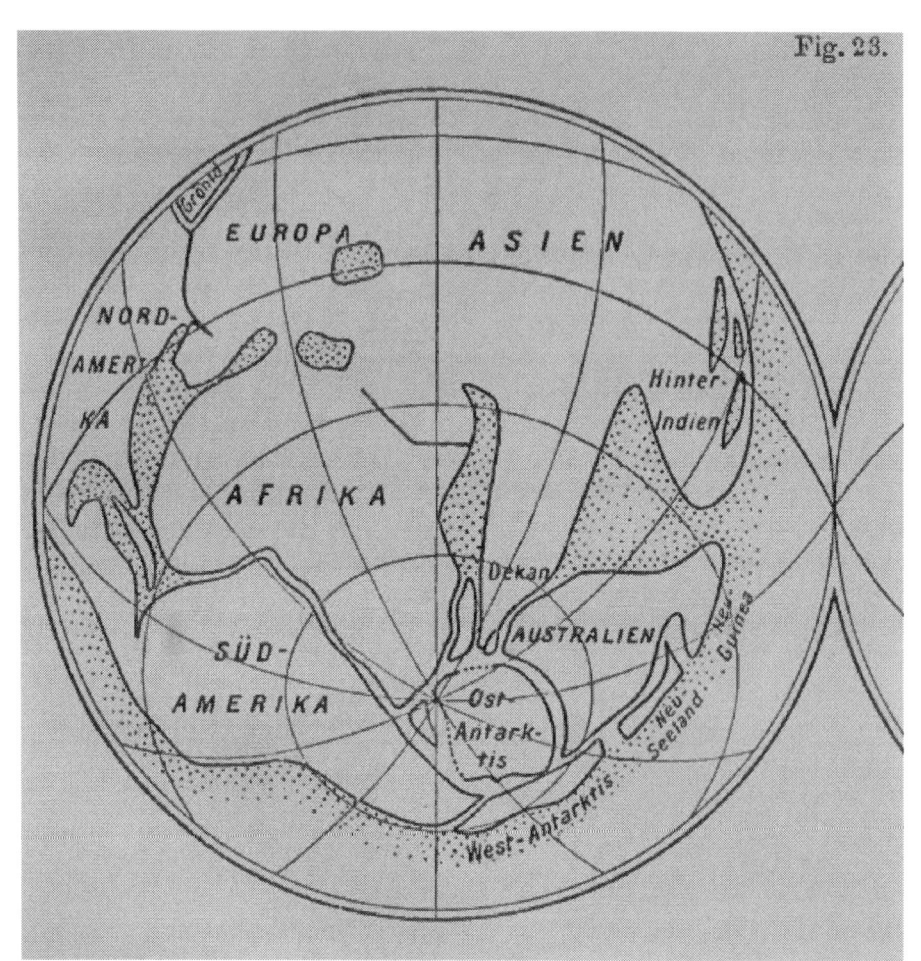

The Earth during the late Carboniferous, as Alfred Wegener imagined it in 1920. This map is from *Die Entstehung der Kontinente und Ozeane*, the first publication with a realistic drawing of Pangaea, the supercontinent. It shows the continents in one hemisphere, largely clustered near the south pole. Wegener recognized that this arrangement solved many of the palaeoclimate issues that baffled geologists in the early twentieth century.

12

Lonely Voices

Selling vacuum cleaners was not Arthur Holmes's original career choice. He was born in Gateshead, England, a gloomy and remote coastal town in the northeast, a spare place of windswept barren bogs and heaths. Reverend John Wesley, the first Methodist, arrived there in a blizzard in 1785, and described it as a pathless waste inhabited mainly by "tinkers, gypsies, pitmen, and quarrymen."[217] In bringing his good news to Gateshead, Wesley must have been at least partly successful – Holmes grew up in a Methodist family. Of that, Holmes says he was puzzled, as a child, to read the Earth's Creation date printed in the family Bible. It appeared at the beginning of the book's summary of the world's chronology – 4004 BC – the year Bishop Ussher had calculated. "I was puzzled by the odd '4.' Why not a nice round 4000? And how could anyone know?"[218] Holmes would prove the Earth's age was in the billions, not thousands, of years. At 17, he escaped dreary Gateshead and studied physics and geology at Imperial College in London. For food, lodging, and class fees, he survived on a scholarship of £5 per month – about $300 today. Holmes would be plagued by poverty all his life.

Upon graduating, Holmes joined a mining company and began prospecting in what is now Mozambique. He spent six months diligently discovering nothing, then became so sick from malaria that his employer rushed his death notice home. But Holmes survived the malaria, left Africa, and was given a job at Imperial College. There he earned his doctorate and at age 30 was appointed chief geologist of a British oil company exploring in Burma. With a wife and the chance to earn his fortune on the other side of the world, Holmes settled into Rangoon and began developing his employer's oil field. Within four years, the company was bankrupt. Unpaid for his work, luggage packed, and awaiting their return ship to England, Holmes and his wife lost their three-year-old son to dysentery just before leaving Asia. After their loss, the company's failure, and his unproductive years in Asia as an oil exploration geologist, Holmes was back in London, penniless. Holmes took a variety of jobs. He lectured whenever the university had an opening for him. When it didn't, he sold vacuum cleaners door-to-door. Within a few years, his young wife died.

Arthur Holmes, 1912

Through the tragedies, Holmes had been – and remained – a persistent scientist. He was arguably the most prodigious of England's twentieth-century geologists. Arriving on the scene in 1910, still an undergraduate, Holmes figured out the true antiquity of the Earth and put an end to the bickering centred on Lord Kelvin's enormous ego and enormous underestimate of the planet's age. At 21, Holmes performed the world's first uranium radiometric dating. The rock he tested was an astounding billion years old. He published his results in 1911, as "The association of lead with uranium in rock-minerals, and its application to the measurement of geologic time."[219] It was revolutionary. First, it thoroughly rejected Kelvin's 20-million-year estimate; second, it made unexpected use of newly discovered radioactivity. Barely past his teenage years, Holmes created a brilliant geological technique still used a hundred years later.

Within two years, Holmes published *The Age of the Earth*. He demonstrated a wry sense of humour by beginning his book, "It is perhaps a little indelicate to ask of our Mother Earth her age, but Science acknowledges no shame and from time to time has boldly attempted to wrest from her a secret which is proverbially well guarded."[220] Most geologists cheered the results because the processes they had encountered required the huge passage of time which Holmes had measured.

Not everyone welcomed the news. The discovery pitted Holmes against the diminishing gaggle of Lord Kelvin's followers. They maintained the Earth had existed for less than 100 million years, though Rutherford's recent suggestion that radioactive heat had been skewing everyone's estimates was gaining acceptance. However, until Holmes's experiments, few guessed the planet might be over a thousand million years old. Nor had they ever had an actual number, a nearly precise value, rather than a vague notion of timelessness. Holmes continued to refine the way radioactive decay revealed deep time. By 1927, he revised his estimate to three billion years. By the early 1940s, Holmes figured the Earth had had a solid crust for about 4.5 billion years. With a minor correction, this is still considered valid. The reason for Holmes' revisions is related to the way radioactive isotopes were becoming understood and the way measuring techniques and equipment became increasingly accurate.

In the early days, radioactive dating was like using an ancient grandfather's clock, tinkering to get the time right, knowing it will be somewhat reliable, but never perfectly accurate. The nuclear clock has been steadily ticking away ever since elements coalesced. Physicists found it works like this: Radioactive materials change with time at a steady, predictable rate. They lose their radioactivity as their unstable bits shed energy and become stable forms of matter. For example, uranium can be dangerously radioactive, emitting cancer-causing rays or exploding in nuclear bombs. Inside the Earth, radiation releases enormous amounts of energy, keeping our planet's interior hot with energetic material slowly transforming into permanent forms. Once uranium has shed some of its mass as energy and settled into a life of lead, it is stable and non-radioactive. In a rock sample, a comparison of the amount of uranium with the amount of lead yields its age. A specimen with quite a bit of uranium and virtually no lead is young. As it ages, there is more and more lead.

Given enough time, a considerable portion of the unstable uranium ends up as the stable element lead. Physicists cannot predict when any individual atom will spontaneously transform, but a lump of granite has billions of unstable atoms, so scientists predict the percentage of material that will undergo the change each second. Just as a theatre manager usually cannot tell which patron will be next to head over to the washroom during a three-hour film, the manager knows that a certain percentage will be walking at any moment; given a long enough program, everyone eventually leaves. Even though the transformation of any single uranium atom is unpredictable, the average rate is precisely known. Half of the unstable isotope U-235 becomes lead (Pb-207) in 704 million years. In twice the time period, 1,408 million years, three-quarters of the original U-235 is Pb-207. For this particular kind of uranium, geologists simply compare the proportions of lead and uranium. You may wonder how they account for any lead already in the lump of granite when the rock first formed. It is easy – there wasn't any.

Uranium is sometimes found in the mineral zircon. Zircon is a common mineral, plenty is found in ancient crustal rocks. Pure zircon has one zirconium, one silicon, and four oxygen atoms stitched together in a molecule. When zircon molecules form, uranium frequently sneaks in, replacing the zirconium atom. Then you have a molecule with one uranium, one silicon, and four oxygen atoms. Lead never combines this way during zircon's formation – there is never a newly formed zircon molecule with a lead atom replacing the zirconium atom. Geologists inspecting zircon minerals in igneous rocks realize every bit of lead in the mix is there only because in its former life, the lead was uranium, and then radioactively decayed into lead. They can be certain that a bit on zircon with equal amounts U-235 and Pb-207 is 704 million years old.

There are other isotopes of uranium – U-238, for example, also ends up leaden, as Pb-206, but at a much slower pace. Using two varieties of uranium decaying into two different types of lead improves age-dating accuracy. It is like having someone check your math when you examine a cafe tab. Geophysicists plot the ratio of one against the other on a concordia graph and the rock's age jumps out.

In addition to uranium, various other radioactive isotopes also give good age-dating results. The extremely common element potassium has an isotope that becomes argon gas. Igneous rocks, of course, start as melted rock. During melting, any preexisting argon gas is released from the rock into the atmosphere. A newly cooled solid igneous rock has no argon residing in it. As time goes on, some of the potassium in the rock decays into argon which is trapped in the solid matrix. In a lab's vacuum chamber, the rock is heated, the argon liberated, and its volume is calculated. From the ratio between the remaining amount of potassium atoms and the liberated argon atoms, scientists can calculate the sample's age. Arthur Holmes, working in his 1911 undergraduate laboratory, started with uranium and created the world's first calibrated geologic timescale. Before Holmes, geologists had no idea that the Jurassic period started 200 million years ago. They only knew it was rather old. They had figured the Jurassic, with its dinosaurs, was older than the Tertiary, with its sabre-toothed cats, but Holmes was the first to assign actual years to the previous relative scale. He quantified the broad generalizations made by earlier generations of geologists.

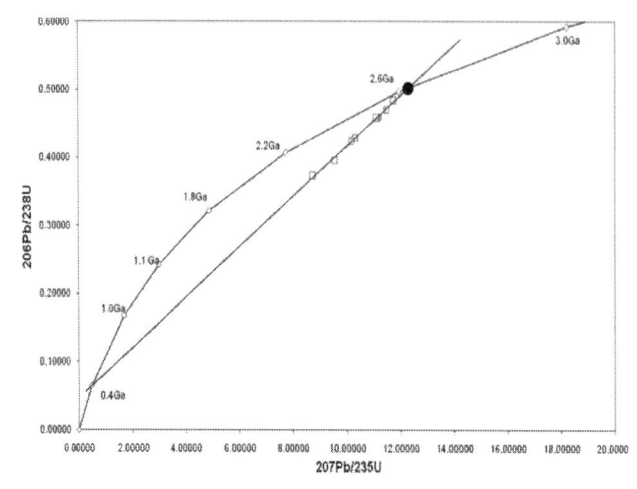

Concordia graph: This rock's age is 2.6 billion years.

"Every radioactive mineral can be regarded as a chronometer registering its own age with exquisite accuracy," said Holmes.[221] Every rock is a clock. With his method, Holmes, in 1913, found the oldest rocks in his collection dated back 1,600 million years. Arthur Holmes proved that the geologists had been right in their decades of battle against Lord Kelvin, T.C. Chamberlin, and the many others who steadfastly disapproved of suggestions of extreme antiquity. Based strictly on the geologic record, with "miles and miles of stacked layers" of stratigraphy, Holmes realized, "Earth history could not comfortably be squeezed into less than 100 million years."[222] Nor need it be. The rocks' natural radioactivity spoke the truth. Still in his early twenties, Holmes had secured his place as one the century's most important geologists. But his creative mind was not finished solving the great mysteries of the Earth.

At an early date, Holmes accepted Wegener's theory of continental drift. In 1931, he published "Radioactivity and Earth Movements,"[223] which described his

Lonely Voices

idea of drift's power source. It was the first real description of the mechanism of crustal movements. To Holmes, the evidence was clear. In Burma, exploring for oil, he had seen the aftermath of tectonic activity that had formed deep sedimentary basins. Holmes could even explain how the massive continents were transported along the planet's surface. He argued that convection currents of hot plasticized rock carried the mantle's heat up to the lithosphere where it cooled, crept along the surface, then sank back into the mantle, a process known as convection. We park our furnaces in the basement because air expands and rises when heated while cooler air sinks back to the cellar, creating a useful convection current. That's what happens on a grand scale inside the Earth. Currents of molten rock, hot from the lower mantle, rise, spread out, and push the continents into each other, forming mountain ranges. Or they drift apart from each other, creating seafloors, according to Arthur Holmes.

His 1944 book, *Principles of Physical Geology*, concludes with a chapter on continental drift. This was the first college text to include such material. Holmes wrote, "Currents flowing horizontally beneath the crust would inevitably carry the continents along with them."[224]

In the preface to his book, Holmes writes, "While I have not hesitated to introduce current views, since these reveal the active growth of the subject, it should be clearly realized that topics such as the cause of mountain building, the source of volcanic activity and the possibility of continental drift remain controversial just because the guiding facts are still too few to provide a formation to more than tentative hypotheses. It is my hope that recognition of some of the outstanding problems may stimulate at least a few of my readers to cooperate in the attempt to solve them."[225] Eventually they did, but it took decades. Until then, most geologists were unmoved.

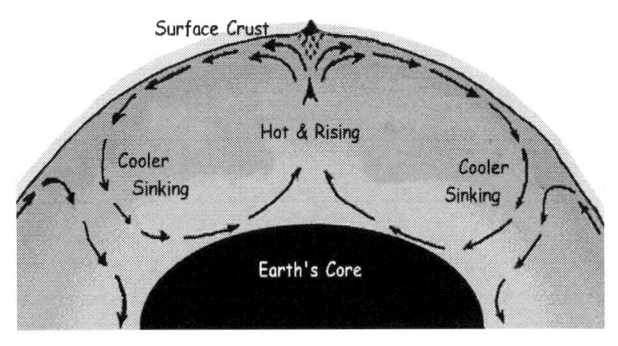

Earth's convection currents (shown as black arrows) as imagined by Holmes

Another Voice

Among the few who agreed with Wegener and Holmes was South Africa's Alexander du Toit[i]. Du Toit was particularly convinced of Wegener's theory – he had studied rock outcrops and fossils in both South Africa and South America and

---
i Alexander du Toit (1878-1948), South African geologist.

considered them separated siblings. His studies assured him the two places had previously been connected. The Cape Town geologist was descended of French Huguenot immigrants, earned his mining engineering degree in Scotland, studied geology in London, then went north again to lecture in mining and geology at the University of Glasgow. Du Toit finally returned to Africa when he was hired by the Geological Commission of the Cape of Good Hope. From 1903 to 1910, he travelled the whole of southern Africa on foot, ox cart, and ordinary bicycle with a mapping table slung over his shoulders, unravelling Africa's geology.[226]

Du Toit became the world's expert at understanding the complicated system of rocks that cover two-thirds of South Africa. The Karoo Supergroup is a sequence of sedimentary and igneous layers of relatively recent age that extend an enormous distance beneath the surface – they are up to 12,000 metres deep in places. The volume of material in the Karoo is incredible – all of it deposited or intruded within a geologically short period. It creates a strange mish-mash of layers that include one the world's largest deposits of coal. Du Toit spent years mapping and deciphering the whole convoluted mess.

By 1923, the Carnegie Institute in Washington recognized du Toit's talent and granted him finances to look, ponder, and work abroad. A geologist's sabbatical. With wife and young son, he went to Brazil, Uruguay, and Argentina for five months. It was an inspired choice – du Toit's previous mapping had led him to the edge of Africa, where the rocks were broken off at the Atlantic coast. He wanted to see if he could continue his map, skipping the Atlantic and picking up in Brazil. In South America, he found those rocks. And the same animal and plant fossils as he had observed in Africa. Stunningly, he also found the same complex rock sequences – stacked layer on layer – exactly as he had mapped at home. About South America, he wrote: "I had great difficulty in realising that this was another continent and not some portion of one of the southern districts of the Cape."[227] He was convinced of Wegener's theory, believing it was the simplest and most logical explanation for similarities between the southern continents. It was more convincing to him than mysteriously disappearing land bridges. Besides, although some animals might amble across a 3,000-kilometre land bridge, geologists were reluctant to suggest rocks would make the same trip. Forget land bridges, said du Toit. Pangaea had broken and drifted.

Alexander du Toit, 1898

Lonely Voices

The immediate result of the Carnegie Institute grant was du Toit's 1927 book, *A Geological Comparison of South America with Africa*. It showed fossil, stratigraphic and radioisotope evidence that clearly indicated (to du Toit, at least) these were separated from the same original supercontinent. In his book, du Toit wrote that the match between the continents "has consistently been extended by each fresh geological observation until at present the amount of agreement is nothing short of marvellous."[228]

Du Toit discovered a trove of high-quality evidence supporting continental drift, then refined his ideas in a second book, *Our Wandering Continents*. It was published in 1937, a few years after Wegener's death and was dedicated to that great scientist. Alexander du Toit was convinced he had proof and was irritated that most American geologists remained fixed in their disbelief. But the parting of continents in the southern hemisphere is more clearly obvious than it is in the northern. In the south, the split caused by the Atlantic's birth cleanly cuts across the tectonic features and geological outcrops of South America and Africa; in the north, the distinction is smeared. Caledonian mountain-building runs parallel to the Ocean, making continuities ambiguous. This is not to imply all geology is local, nor does it excuse the attacks levelled against Alexander du Toit by his northern colleagues.

Du Toit continued to be a respected geologist on the world stage, even after his unequivocal acceptance of continental mobility. He was by all accounts likeable

Du Toit, his wife and son, finding clues of Pangaea in South America, 1923.

and possessed a great sense of humour. During some of the darkest days for the continental displacement theory – the late 1940s – a group of American geologists appeared in South Africa, intent upon exploring palaeontological and geological sites. They specifically asked that the country's greatest geologist, Alexander du Toit, join them for part of their tour. South African Sydney Brenner, who was studying geology at the time, but was later awarded the 2002 Nobel Prize in Medicine for genetic decoding, tells of meeting Alexander du Toit on that tour. Brenner led the 1947 expedition of University of California geologists. Alexander du Toit, who was approaching 70, turned up "dressed like a Dutch Reform Church preacher in a black suit with a waistcoat, and a tie, and a hat. All the Americans had turned out in expedition outfits – with bush hats with little leopard skins. I had come out in my thing – shorts and boots and a shirt and hat – but du Toit arrived dressed in this beautiful suit."[229] Brenner continues to describe the tour, riding in a big truck, where talk unexpectedly turned to water divination, the pseudo-science of wandering about the countryside with a stick, expecting it to point downwards when underground streams were crossed. The Americans were aghast when the world-famous geologist told them that he believed it worked.

Du Toit told the driver to stop. They parked in the middle of a large featureless section of the Kalahari. Du Toit rustled around and cut a branch from some bramble in the desert, and with much formality and ritual, prepared to find underground water. As du Toit walked along the road, the stick suddenly pointed down. "Aha," he said. "There must be a fault below us and water has been flowing through it. Let's follow it downslope and see where it leads." In a short while they found a small feeble spring had ruptured through the hillside, exactly where du Toit's divining rod predicted. The American visitors, according to Brenner, were astonished and impressed. They came away with a new respect for both du Toit and water witching. Back in the truck, speaking privately in their Afrikaans language, du Toit told Sydney Brenner that fifty years earlier he had passed the same spot with his ox and wagon, doing a geological survey. He hadn't discovered underground water with a twisted stick, he had remembered the spring from his early mapping days.

The famous field geologist was greatly respected, but his monumental book about wandering continents was not well received. Although packed with detail, it met a barrage of criticism. Chamberlin, still head of geology at the University of Chicago, dismissed it, saying "Certain advantages of the hypothesis are at once apparent. If a landmass can be drifted for a desired distance and can, in addition to drifting, be rotated either clockwise or counterclockwise at the wish of the manipulator, the possibilities of producing a given fit, pattern, or arrangement with other similarly movable landmasses must be very considerable. Such maneuvering has few restrictions."[230] Of course, this was pure sarcasm. Chamberlin derided the methods of Wegener and du Toit as sloppy, undisciplined, and unscientific. So it was perhaps Chamberlin whom du Toit had in mind when he started his book with the lines "Looking back dispassionately into the history of geology, it is interesting to observe how deeply conservatism appears to have become entrenched. Particular theories have come to be so widely accepted that any doubts regarding their validity are apt to be overlooked."[231] Du Toit could not know it, but such conservatism would remain entrenched at least another 30 years, though he did much to persuade change. However, resistance from reactionary geologists was relentless. The Earth was cooling and contracting. Or the Earth was inherently designed stable and permanent. Either way, the continents were not drifting. Du Toit's book, *Our Wandering Continents,* also suffered from Alexander du Toit's unfortunate writing style. His sentences were too long. Too ornate. Too dramatic. Like this one:

> "The dumbfounding spectacle of the present continental masses, firmly anchored to a plastic foundation yet remaining fixed in space; set thousands of kilometers apart, it may be, yet behaving in almost identical fashion from epoch to epoch and stage to stage like soldiers at a drill; widely stretched in some quarters at various times and astoundingly compressed in others, yet retaining their general shapes, positions and orientations; remote from one another throughout history, yet showing in their fossil remains common or allied forms of terrestrial life; possessed during certain epochs of climates that may have ranged from glacial to torrid or pluvial to arid, though contrary to meteorological principles when their existing geographic positions are considered – to mention but a few such paradoxes!"[232]

His expanded thesis was slightly different from Wegener's, but fully supported the idea of drifting continents. Du Toit deduced there were originally two supercontinents (Wegener envisioned one.) and these two very ancient landmasses were separated by the Tethys Ocean, a sea that does not exist today, and perhaps never did. The Tethys would have stood between ancient northern Laurasia and southern Gondwana. Combined, they created Pangaea. The massive empirical evidence generated by the renowned South African explorer made it difficult for mainstream geologists to brush aside all evidence of continental drift. Difficult, but not impossible.

Two years after du Toit died in 1948, T.W. Gevers, du Toit's close friend for 25 years, praised du Toit as one of the world's greatest geologists, but still couldn't resist adding a disclaimer in his obituary, writing, "notwithstanding the zealous and valiant efforts of du Toit and others, there has in recent years been a marked regression of opinion away from continental drift."[233] Although nothing had been discovered which weakened continental drift theory, geologists had clearly distanced themselves from any appearance of interest in the idea. With the death of du Toit, drift became a truly orphaned theory.

Despite the combined efforts of Holmes and du Toit building on Wegener's work, throughout the 1940s the idea of continental displacement had become almost universally condemned and ridiculed. Gevers attributes to some unnamed geologist the idea that drift theory is "only a drunken sialic upper crust hopelessly floundering on the sober sima."[234] Sialic being the vogue term for silicon-aluminum continental crust and sima the silicon-magnesium ocean crust. Even without these definitions, you can taste the disgust. At the same time, American geologist Bailey Willis added: "I confess that my reason refuses to consider 'continental drift' possible. . . that hypothesis should, in my judgment, be placed in the discard."[235]

With displacement discarded and contraction theory fading in the presence of contradictory evidence, there arose a few novel ideas to account for the things Wegener, Holmes, and du Toit observed. Some achieved a level of short-lived notoriety. Even Einstein endorsed one such fanciful notion.

13

Slipping and Sliding

We have seen that some of the most interesting – and ultimately correct – ideas on the origin of mountains came from outsiders: Wegener, the weatherman; Mantovani, the violinist; Taylor, the hobbyist. So we might hope Hapgood, a history professor, would also contribute some useful insight into the Earth's restless transformations. Especially when Einstein reviewed his work and gave it his reserved endorsement. But our expectations for anything more than drivel from Charles Hapgood[i] are unrewarded. In his first book, *The Earth's Shifting Crust*, published in 1954, Hapgood explained that the entire crust of the Earth can slip around in one massive, unbroken chunk, changing climates and creating new geography. The surface upon which we stand occasionally becomes top-heavy and spins wildly, like an unbalanced top, he wrote.

Hapgood speculated that every twenty or thirty thousand years, due to shifts in weight caused by melting polar ice and the spinning planet's centrifugal force, the skin of the Earth suffers a catastrophic skid. The ultimate carnival ride. "Some will be shifted nearer the equator, and others farther away. Points on opposite sides of the Earth will move in opposite directions. For example, if New York should move 2,000 miles south, the Indian Ocean, diametrically opposite, would have to be shifted 2,000 miles north."[236] Hapgood imagined the solid crust's great leap was enabled by a slippery zone which was "immediately under the crust and thought to be extremely weak. . . extremely plastic fluid, but very stiff, as tar might be."[237]

Hapgood did not originate the theory he promoted. It was a rehash of an old discredited notion, originally proposed in 1863. Charles Lyell reported that a scientist he identified simply as J. Evans had described a slipping crust in a paper presented to the British Royal Society. Evans was attempting to explain cycles in the Earth's climate and show how it might be possible for glaciers to have once formed in places now along the equator. He proposed "former changes of climate might be connected with the sliding of a solid shell over an internal fluid nucleus."[238] Hapgood rode this dead horse far across the field of geology.

---

i   Charles Hapgood (1904-1982), American history lecturer.

Mechanically, Hapgood's resurrection of the old conjecture falls into the realm of plausible. However, the idea of a skidding crust is very different from Wegener's continental drift theory. In Hapgood's book, upon considering continental drift, he mentioned Caster and Mendes, two geologists who wanted to test Wegener's theory. Hapgood reported the pair spent months in South America, travelled 25,000 miles making field investigations, and compared details of the geology of South America with Africa. They traced the steps of du Toit. Hapgood wrote, "Their conclusion was that the rock formations did not prove [Wegener's] theory. Neither, however did the evidence disprove it. They added, 'Only time and more facts can settle the issue'."[239] With this, Hapgood then suggested a better theory – his.

The main reason Hapgood felt his system was more viable than drift was he included a driving force, a power strong enough to move continents – namely the unbalanced slipping of the entire crust. Hapgood wrote "the rock under the oceans, which Wegener had thought to be plastic enough for the continents to drift over it, is in fact very rigid. . . It is therefore impossible for the continents to drift."[240] His statement is partly correct – continents are not plows. Hapgood was restating the most common objection to Wegener's theory. Yet, to account for glacial remains in the Sahara, Hapgood needed to move the crust.

Albert Einstein's public endorsement of Hapgood's book – Einstein even wrote its foreword – drew some attention. The two had met at least once. They exchanged a few letters. When Charles Hapgood completed *The Earth's Shifting Crust* in 1954, a few months before Einstein's death, he sent the manuscript to Einstein's Princeton office. Einstein enthused that when he first heard of Hapgood's idea about the Earth's geological history, the concept "electrified" him.

"His idea is original, of great simplicity, and – if it continues to prove itself – of great importance to everything related to the history of the Earth's surface," wrote Einstein in the foreword. But Einstein cautioned that if Hapgood's theory was right – if the crust could slip because of polar ice melting, then perhaps other imbalances (mountain building or earthquakes) could also cause a slippage – which Einstein thought unlikely. Nevertheless, Einstein wrote that the constantly increasing pressure from polar ice "will, when it has reached a certain point, produce a movement of the Earth's crust over the rest of the Earth's body, and this will displace the polar regions toward the equator."[241] From his introduction to Hapgood's book, we can safely say Einstein accepted Hapgood's idea of the Earth's skin slipping around. At least tentatively.

We don't know if Einstein would have supported the sliding unbalanced crust theory after plate tectonics was developed. However, we now recognize a very basic weakness in Hapgood's theory. Glacial ice on a continent depresses it. When the ice melts, there is a slow glacial rebound, a restoring of balance which destroys Hapgood's imbalance theory. Further, if such events had resulted in the entire crust of the planet slipping massively, there would have been observations

Slipping and Sliding

in the rock and climate record confirming it. These are sketchy, at best.

Charles Hapgood's name appears in a variety of interesting places. His unfinished doctoral thesis was a study of the French Revolution. He lectured history, economics, and anthropology – but not science. Among his wilder claims, Hapgood advocated the authenticity of a huge trove of 32,000 ceramic dinosaur figurines found in Mexico in the early 1940s. He mistakenly declared that the figures of dinosaurs were crafted by an ancient civilization, suggesting that the *Triceratops* and man co-existed. The clay figures (which included horned humans and bearded Caucasians) were credited by Hapgood to a lost culture that lived thousands of years ago in central Mexico. He became a darling of some religious groups who believed the figurines confirmed a young-Earth and supernatural creation with dinosaurs and humans living side-by-side. But later testing revealed the ceramics were fired in a kiln during the 1930s. An investigative archaeologist, Charles Di Peso, reported they were fashioned by a local family with a furnace. Their models were taken from comic books.[242]

Hapgood had other interests. He spent ten years helping occultist Elwood Babbitt contact Mark Twain, Jesus Christ, and the Hindu god Vishnu. According to Hapgood's books, they were successful on all three accounts. But among geologists, Charles Hapgood is best known for his resurrection of the hundred-year-old theory about the Earth's crust, which was once predicted to slip violently every few millennia. Possibly because of Hapgood's books on Atlantis and the Mexican dinosaur figurines, scientists rarely glance at his peculiar stab at geology. The theory, once good enough for Einstein to consider, now has only a small following among advocates who suspiciously feel mainstream science is neglecting some rather clever ideas. One example of a revamp of Hapgood's geology is the 1997 Richard Noone book, *5/5/2000 ICE: The Ultimate Disaster*. The book predicted a massive shift in the Antarctic ice cap concurring with solar storms and planetary alignment, causing a cataclysmic crustal displacement of the Hapgood kind. Noone prophesied it would happen on May 5, 2000.[243] It didn't.[i]

Hapgood's bold assertion aside, a key reason the theory of continental drift remained rejected for decades was the lack of a believable mechanism that could power the movement of continents. Arthur Holmes had published and promoted the convection theory, but contemporary geologists saw continents as huge, heavy, immobile structures, while the crust was clearly too rigid and solid to break apart. Intuitively, the continents were not going anywhere, even with the force of convecting mantle currents – which the majority of geologists did not even believe existed. As recently as 1954, when Hapgood promoted his notion, scientists were still a long way from understanding the origin of mountains.

---

i   The author's name, Richard W. Noone, may be a pseudonym and the book may have been a hoax as Noone hasn't been found since the book's publication. Of course, this minor detail has not kept his tale of death and destruction off conspiracy theorists' websites where the science is considered good and the predicted doomsday date was simply miscalculated.

The Mountain Mystery

A Bankruptcy of Ideas

Less than a hundred years ago, one of the world's best geologists lamented to a friend that no clear idea existed that explained the planet's mountains. In 1921, Reginald Daly[i], head of geology at Harvard, told mathematician Walter Lambert, there was "a bankruptcy in decent theories of mountain building."[244] The most widely accepted idea was still one or another form of a contracting-earth.

As we have seen, contraction was first proposed centuries earlier by Bruno, then refined in Suess's dried-apple model. We have also already noted the various problems with this scheme: contraction would have created regularly spaced mountain ranges, which the Earth lacks; and, the amount of shrinking based on Lord Kelvin's physics was not enough to create the phenomenal crustal-compaction found in the Alps. With the discovery of radioactive heat, the rate of cooling was realized to be slower than early scientists expected. This, too, was challenging the idea that mountains arose because the planet was cooling and shrinking.

Thus, rather than cooling, a few geologists speculated contraction might happen by other means. They suggested molten rock squeezed from the Earth shrinks the globe's volume in some places while padding mountains in others. However, the amount of extrusive rock on the planet's surface is not enough to explain the required crustal shortening. To make up the difference, still others proposed an idea called chemical shrinkage. Radioactive decay showed that some forms of mass turn into others. It was proposed that perhaps some rocks decompose into helium, a gas which eventually drifts into space. This would leave the Earth smaller. But there is a bit of a problem with this sort of shrinkage – there are no rocks that actually decompose into helium (although alpha particles, which are similar, may be emitted). Even if we haven't yet discovered how nature might reduce entire stones to helium, the numbers don't add up. Assuming helium has been disappearing at a steady rate, the Earth has lost six billion tonnes of helium over the past 4 billion years – this seems to be a huge quantity, but the Earth is a trillion times more massive. So the lost gas is not making a mountain of a difference, but instead amounts to a paper's thickness of global shrinkage.

Not every geologist explained mountains and oceans in terms of a shrinking, wrinkling Earth. Reginald Daly, the Harvard geologist who noted the bankruptcy of ideas about mountain building, thought a type of drift theory be right. Daly published furiously – his first paper, in 1896, was followed by 150 more. His last came in 1957, the year of his death, when he was 86 years old. Between those bookends were ground-breaking papers about rocks (how they form), the moon (how it formed), and the Earth (how it was constantly re-forming).

Daly was the youngest of nine children born on a remote Ontario farm. His grandfather had immigrated from Ireland, arriving in Ontario just before the Irish

---

i   Reginald Aldworth Daly (1871-1957), Canadian geologist, Harvard professor.

Slipping and Sliding

Potato Famine. In Canada, the family first settled as farmers, but later moved to a town along the Saint Lawrence River. At Victoria College (now part of the University of Toronto), Daly excelled at everything. He won prizes in subjects ranging from English literature to astronomy. The college recognized his broad talents with a gold medal in general proficiency.[245] After completing a maths program in Toronto, Daly began to teach. According to his biographer, Francis Birch, Daly switched from math to geology the moment Professor A.P. Coleman held up a piece of granite and remarked, "This is made of crystals."[246] Until then, Daly knew almost nothing at all about geology. Having already earned two degrees, Daly went to Harvard where he quickly completed a geology doctorate. In 1896, he travelled to Germany and France where he learned how to slice rock into thin translucent wafers, like tiny stained-glass windows. Viewed through a microscope, sometimes with polarized light, thin-sections are a reliable way to identify the minerals of a rock. It was a novel idea, brought to perfection by Harry Rosenbusch in Heidelberg. Daly was one of the first North Americans to master the technique of slicing rocks with diamond blades and polishing the slivers to 30 millionths of a metre – slightly less than the thickness of a hair. This technique proved especially indispensable for identifying minerals in igneous rocks where heat had melted and mixed components into an indistinct jumble.

Daly returned to teach geology at Harvard. But he soon agreed to spend six years battling mosquitoes, heat, wind, and miserable cold along the western border between Canada and the United States. He became the field geologist for the Canadian International Boundary Commission, which was established to bolster Canadian sovereignty – and to discover mineral riches hidden along the international border. Daly took the job because he wanted to be in the field doing original research. Ten years later he would return to Harvard yet again, this time as department head. But during his six field seasons on the frontier, Daly mapped the border from the Pacific Ocean to the Great Plains, a rugged swath 400 miles long and 10 miles wide. The project ultimately included three thousand rock samples, thin sections, and photographs. Daly produced dozens of lake soundings, stratigraphic and structural maps, and studies of petrology and morphology. In 1912, he filed his final report with the Geological Survey of Canada – a massive 3-volume tome he called *North America Cordillera: Forty-Ninth Parallel*.

When Daly's work with the Geological Survey of Canada was nearly finished, he was invited to teach at Massachusetts Institute of Technology, but still spent two more summers in the Rockies, mapping and collecting samples along the Canadian Pacific Railway. He wanted to discover precisely how igneous rocks formed and why mineral content varied so much in the thousands of different melted, cooled, crystallized stones found around the world. There was no systematic catalogue for igneous rocks until Daly pursued the task.

During his many years of frontier research, Daly did nearly all his field work alone or with just a temporary assistant. One such field hand was fellow Canadian

The Mountain Mystery

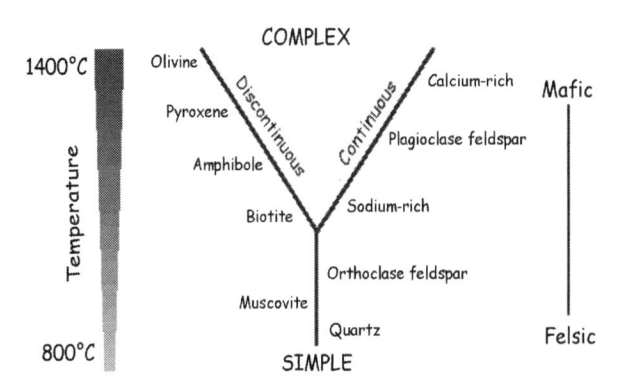

Bowen's Reactive Series with igneous rocks grouped by temperature and chemical composition.

Norman Bowen, a geophysicist who also had a keen interest in the way the igneous rocks of the west had formed. Bowen invented a chart, the reaction series, that relates temperature and pressure to the way minerals group themselves into rocks. His simple chart, growing out of that frontier research, is an elegant diagram that explains why certain minerals frequently occur together while others never do, even when those minerals contain many of the same elements. Bowen's Reactive Series became a key concept in understanding the basic building blocks of the Earth and identifying the depth and conditions required to form igneous rocks. His chart is still a basic memorization requirement for first-year geology students. Daly and Bowen would later bitterly spar over what appears to be an arcane aspect of the chart's implication – the role limestone plays during basalt's creation.

Daly felt basaltic rocks were the most critical material on the planet. No rock type was more important than Daly's "bringer of heat,"[247] the magmatic material that spewed up from below, affecting every landmass, building every ocean floor. Basalt is usually grey, dull, and monotonous, but Daly saw beauty in it. To him, basalt built the planet. We will see he was very nearly right.

From the Canadian perspective, part of the goal of the frontier Boundary Commission was to reinforce independence from the United States. Alberta was just becoming a Canadian province. It was apparent the vast region was awash in minerals. Governments on both sides of the border wanted to catalogue their resources, to tally the oil, gas, coal, copper, silver, and gold in the western mountains. Daly was a Canadian who felt strongly that economic geology, as mining and oil exploration was called in those days, was crucial to his young nation's progress. A proper inventory of the country's riches, especially in the Rocky Mountain border zone, depended on regional geological studies. In trying to understand the igneous rocks he encountered on the boundary study, Daly arrived at "the realization that geology must be based on geophysics."[248] This basic truth – the role of the geophysical manipulation of rocks and earth materials – would ultimately be the key to understanding all the planet's landforms.

But geology encompasses much more than broad studies of the Earth and it can explain a lot more than the origin of igneous rocks – or the origin of fossils

Slipping and Sliding

and mountains. Sometimes geology is used to figure out if manslaughter has taken place. It was Daly's experience in the Rocky Mountains that thrust him into one of his most controversial studies: a report he helped write about a coal mine blamed for destroying half of a town. Today, Frank, Alberta, is an eerie site.

Frank's Slide

One morning I walked among Frank's gravel, rocks, and huge boulders – some larger than the houses buried in the rubble below my feet. I tried to imagine the sound of the landslide that began on the face of Turtle Mountain, then crashed into the village. At four o'clock one cool wet April morning in 1903, the 2,200-metre mountain lost a broad slice of its shell. The previous day, townsfolk had heard creaking sounds from the slopes above them. The ominous groaning continued for a day, but it took just 90 seconds for eighty million tonnes of rock to break loose, race down the ridge, cross the new Canadian Pacific rails, drive through the town of Frank, and slide past it. It was one of the deadliest landslides ever in North America. Of the 90 people killed, only twelve bodies were found, the rest still lie under fallen limestone. Under the mountain's collapsed face, the coal mine entrance was erased, trapping miners inside. They moled their way through metres of coal and limestone to escape. The view from their perch above the village was devastating – much of the town was gone. Almost immediately,

Turtle Mountain and the buried town of Frank, shortly after the 1903 landslide

the huge coal mine that had been burrowed into the side of the mountain was blamed.

Reginald Daly had seen the mountain earlier, during his survey of the borderlands. He was regarded as an expert on the rocks and mountains straddling Montana and Alberta, so he was asked to help a commission explain why Turtle Mountain had turned deadly. The ridge had been named Turtle Mountain twenty-five years earlier by a rancher who thought the mountain's profile resembled a tortoise. But the natives had always referred to Turtle Mountain as the Mountain That Moves. They avoided the place and did not camp below its slopes. They knew that landslides had occurred there in the past.

The coal mine was huge, hastily constructed, and dangerous to the men crawling into it. But activities at the mine were not the main reason the town was obliterated, Daly concluded. Although, of course, without Henry Frank's mine, there would have been neither town nor disaster. The town of Frank, namesake of a 48-year-old businessman from Butte, Montana, had been destined to be the Pittsburgh of the west, said its creator. Henry Frank built Frank, Alberta, with the same hustle and commitment he applied to everything he encountered.

At 26, Frank moved west from Ohio to Montana. His first venture in Butte was a wholesale liquor business, operated out of a log cabin with a dirt floor. He invested his profits in mines in Montana, Idaho, and British Columbia. Henry Frank built a log cabin saloon, founded the Butte Water Company, and started Silver Bow Electric, bringing power to his town.[i] But he made his fortune from a gold mine that he sold for three and a half million dollars in 1899.[249]

Henry Frank had built the town of Frank, Alberta, after geologists pointed out Turtle Mountain, a ridge with thick exposed layers of coal. The Canadian Pacific Railroad had recently opened a line through southern Alberta, alongside the mountain. With easy rail access and exposed coal, the town of Frank was set to thrive. But the mountain destroyed those plans.

Along with miners and their families, the disaster also took a toll on Henry Frank. A few years after the town was destroyed, Frank was visiting his mother in the east. He was only 57, but he was fatigued and feeling nearly as exhausted as his older coal pits. His health may have been weakened by the chemicals and stale air from his mines, plus pollution and smoke from copper smelting near his Montana home. In the lobby of a posh Chicago hotel, Frank was rambling and incoherent. Police were called. Newspapers were not kind, running the headlines "Rich Man Insane"[250] and "H.L. Frank Demented."[251] Friends reported he was depressed. A few months later, Henry Frank died a suspicious death, as police were wont to describe suicides of the well-placed. Everything he had was bequeathed to charities.

---

[i] Henry Frank was described as liberal, enterprising, a life-long Democrat, a thirty-third degree Mason, and a member of the Elks and the Knights of Pythias. Frank was mayor of Butte, served two terms in the Montana Legislature, and missed becoming a US Senator by six votes.

Slipping and Sliding

The 1903 Frank Slide wiped out two kilometres of rail track and all the surface buildings of Frank's Canadian American Coal and Coke Company. The mine had just opened in 1900, but at the moment of the collapse, it was the most productive coal mine in Canada, shipping 2,000 tonnes of coal each day. The owner was making a fortune, but working conditions were dismal for the immigrant labourers. Even today, you can see small caves dug into embankments where men slept, summer and winter, unable to afford better shelter. Many of the miners were dismally saving their pay, hoping to eventually bring their wives and children from central Europe.

The immediate blame among the town's survivors was the obvious target – the wealthy mining company's hurried greed to extract coal. Meanwhile, the rail track was rebuilt over the rubble of the landslide and reopened in just 17 days. Within a month, coal production was back to its peak. None of this looked right to the friends of the dead in the town of Frank. But geologists, including Reginald Daly, disagreed with their placement of blame.

The commission found Turtle Mountain's geology had caused its collapse. The mountain had a soft core covered by a limestone sheet. The limestone coating dissolved as acidic water percolated through it. Eventually, it no longer held the side of the mountain together. The commission blamed the shape of Turtle Mountain, heavy spring rains, minor earthquake tremors, and finally the excavation of large coal caverns within the mine. Although mining was mentioned, the 1904 summary stated, "The steep slopes, the shattered and fractured nature of the rocks, coupled with unusually heavy precipitation are causes which in themselves are quite sufficient to have produced the slide."[252] The face of Turtle Mountain was ready to collapse. The mine and its owner were not considered culpable, except in the sense that it was the mine which had lured the people and had built the town. Although half the town was gone, mining continued. However, Frank, Alberta, with a population of just 112 today, never became the west's new Pittsburgh.

Daly Opinions

After his work at Frank in Canada, Daly was back in the United States. At Harvard, he was appointed professor of geology and immediately, chair of the department. He stayed for the next thirty years. By examining the samples he hoarded from his 700-kilometre trek along the mountainous Canadian border, he formulated a theory on the origin of igneous rocks. His stash of stones led to *Igneous Rocks and Their Origin* in 1914, a textbook that summarized his years of field work and his lecture notes. Daly showed that all the igneous rocks of the world belong to two types: intrusive and extrusive. Intrusive rocks are usually formed within the crust from slowly cooling melted rock. They become granite, diorite, gabbro. Extrusive rocks – rhyolite, andesite, basalt – are exposed to the surface while forming.

Mountains are largely built of granite materials; ocean floors are basaltic. Granite is colourful, pink and red, with large showy crystals; basalt is fine-grained, dark, dense, and grey; granite is lighter and would heap above the top of a container which the same weight in basalt would barely fill. Basalt is a flat barge of rock floating on the asthenosphere sea; granite is a tall sailboat with a deep stabilizing keel. On average, ocean crust starts four kilometres below sea level; continents float high with peaks topping eight kilometres and roots extending down fifty. For Daly, understanding these differences and deciphering rock origins were the heart of understanding all of geology.

Between his semesters of teaching at Harvard, Reginald Daly conducted field work that took him to Duluth, Butte, the Grand Canyon, and the Adirondacks and abroad to Europe, the Pacific, and Africa. Each summer it was somewhere different, it seems. In Hawaii, he helped establish America's first volcano observatory. In Sweden, he spent a season mapping ore deposits. His wife often worked alongside him. He had met her through their mutual friend, a mechanical engineer who worked at the Harvard physics lab where Daly occasionally commandeered huge presses and furnaces, trying to mimic the Earth's inner conditions of intense heat and pressure. His friend allowed Daly laboratory access and he introduced Reginald Daly to Louise Haskell. Daly is claimed to have had great respect for his wife, who graduated at the top of her Radcliffe class. She was an historian, taught literature at the women's school she owned, helped with Daly's field work, and edited his papers. In fact, Daly once told his wife's sister, "She had the finest mind I ever knew, in a woman."[253]

Daly was usually a sober, serious man – he never drank alcohol. Recognizing liquor was the cause of all manner of evil, he marched in New York City temperance parades and wrote long editorials in New York papers denouncing the foul stuff. In "Beer and Brutality," which he wrote in 1917, Daly pointed out that German war atrocities were the result of alcohol.[254] The Germans, it would seem, were fond of beer, claimed Daly, and that caused the disposition that started the World War.[255] Anti-German sentiment was powerful and popular. Daly, still a Canadian, saw his home country enter the war in 1914. It took almost three years before the Americans joined; by then the mood was feverish. But Reginald Daly's feelings were quite typical for the times.[256]

When the Americans finally joined the Great War, Daly served in France, as a librarian. After the war, the Carnegie Institute supported his mapping expedition to the South Pacific at about the same time it funded Alexander du Toit's field work in South America. Du Toit's trip resulted in compelling evidence that Brazil and Africa were separated twins. Similarly, the Carnegie grant that sent Daly to the Pacific led him to realize there was a huge and significant difference between oceanic and continental rocks, which in turn was vital to unravelling the Earth's evolution. Daly said it was consequential that "not a cubic inch of granite rock has ever been found in the hundreds of volcanic islands which dot the region of the central Pacific. . . These facts suggest that the suboceanic crust is composed of rocks decidedly different from the rocks constituting the continents."[257] Knowing that the ocean crust is fundamentally different from the continental crust enabled much that followed in earth science.

The field work of both du Toit and Daly contributed to solving the puzzle of the problematic mountains. The two men met in South Africa, when Daly was part of an international team focusing on the huge two-billion-year-old volcanic rock exposure in the central Transvaal which du Toit had already mapped during his bicycle trips. Shortly after the two met, Daly declared that du Toit was "the world's greatest field geologist."[258] That was a considerable compliment from someone who was surely the world's second greatest field geologist. Expanding on the compliment, Daly gave a very favourable review of du Toit's 1937 book about continental drift. That book review placed Daly squarely in the centre of the minority who agreed with Wegener and du Toit on the drift hypothesis.[259]

Reginald Aldworth Daly

On his return from visiting du Toit in Africa, Daly stopped at Saint Helena Island – a remote speck of volcanic rock in the south Atlantic Ocean where the British had established a prison. Among the past inmates had been various enemies of the British Empire, including (at various times) 5,000 Boer soldiers, the last king of the Zulus, and Napoleon Bonaparte. Daly spent a month mapping the island, being particularly intrigued by some of the unusual mineral crystals he discovered. "What a pity Napoleon was not a geologist," said Daly. "He would have found Saint Helena much more interesting."[260]

At the conclusion of his Saint Helena and South Africa expedition, Daly's field work abruptly ended. In thirty years, he had crossed the Atlantic 14 times, crossed America 24 times, and visited every American state (except South Dakota) at least once. Back at his Harvard office, he had several large cabinets stuffed with samples of the world's rocks. His travelling days were over. But his career was not. He would spend the next thirty years making sense of what he had seen and collected.

And Yet, It Moves

Daly's expertise was with melted, cooled, crystallized rocks – igneous rocks. He believed "this planet is essentially a body of crystallized and uncrystallized igneous material. The final philosophy of Earth history will therefore be founded on igneous-rock geology."[261] He also recognized the Earth's surface is unstable and constantly changing. In his compelling *Our Mobile Earth,* Daly wrote:

> "Our Earth is very old, an old warrior that has survived many battles. Nevertheless, the face of it is still changing, and science sees no certain limit of time for its stately evolution. Our solid Earth, apparently so stable, inert, and finished, is changing, mobile, and still evolving. . . lava floods and intriguing mountains tell us of the plasticity and mobility of the deep interior of the globe. . . the secret of it all – the secret of the earthquake, the secret of the 'temple of fire,' the secret of the ocean basin, the secret of the highland – is in the heart of the Earth, forever invisible to human eyes."[262]

Daly agreed with Wegener's theory that today's continents were once merged in a single supercontinent. And, of course, the continents later wandered off to their present positions. The cover of his 1926 book, provocatively titled *Our Mobile Earth,* has a sketch of the globe with an alluring subtext, borrowed from Galileo: *E pur si muove!*

Daly's book begins, "Let us run quickly through the history of the globe since its primitive single continent was completed. . . The reader himself has already savored the uncertainty which must attend any efforts to sketch the Earth's development. He may be able once more to forgive apparent dogmatism in the description."[263] Reviewer E.A. Hodgson described the book as an easy read for the average, educated, non-specialist: "The English is clear and forceful. Short sentences predominate. Rhetorical questions receive immediate and satisfactory answers. Startling facts are presented in a manner which serves to drive them home and fix them in the mind of the reader."[264] The ideas were dogmatic, said Daly; startling, said Hodgson. The same qualities that Wegener's accusers disparaged. Although Daly didn't convert many geologists, his contribution to continental drift science was significant. And he included reasonable suggestions about the mechanics of the force which moves the continents.

Daly explains early uncoupling of basalt and granite liquid rocks, due to their density differences, as the reason the Earth has continents. "The continental surfaces now stand so high because the granitic crust was folded and rafted together and therefore thickened. The rocks thus aggregated are especially light rocks. Today each continent floats higher than the denser, heavier basaltic crust under the ocean."[265] Igneous granite rose kilometres above the ocean floors' basalt because of its lighter weight and Daly employs isostasy to strengthen his case.

Slipping and Sliding

But what of all those stacked, thrusted, and folded *sedimentary* rocks heaped atop Daly's igneous crystals? For those layers of sedimentary rocks – sandstone, limestone, shale – Daly describes mountain-building rather counter-intuitively as the result of landslides. Massive accumulations, he wrote in *Our Mobile Earth*, slid into great near-shore chasms created by the weight of eroded sediments, buried deeper and deeper in geosynclines that existed on the single-continent, primordial Earth. Daly said the accumulation of sediments became so thick they sank far down into the hot basement of the subsurface and melted. Eventually everything expanded because of heat, and rose high on the surface in the form of mountains with granite cores under sedimentary rocks.[266]

In a speech in 1932, as President of the Geological Society of America, Daly said the true strength of the Earth was confined almost entirely to its thin outer crust: ". . . more than 97 percent of the Earth is too hot to crystallize, its body is extremely weak. The crust, being so thin, must bend, if, over wide areas, it becomes loaded with glacial ice, ocean water, or deposits of sand and mud. It must bend in the opposite sense if widely extended loads of such material are removed. This accounts for the origin of chains of high mountains. . . and the rise of lava to the Earth's surface."[267]

Although some of his theories were contested, his book was well-received. It was thoughtful, provocative, and entertaining. Daly describes many geological disasters: the wicked Lisbon earthquake of 1755, the still mysterious Charleston, South Carolina, quake that killed 27,[i] and the Gansu, China, earthquake where "the mountains walked in the night"[268] and trampled nearly 300,000. Dramatic and devastating as they certainly are, Daly puts earthquakes in perspective in his book, pointing out that an estimate of deaths from earthquakes since the dawn of written history was about 13,000,000 – "only about one-sixth of the rate at which the automobile is destroying lives in the United States alone [in the 1920s]."[269] Hodgson, at the Dominion Observatory in Ottawa, wrote, "Though one may disagree with Prof. Daly in regard to some of his theories, it must be granted that *Our Mobile Earth* is a veritable mine of accurate geological facts."[270]

Daly was comfortable with controversy. He knew some of his ideas were speculative. In describing mountain building, he reminds the reader, "Speculation is not science or knowledge. Speculation, even of the happiest kind, can do no more than point the way to possible future knowledge."[271] Later, addressing the Geological Society of America, he explained why geologists must hypothesize

---

i The Charleston earthquake, August 31, 1886, destroyed a quarter of all the buildings in the city. Its cause is still unknown – the area does not sit on a seismically active zone. Because of its unusual and unexpected occurrence, many assumed the destruction was the act of an angry God. Canadian Ezekiel Wiggins (known as the Ottawa Prophet) was told by an angel that a second, much more devastating, earthquake would strike the eastern seaboard one month later, on September 29. His prophecy created widespread panic. Believers quit their jobs and donned fashionable "ascension robes" in preparation for the Second Coming. Most were back at work the following Monday.

and speculate: The majority of the Earth is hidden from us, deep beneath the surface: "Ninety-nine and nine-tenths of its volume must forever remain invisible and untouchable."[272]

Although Daly understood the power of the mantle's convection currents and how they might propel continents, his 1926 book on crustal mobility ended by invoking the creation of the Moon as the main cause of the horizontal drift of the continents. On this point, Daly built upon the commonly held idea that the Moon was expelled shortly after the Earth formed. This left a gaping hole in the planet into which parts of the shattered supercontinent slid.

In explaining continental movement as an adjustment to the Moon's birth, Daly suggested the Earth has been off-balance ever since the lunar expulsion. The spinning Earth rounded its rough edges, slowly filling the hole created when the Moon departed. The crust slid until things were balanced again. His idea caught the public's imagination and made a fine rival to contraction theory. In his review of *Our Mobile Earth,* the science editor of *The New York Times* cleverly simplified Daly's theme as "Earth Shakes to Gain Symmetry."[273] It was a reasonable notion, but later, Daly himself would challenge his own idea.

Daly's moon-birth scheme was different from most of the lunar theories of his day. Daly opposed the common idea that the Moon was a captured wayward planet. A passing space rock would almost certainly be slung far away after encountering Earth's gravity or would come crashing into the Earth. The chances it would benignly enter orbit are slim. If a passing planetoid had been caught in the Earth's gravitational field, we and the man on the moon would not be related. With the Apollo landings, we discovered that our rocks and lunar rocks are indeed kith and kin. The moon was born at least partly of Earth. Using simple Newtonian physics, George Darwin once calculated that the Moon originally orbited much closer but has been drifting away, hence also supporting the idea that the Moon originated from within the Earth and was not some imprisoned passerby.

Reginald Daly differed with George Darwin's proposal that centrifugal force on the rapidly spinning primordial Earth ejected a molten satellite, even though this was a valid academic explanation for the Moon's origin at the time.[274] Daly was in partial agreement. His calculus confirmed that the lunar tidal drag on the Earth slows rotation, continually increasing the length of our day. Without the Moon's drag, the Earth would spin much faster. Your day would now last just 15 hours. The tidal effect results in a slow wayward drifting of the Moon, which was confirmed when Apollo astronauts put laser reflectors in a crater, and earthlings began measuring the lingering lunar farewell. Much of Darwin's speculation was correct, said Daly. George Darwin showed that the Moon originated from the Earth, but Darwin's calculations could not trace the Moon all the way back to the surface of the Earth. The primitive Earth could not have been rotating fast enough for a moon-sized blob to simply burst out of it. Daly also showed that the physics does not work – in fact, the Earth would have to spin ten times faster than it does

Slipping and Sliding

now, and no one can explain how that might have occurred without the whole planet ripping apart. Others suggested a passing star's gravitational field exorcised Moon from Earth – but this idea fails because we now know that stars rarely "pass by" the Earth on their way to wherever stars are going.

Daly suggested instead that the creation of the Moon was caused by an impact rather than centrifugal forces ejecting the satellite.[275] In the early days of the solar system, there were a lot of large hunks of space junk orbiting the sun. Much of creation's debris coalesced into planets. Daly thought that the same gravitational attraction that brought space debris together to build the Earth may have resulted in something huge striking our planet. He suggested it had ricocheted off the Earth and ripped the Moon loose. All of this would have occurred when our planet was perhaps only 50 million years old, hot and unsolidified. The mix of Earth with the impacting smaller planet spewed material which gravity later smoothed into our round lunar companion.

The birth of the Moon, by the way, is still not confidently understood, even today. However, since the 1980s, lunar scientists have mostly favoured a form of Daly's *Great Impact Theory,* otherwise known as the Big Splat. Our current Moon creation story is that during Earth's formation, more and more dust and stones gathered together into a larger and larger ball, forming our planet. Finally, a huge "Mars-sized" rock slammed into the Earth. The impact was rather devastating – the entire surface of the Earth melted. The ensuing smokey cloud of evaporated earth-bits reached 10,000 degrees. The mischievous rocky planetoid melded with the melted Earth, and a cloud of moon rock exploded off the surface. Some scientists think two moons formed – the second, somewhat smaller one, re-splatted. This, they say is why the two sides of the Moon are so different.[276] The face we see is largely flat and smooth, while the back side of the lunar head – the hidden far side of the Moon – has mostly rugged mountains. It is as if the man on the Moon needs a hair cut. The Big Splat (or perhaps Double-Splat) was the start of the earth-moon odyssey. The result is both a heftier Earth and a big orbiting satellite. The physics makes sense and the geology of the moon does not disagree with this scenario.

Thus Daly, the head of Harvard's geology department, attributed earthquakes, volcanoes, and the building of new mountain ranges to the readjustment, or symmetric correction, occurring because an impact melted the Earth's surface, and spun off a moon-sized glob of melted basalt. This event also generated tremendous mile-high tidal waves – not of water, but of molten earth-rock – that rippled across the surface as the new Moon passed closely overhead. This, Daly figured, tore at the planet, creating a deep wound in the side of the Earth, about the size of the Pacific. In fact, he thought it may have been the origin of the Pacific. Ever since, our planet has been in a healing phase with continents sliding apart to fill the void while gravity and centrifugal forces bring the planet's shape back to a more perfect oblate. In some ways, the mechanics of the movement are similar to the idea which Einstein endorsed – the idea that huge unbalanced segments of crust can slide to restore equilibrium.

Daly suggested the rebalancing and infilling of the lunar hole can be seen in the westward drifting of the Americas. He said evidence includes the long coastal mountain ranges (Rockies and Andes) which form the western spine of the moving Americas, crumpled by their resistance to sliding. The east coasts of the two continents are stretched out, meeting no resistance, and are mostly flat plains. It is a clever solution for the movement of the continents and several key points are basically correct, but the scheme is no longer considered feasible.

Twenty years after proposing that continents drift because of gravitational rebalancing following the Moon's creation, Daly changed his mind. He realized a different mechanism was likely responsible for the movement of continents. He tentatively endorsed the convection hypothesis of Arthur Holmes. Shortly before retiring from Harvard in 1942, Daly released *Strength and Structure of the Earth*, a book in which he anticipated parts of plate tectonics. He introduced a mesospheric shell, which he described as the part of the mantle sandwiched between the plasticized asthenosphere and the tough outer core. He also described a slippery zone, a glassy layer just below the asthenosphere coinciding with Mohorovičić's velocity discontinuity about 50 kilometres below the surface. The Croatian geophysicist had earlier shown that a major change took place at that depth. Daly expanded on the nature of the disruption, suggesting it could easily allow continental movement.

The geologist Reginald Daly lived a long and rewarding life. He had public recognition and interesting adventurous work. Daly was so much appreciated as a lecturer at Harvard he was frequently applauded by the hundred or so students attending his introductory geology classes. Fittingly, in addition to having craters named after him on both the Moon and Mars, the brilliant geologist also shares his name with dalyite, a rare, sparkly, zirconium-rich mineral found on St. Helena Island and in his native Canada.

Reginald Daly had advanced the science of the Earth immeasurably. But at the time of his death, in 1957, we were only slightly closer to understanding the creation of mountains and their ingrained fish fossils than Palaios was 3,000 years earlier. Ultimately, answers would be found at the bottom of the sea. But to arrive there, we would need to see the Earth in a new way. Surprisingly, the new approach came with the search for oil and diamonds.

14

Looking into the Earth

As Daly knew, and pointed out repeatedly, understanding the difference between continental crust and ocean basin is essential to any understanding of our planet's history. But it would take another generation before scientists discovered that ocean basins form because lava bleeds through enormous crustal scars, causing intense pressure that shoves seafloors and displaces continents. By this action, landmasses variously split apart, opening oceans, or push together, creating mountains. As we have seen, most geologists in the early twentieth century still felt the Earth was cooling and shrinking, resulting in mountainous ridges separated by canyons. For these scientists, ridges became continents; canyons became oceans. We have noted flaws in this idea – Earth's mountain ranges are not distributed the way a drying apple's bumps are, and the contraction system can not differentiate continental granite from seafloor basalt. Seeing flaws in the cooling theory, other geologists were convinced the expansive heat of volcanic action created continents. The surface swelled across vast areas because of heat and pressure below the surface. The occasional volcano erupted when local pressure was stronger than the crust. This is occasionally true, but there is more to the story of mountains than hot pressure points.

Over 80 percent of the Earth's surface has been formed by volcanic activity – either slow basaltic flows, explosive volcanic eruptions, or granite seeping between rock layers. There are lava fields, spreading seafloors, exposed intrusions and extrusions. Volcanics affected most of the Earth's surface, but today's igneous activity is less frequent than in past ages. Scientists now keep track of about one thousand active volcanoes. This doesn't include the seafloor's rifts, which some geologists count as many thousands, while others see them as a single skinny 70,000-kilometre-long volcano. Above sea level, fewer than a hundred volcanoes show any signs of restlessness during any given year. People who count such things figure the past 10,000 years saw 170 eruptions in Russia, followed closely by the USA (160), Indonesia (140), Chile (110), and Japan (100). Ethiopia, Papua New Guinea, the Philippine Islands, Ecuador, Iceland, and Mexico each have had between 25 and 50 active volcanoes during those past ten millennia.[277]

The Mountain Mystery

A few hundred big events is not an overwhelming number of pyroclastics for a period which saw man advance from throwing stones to launching rockets. Compared with ancient geological records, modern volcanic activity is far less common – although the chance of an earth-altering catastrophic eruption is still with us. However, most volcanoes are a good thing.

Some of our most coveted treasure comes directly from the throats of volcanoes. Especially diamonds. But gold, silver, copper, lead, and zinc are also among the minerals associated with magmatic spews. It's not that magma necessarily contains precious metals, but those elements become concentrated by hot water circulating through volcanic action. The rich warm water oozes through fissures and fractures, cooling and releasing prizes that grizzled prospectors may someday find. Diamonds, however, are tossed up directly from the throats of volcanoes.

Most diamonds are formed a hundred kilometres below the surface where temperatures hover around one thousand degrees. Under enormous pressure, disorganized carbon molecules (resembling pencil graphite or lumps of chimney soot) are squeezed into the diamond's lattice pattern, though it takes time. A billion years, some geologists think. If not for diamond elevators in the form of volcanoes, those diamonds would remain out of reach, deep below our feet. But occasionally magma bursts out of the mantle from great depth, races to the Earth's surface, and forms an igneous rock called kimberlite. The journey, once begun, is fast, and diamonds ride along within the kimberlite. Kimberlite is probably the deepest of any earth material to arrive at the surface. Its name comes from the South African town of Kimberley where kimberlite was first noticed. And it was there that a group of children started Africa's diamond industry.

The first commercial diamond was picked among stones cleared by a fifteen-year-old farmer's son, Erasmus Jacob, in 1866. Four years later, Cecil Rhodes, 17, arrived from England and began selling ice and water pumps to diamond prospectors. Three years after that, 18-year-old Barney Barnato started digging on a farm owned by the De Beers brothers. That became the first kimberlite diamond mine in the world when Barnato and his brother Harry Isaacs[i] began excavating it in 1871. This led to the largest hand-dug excavation in history, the aptly named Big Hole. In forty years, up to 50,000 miners worked with picks, shovels, and bare hands, emptying Big Hole of 20 million tonnes of kimberlite. And 3,000 kilograms (14 million carats) of diamonds. Hundreds died on the job as the open-pit mine reached a depth of 240 metres. Rhodes, the young English entrepreneur,

---

i Barney and his brother Harry had different surnames. Barnato was born Barnet Isaacs, in London. He scrambled to make a living, scalping theatre tickets and prize-fighting for cash in England. His brother Harry was a magician and started a comedy act in South Africa. Barnet joined him; they billed themselves Harry Isaac and Barney, too. *Barney, too* became *Barnato,* a name that stuck with Barney. The brothers started in diamonds by buying cheap exhausted leases, then sifting through the rubble, finding missed gems. From their profits, they bought new leases. After diamonds, Barnato made a second fortune investing in African gold mines.

bought the Barnato leases for today's equivalent of two billion dollars. Cecil Rhodes consolidated those with other leases he himself had claimed, founding the De Beers Mining Company. The new enterprise continued Big Hole's excavation, going underground to a depth of over 1,000 metres. The effort made Cecil Rhodes fabulously wealthy. A bachelor, he left a huge part of his estate to establish the Rhodes Scholarship which has sponsored studies at Oxford for over 7,000 international students. Since 2002, the Rhodes Trust has also operated the Mandela Rhodes Scholarship which awards the brightest African students with funding for two years of postgraduate studies – assisting the descendants of the men with picks and shovels who dug Big Hole at Kimberley a hundred years earlier.

When a kimberlite volcano erupts, it doesn't form a tall volcanic structure on the surface. Instead, any material lying atop the upwelling magma is pushed into a ring around a sunken bowl-shaped depression, as if all the explosive energy was spent on the long trip up. Frequently, the volcano doesn't quite make surface, stopping hundreds of metres short of sunlight. At other times, a surface mound is formed with diamonds and garnets sprinkled in a ring around its edge. The design is a definite circular shape, as you would expect for the throat of an extinct volcano. The rocky core is a kimberlite pipe, blasted out of the deep diamond-moulding layer. Geologists believe the diamond-transporting volcanoes were most active in the late Cretaceous period – around the time of the great dinosaur extinction. No one has ever seen these odd eruptions occur – the youngest yet discovered are in Tanzania's Igwisi Hills where a small group of pipes swarmed to the surface 100,000 years ago. But it is near the old ones, long since eroded and now level with the surface, that fortune-seekers hammer their claim stakes.

I was in my last year of university when I met a grey-bearded prospector exiled from central Europe. Thirty years earlier, the adventurous prospector had discovered diamonds in Siberia. He invited my friend Lorne and me to survey a magnetic bullseye the Canadian Geological Survey had mapped over a frozen lake in northern Saskatchewan. He thought the landscape, the rocks, and especially the magnetic circles on the government maps were promising signs of diamond prosperity. He pointed to eleven tight magnetic rings on his map – a swarm of kimberlite pipes, he thought. Lorne and I disappeared from classes for a few days and headed into the northern bush, magnetometers in tow. It was mid-winter and cold, but not bitterly so. In parkas and insulated boots, we trudged equipment through metre-deep snow over the promising feature that formed a perfect circle on our client's maps. Our work wasn't far from the place where a released prisoner from the Prince Albert Penitentiary claimed he had found a handful of rough diamonds in a stream bed. His find included small yellowish-white nuggets. It was one of Canada's first diamond discoveries, although farther east the occasional diamond had already been lifted from glacial debris plowed southward from some undiscovered North Ontario kimberlite pipe.

The Mountain Mystery

For several days we tramped across the frozen lake, taking measurements on intersecting perpendicular traverses. The strength of Earth's magnetic field modulates constantly at that far-north latitude, so we kept a permanent base that recorded the background field every few minutes. Although the ionosphere danced above us and challenged our data, we were able to later remove its noisy variable from the recordings we made every fifty metres as we slogged, surveyed, sampled, and recorded for seven kilometres across the frozen Saskatchewan lake.

We watched the magnetometer readings increase, dip, increase, then finally dip again past the far edge of the circular lake. Our magnetic survey gave a rough estimate of the depth to the top of the kimberlite pipe which we now suspected was under the lake. Unfortunately, it was much too deep. There were perhaps two hundred metres of non-kimberlite rocks and surface debris over the hidden pipe. It would take an expensive borehole to sample the rocks below the over-burden to be certain the magnetic anomaly really indicated a kimberlite pipe. Then several more drilled cores before enough samples were recovered to see if diamonds were also there. If they were, excavators would need to dig out all the overlying rock to expose the kimberlite and start a proper mine.

Our friend the prospector was hoping for a near-surface pipe, not a suspicious anomaly hundreds of feet below the ground. That was Saskatchewan, February 1991. During that same season, two geologists found exposed kimberlite pipes farther west, and a bit north, in Canada's Northwest Territories. They had discovered a billion-dollar mine, the Ekati, which has produced 40 million carats of diamonds.

Digging for diamonds at Ekati Mine, Northwest Territories. The first diamonds were found on the surface. Each pit was dug into a volcano throat, or kimberlite pipe.

Looking into the Earth

Diamonds are born of volcanoes, but as we have already seen, most mountains are not. A hundred years ago, geologists mapped volcano locations around the world and found that most encircle the Pacific Ocean. Geologists couldn't explain their abundance along the ring of fire, nor the absence of volcanoes in mountain ranges such as the Alps and Himalayas. Meanwhile, scientists still had all those other pesky geological issues to unravel, questions for which volcanoes held no solutions at all. There remained unexplained fossil distributions, glacial remains in deserts, and coal beds in the Arctic. There were mountain ranges extending from Scotland to Newfoundland (interrupted by an ocean), and river beds and rock formations that leap from Africa to South America.[278] These were all explained by Alfred Wegener's model, but during the 1930s and 1940s, conventional wisdom – especially in the United States – opposed continental drift. Geologists might occasionally hiss at each other about this subject during conference presentations, but most of the smart young geoscientists looking for funding from the offices of old timers knew their seniors had written enough material against drift theory to paper a science hall. Alexander du Toit and Reginald Daly were the only well-known geologists championing the cause. It would take radically new information to force the revolution that was coming.

The Salesman

Over the previous centuries, geologists had learned a great deal about rocks. They reached relatively accurate notions of how rocks form – igneous from material melted in the bowels of the Earth; sedimentary from river, lake, and especially sea bottom deposits; and, if either of those two types found themselves buried, heated, and sufficiently compressed, metamorphic rocks arose. Granite and basalt are common examples of rocks born of fire; sandstone, limestone, and shale are from sediments of beach, coral, and mud. Metamorphic rocks include slate, from transformed shale, while limestone presses into marble. A hundred years ago, geologists could figure out what minerals were in any chunk of rock, and the proportion of elements within those minerals. They even derived estimates of the age of their billion-year-old specimens. They measured magnetic and thermal properties. They were also fairly good at finding oil fields and veins of gold. But they argued bitterly over the origin of mountains. In 1936, Princeton's Richard Field[i] thought he knew why there was so little agreement on that fundamental subject.

Field, the founder of marine geology, told British scientist Edward Bullard that the problem with geology was it studied only the dry land – you could not expect to understand the Earth if you studied only a third of its surface, if you didn't know about the rocks under the sea.[279] Dr. Field's simple solution was to start at land's end and continue surveying out into the ocean, using any means of

---
i   Richard Montgomery Field (1885-1961), American geologist.

investigation possible. The tools available were almost entirely geophysical – heat flow from the seafloors, magnetic patterns, gravity highs and lows. Marine seismic was an infant science, but it soon added to the geophysical studies and then quickly dominated it. Richard Field was aggressive, optimistic, and sometimes wild-eyed, insisting seafloor science must be done, and done correctly.[280]

At least, that was Edward Bullard's assessment when he wrote that Field had "the burning zeal of an Old Testament prophet. He would not take no for an answer, he would not stop talking, he had no doubts, he was embarrassing and sometimes a nuisance, and yet . . ."[281] Canadian Tuzo Wilson had similar feelings. For his PhD studies, Wilson was accepted simultaneously by Harvard, MIT, and Princeton. He chose Princeton partly because the school enticed him with some money, but mostly because of the persuasive powers of Professor Field. The professor "bubbled with energy, enthusiasm, and powers of persuasion. The other professors wished that he had become a salesman, for which he would have been well suited, and they disparaged his ability as a professor. It's true that he did no research, but he held freshmen classes spellbound, he organized great field trips and he had a sweep of imagination lacking in his research-conscious colleagues."[282] Less kind cohorts likened Field to a pitch-man, a carnival barker who exaggerated claims to drag in an audience from the midway. And yet, it was Field who gleaned every source for funding while cajoling brains like Tuzo Wilson, Maurice Ewing, and Princeton professor Harry Hess to participate in his schemes – especially his dubious surveys of the rocks beneath the waves.

Professor Field believed an understanding of the seafloor was the only way scientists would ever decipher the Earth's history. He was right. There was a paucity of oceanographic data – it had been a long time since a serious voyage of discovery had taken place on the high seas – sixty years, in fact. R.M. Field wanted to continue where marine science had ended a half-century earlier.

Although the birth of mountains is inexorably tied to the sea floor – for reasons not apparent at the time – scientists in the 1930s had scant marine data, and most didn't recognize the importance of what was missing. Very little had been accomplished in oceanography since the voyage of the British research ship *HMS Challenger*, which set off as a science ship when Victoria was Queen of England. Originally a sail and steam battleship that served in Mexico and Fiji, she was refitted, her big guns and ammo rooms replaced by laboratories. During her three-year scientific expedition, the *Challenger* covered 125,000 kilometres exploring seafloor deposits, water salinity, and ocean currents. The expedition catalogued 4,700 new species of creatures. It took 23 years and 30,000 pages to write up the final 50-volume report. But it was a bloody awful sea voyage. The winch used to drag up scum from the bottom of the ocean was designed by Lord Kelvin himself. It collapsed the first time it was used. The original 216 members of crew and scientists were reduced to 144 by the time the *Challenger* returned to

Looking into the Earth

England in 1875. The missing had died, deserted, or simply disappeared. Someone among the bunch was lost every two weeks. We can speculate on whether the expedition was so nauseating that another wasn't planned again for over fifty years, or whether the scientists were content with the trove of data harvested from the seas. Either way, the next big ocean mission was planned and paid for by an American team, not British.

Although the *Challenger* had some exciting creature discoveries, the ocean bottom itself proved disquietingly dull. The *Challenger* did not discover the cracks in the sea where lava emerged as continents drifted apart. Nor did she find the hot vents that spewed noxious chemicals while harbouring animal life so bizarre they appear alien to our planet. Although the *Challenger* had been at sea for nearly three years and had collected the first samples of scrapings from the bottoms of really deep parts of the seafloor, most of the results showed the ocean basins were monotonous – a smooth, gently sloping layer of muddy ooze.

No one realized those muddy films were a thin veneer hiding crust younger than almost any rocks above water. Most geologists believed that the sea floors were polished extensions of the continents themselves; not notably different in genesis. Few geologists wished to spend a career studying something so lacking in excitement. But Princeton professor Richard Field felt that the submerged two-thirds of the Earth could not be ignored. He was more correct in this than even his own grand imagination allowed. Later oceanographers would discover that the seemingly monotonous seafloor was not made of sunken continental rocks. Hidden beneath the waves was the Earth's longest mountain range, its tallest mountains, and proof that the continents actually do slide around the surface of the Earth.

Field needed scientists and an exploration ship. He found some of the men, but none would prove better at exploring the oceans than a former farm boy from the Texas desert panhandle. At the annual meeting of the American Geophysical Union in 1931, among the handful of speakers was Maurice Ewing[i], a physics professor at Lehigh University in Bethlehem, Pennsylvania. Field introduced himself, then asked if Ewing's seismic science could help geologists figure out what was going on under the waves. Ewing was a few steps ahead of Field – he had already been setting off illicit sticks of dynamite near the ocean, along the New Jersey shores. It was the recorded echoes of his dynamite which he presented at the 1931 meeting where Ewing and Field first met. Professor Field had found his man. If he knew about Ewing's long struggle to arrive at this point in his life, he would have been even more impressed.

Doc, as Maurice Ewing became known, was the eldest surviving child of an impoverished Texas farm family. Three older children died young, but Maurice and the next six survived. They were all brilliant, excelled at university, and eventually all but one left the ranch for professional careers. Their father loved

---

i William Maurice Ewing (1906-1974), American seismology geophysicist.

music and literature more than cow-poking and horse-trading, so it was tough going for the big family of future scholars.[283] Maurice Ewing grew up extremely isolated, extremely poor. But at 16 he won a scholarship to Rice, in Houston, to study maths and physics. It was 1922. Petroleum would loom large in Ewing's future and would fund much of his work. To understand its importance, we need to know a bit about the oil business and its role in Houston when Ewing arrived there to study. The great Texas oil boom was underway. With oil, Texas had transformed itself from a cattle and cotton state to the keeper of the most productive oil fields in the world. The greatest oil well ever drilled had been found on a small hill an hour's drive east of Houston. The discovery was made by an industrious American immigrant from Croatia's coast, a navy captain with an engineering degree who became a petroleum geologist.

An Oil Interlude

The Croatian engineer, Tony Lucas[i], used a combination of mechanical engineering and salt mining experience to find the right place to drill for oil. His early background had not prepared him for anything like this. Lucas grew up on the shores of the Adriatic Sea, in southern Europe. His childhood home was near the huge retirement palace that Roman Emperor Diocletian had built in Split, along the Dalmatian coast. In Split, the Lucas family business was shipbuilding. They were wealthy enough to send Antun to Graz, Austria, where he studied with another prodigious youngster from a village near Lucas's Croatian home – the eccentric genius Nikola Tesla. Together, it was maths and physics in Graz for the young men, and separately, emigration to America. But before America, and after school, Lucas returned to the Croatian coast as a navy officer. He was searching for a career when he travelled to the United States to visit his uncle in Michigan. The visit became a permanent stay when the 24-year-old was offered a job designing a Saginaw lumber mill.

With an Americanized name and an American wife (Caroline Fitzgerald), Lucas scrambled around the country

Anthony Lucas

---
i   Anthony Francis Lucas, born Antun Lučić (1855-1921), Croatian-American engineer.

Looking into the Earth

prospecting for gold, then exploring for rock salt for a New Orleans company. Oil was often found alongside Gulf Coast salt. Lucas soon knew more about salt domes than anyone else in the world. He realized they originated as deposits in flat sheets when ancient shallow seas evaporated. Later, mud, reef material, and sand buried the salty evaporites. The weight of the new overlying sediments, packed and hardened into shale and thin layers of limestone and sandstone, was heavier than the salt below, which began to bulge up through weak spots. Meanwhile, the shale that buried the salt held trillions of ancient decomposed sea denizens, mostly single-celled fatty bacteria. Pressure and heat squashed and cooked them until oily droplets of life drained from their dead bodies. The salt domes gained size as they plumed upwards and the oil from the shale creatures became trapped around the edges of those giant bulbs of salt. This was Lucas's theory. It was an unbelievable stretch of creative thinking, according to experienced geologists wildcatting oil wells. They unanimously rejected Lucas's theory that oil could be found near salt domes. This isolated Lucas until he met an interesting Sunday School teacher from Beaumont, Texas.

When the weather was pleasant, Pattillo Higgins[i], a small-town businessman, would take his young Bible students – at least the young ladies of the group – for a picnic at a mound just south of town. They would eat their sandwiches at the site, a good five metres above the flat expansive prairie. For entertainment, forty-year-old Higgins would sometimes poke his walking stick into the ground, then light the gas that escaped from the hole he'd made. It seems a bit of an odd activity for a Sunday School picnic, but Higgins had been an odd man all his life.

After he'd dropped out of school at age nine, his father apprenticed him to a gunsmith. As a teenager, he was often in fights, and once, with a rifle, he threatened a black family heading off to their church on a Sunday morning. The sheriff was called. Higgins shot – and killed – a deputy. Higgins himself was wounded in his left arm, which was amputated. He was seventeen years old and charged with murder. He pleaded self-defence. A jury of his peers believed him.

One can almost feel the congratulatory slaps on Higgins's back from the white jurors who found him innocent. The judge, however, ordered the violent teenager to leave town. In north Louisiana, Higgins worked as a hard-fighting, one-armed lumberjack. Someone who worried about the young man's explosive temper persuaded him to attend an evangelical tent revival. Higgins immediately converted himself into a pious Baptist. He built ties within his new church, using fellow worshipers as business allies while teaching Bible lessons to their daughters.[284] Using his connections, Higgins shrewdly built a real estate business, a brick-firing plant, and an oil company. It was the smelly gaseous hill in Beaumont that got him interested in petroleum.

Higgins knew that he and his troupe of young lady students were sitting on a dome of gas, maybe oil. Nearby, the grass was waxy from paraffin. For hundreds

---
i   Pattillo Higgins (1865-1955), American entrepreneur.

of years, local seeps had supplied natives, and then French explorers, with sealants and lubricants.[285] A few marginal oil wells had been drilled in south Texas, but nothing profitable came from them. Higgins guessed that Spindletop, as his Sunday School mound was called, rose above the plains because somewhere far below the surface, hard capstone arched into a bowl-shaped trap that held oil.[286] But Higgins was no geologist, and he couldn't explain why rocks in the Texas coastal flats would bow upwards. So he advertised for an expert.

Lucas answered the newspaper ad placed by Higgins and his partners. Those partners had formed an oil company – Gladys City Oil. The small drilling outfit was named after Gladys Bingham, a sixteen-year-old, one of Higgins's favourite Sunday School students. The oil company drilled several wells, all abandoned too early, too shallow, too dry. As soon as Anthony Lucas saw the paraffin-stained Spindletop mound, he put the whole geological story together.

Lucas took over the operation. We was a mechanical engineer, a geologist, and a disciplined former navy officer. He elevated the enterprise to a more professional level. He bought the Gladys lease and gave Higgins a ten percent interest. One thing Lucas did not have was a fortune to squander. He sunk all his money into the enterprise, but halfway to his target, after battling quicksand for weeks, he was in trouble. Lucas found investors in Pittsburgh, but the deal left him with only a small part of the stake, and Higgins with virtually nothing. For Lucas, though, it was less about money than about stubborn persistence. With money to vindicate his reputation, sullied for months by Texas oil men who scoffed at his salt dome theory, drilling continued. When Lucas reached a depth of 347 metres on January 10, 1901, everything in the world changed.

The Lucas Gusher shook the ground like an earthquake. Startled rig hands fled for their lives, just ahead of six tonnes of four-inch drilling pipe that came blasting out of the hole. Oil exploded 50 metres into the air. It took nine days to take control of the runaway well. By then, a million barrels of oil were on the ground, drenching Texas with a black ooze that still stains the ground, a hundred years later. Within months, 500 registered oil companies were drilling in Beaumont, which had grown from 10,000 to 30,000 people in a hundred days. Each day the new Beaumont wells produced more oil than the rest of the world combined. Unbridled greed and ambition wrecked the unregulated field as dodgy oil men competed with their rigs, pumping the reservoir so fast the oil field collapsed and depressurized, leaving most of the oil in the ground, unrecoverable. But similar fields were quickly discovered along the coast. Oil was suddenly in such surplus its price plunged to three cents a barrel – cheaper than water in some places. Cheap oil meant cheap fuel for the automobile, which had just driven onto the scene. The United States was booming with cars, steel, oil. Within weeks, America was the world's largest oil producer; Texas, the world's oil centre; and Houston, lying just west of Spindletop, was the world's most important oil city.

Today, the site of that spectacular well, located near Sulphur Drive, at the end

Looking into the Earth

of Spindletop Avenue, is a dead grease spot 500 metres in diameter. Never again would Pattillo Higgins take Sunday School girls for a picnic on the mound. The landscape was so ruined – the soil dead and the hill deflated and sunken – that the original granite memorial marking the spot where the world's oil revolution began was moved to a lovelier site. A replica of Spindletop stands in a park next to the kitschy Gladys City Boomtown which houses a neat collection of oil-boom artifacts. Left out of the riches, Higgins sued Lucas and others, then scattered money into a variety of unsuccessful oil deals. At age 42, Higgins adopted a fifteen-year-old girl named Annie Johns. He married his adoptee a few years later. Higgins, by the way, never wavered in his staunch religious conviction. Meanwhile, Lucas made some money, but not much, from Spindletop. He moved to Washington, D.C., his base as a geological engineer that took him to Romania, Russia, Algeria, Mexico, and new oil fields across the United States. Considered the founder of modern petroleum engineering, he also had a hand in developing blowout preventers – the sort of device that might have checked the million barrels of oil that desecrated Spindletop's grassy knoll.

Spindletop, October 1902: Unbridled greed and ambition.

The Explorer

By 1922, the city of Houston was twenty years into the oil boom that Anthony Lucas started and the young farm lad, Maurice Ewing, had just arrived from the Texas panhandle. The final leg of his trip to university was by police escort. The police lift was the end of a long string of disasters that started when the teenager set off from the family ranch on an old motorcycle that quickly disintegrated. He hopped a railroad boxcar, but was punted and robbed by the brakeman, then by the two hobos who had been travelling with him. A sympathetic farmer and a salesman gave the boy rides, finally getting him into Houston. Broke and embarrassed by his rough appearance from the trip, he stopped a friendly constable who drove him to the Rice University campus.[287] That was Maurice

The Mountain Mystery

Ewing's auspicious arrival into the wild and bustling oil city.

Ewing found Houston's streets lit by cheap natural gas. Cars clogged narrow roads designed for wagons. People were surging in from all over the world, bloating the city from 150,000 residents when Ewing arrived to 300,000 by the time he finished his PhD, eight years later. All of Ewing's education was at the same school – Rice University – because he couldn't afford to be anywhere else. As a student, he tutored classmates, worked shifts at an all-night pharmacy (not mixing drugs, he noted, but delivering late-night sandwiches to the young girls cruising the nearby hotels).[288] Summers, Ewing shovelled grain from prairie elevators, and took gravity readings as a prospector for Texas oil companies. Exploring with gravity and rudimentary seismic – bouncing sound waves from the flanks of salt domes and recording the echoes – led to the discovery of swarms of oil-rich traps such as Lucas had drilled. Maurice Ewing was on his way to becoming a practical geophysicist with a knack for research and for improvising scientific equipment. His PhD thesis analyzed the way sound travels through rock layers, bending and bouncing back as it echoes, but also travelling along the boundary between different rock types. The latter, the essence of refraction geophysics, would later help him understand the ocean's seafloor – and also eventually devise a noise-bomb that could save the lives of airmen shot down over the Atlantic during the next world war.

By 1930, Ewing had moved north to Pennsylvania, taking a job teaching physics at Lehigh University in Bethlehem. Neither he nor his new school had much money for research. It was the Depression, a dismal time for the region's steel mills, many of which had closed. Ewing was nevertheless eager to experiment and advance seismic methods, so he improvised, taking advantage of scarce resources as he would his entire career. He asked the local stone quarries for their schedules. When the quarry bosses were blasting dynamite in local limestone pits, young Ewing poked microphones into the ground and listened to echoes arriving from deep within the Pennsylvania hills. He improved a system that amplified the ground rumble picked up by his geophones, using mirrors that wobbled a narrow beam of light across a roll of photographic paper that he moved by cranking a handle. The result was a trail of wiggles, a seismogram recording the blast as it bounced among layers of rock deep beneath Ewing's feet. Summers, he worked in the oil patch, which gave him a little cash, added to his seismic skills, and sometimes allowed him to drive home with sticks of dynamite in his car for his own research.

Several of Ewing's physics students became geophysicists. Lamar Worzel[i] accompanied Ewing during his early experiments with explosives and geophones. Most weekends they loaded their equipment into a rough old Model A Ford they called Floosey Belle. "It was a product of the depression," in the words of Worzel. The back seat could be pulled out, back door removed, and Ewing's family car

---

i   J. Lamar Worzel (1919-2008), American geophysicist.

Looking into the Earth

was suddenly a delivery truck. They would load their oscillograph, amplifiers, cables, and geophones on a Friday evening, then drive off at four the following morning. The crew tried to arrive in the field at daybreak and conduct their experiments until dark, munching on a peanut butter sandwich for their lunch. In the evening, when it became too dark to work, they would enter a nearby town. After dinner, they'd work into the night developing the photographic paper they'd recorded during the day. "We'd find a rooming house, get some supper, wash the records in the bathtub, plot the data on graph paper, then go to bed at midnight," said Worzel.[289] It sounds charming. The development of one of the most important tools for understanding the Earth included a phase when poorly-paid geophysicists toiled without a break in the field, then spent their evenings washing photographic records in bathtubs in boarding houses.

In November, 1934, Richard Field of Princeton and William Bowie[i] of the US Coast and Geodetic Survey met Maurice Ewing a second time. He was sitting in Floosey Belle at Lehigh when they arrived. The visitors wanted to know if Ewing would like to try a really big project, rather than the small-scale experiments he'd been cobbling together. Would he like to map the oceans? Field could get lots of money for the work. Field presented himself as the founder and chairman of the American Geophysical Union's Committee on the Geophysical Study of the Ocean Basins. He was likely its entire committee. His talkative, persuasive style was balanced by Bowie's calm southern charms.[290] As a result of Field's persistence, in the summer of 1935, Ewing acquired seismic data far off the Virginia coast, in 100 metres of water. It was the world's first marine seismic data. Ewing sank a heavy seismic receiver, his microphone, which was attached to his ship by an insulated electric wire. Someone in another boat, 8 kilometres away, blasted dynamite in the water. At the same moment, a radio signal was broadcast to Ewing on his recording ship. The radio signal arrived immediately, but sound waves from the dynamite travelled more slowly through water, then various rock layers, working their way along rock interfaces until they echoed up into the submerged microphone. Ewing recorded the sounds on a magnetic wire when they arrived. The time delay gave an idea of the depths of the various rock layers below the seafloor. It was a lot to figure out and Ewing was nervous. Not so much about the quality of his work, but about the integrity of some of his colleagues.

Ewing was anxious to get credit for being the first to succeed at marine seismic acquisition. He wrote that he had the experiments "so arranged that I see no possibility of anyone stealing the credit from me."[291] Four years earlier, publication of his groundbreaking PhD dealing with seismic waves had been clouded by issues that led to a permanent and bitter estrangement from Don Leet, an established Harvard seismologist with whom he co-authored two papers. Ewing was anxious that this might happen again. Ewing's enormous drive for recognition as a scientist was partly responsible for his habit of working eighteen

---
i William Bowie (1872-1940), American engineer.

hours a day, mostly alone. Despite Ewing's efforts, the initial results of the first marine seismic studies would glorify the flamboyant Professor Field.

As we've seen, Field's talent was in grabbing funding for research, not in actually struggling with data. In 1931, he persuaded the American government to buy a vessel for seismic research. The *Atlantis* became the first ship Maurice Ewing used. During the depths of the Great Depression, through the efforts of Field and a few others, the Woods Hole Oceanographic Institute[i] was able to gather over a quarter of a million dollars in government funding to commission the *Atlantis*. A princely sum for a boat conducting pure scientific research, especially when unemployment was approaching 20 percent in the United States. But the vessel returned huge dividends from the seas. She would sail 299 expeditions, survey over a million kilometres, serve for 35 years, then be sold to the Argentine government for South American research.

It was expected that Ewing's work would help geologists understand the transition from continent to ocean basin. When his work began, no one knew how far the continental shelf extended. Was there a quick transition (as geologists expected if the continents were mobile) or was there a long slope of sediments as might occur with stationary continents? A long trail of continental run-off might mean that below the sediments there was a sunken continent that had become the ocean basin, suggesting the North American continent had been sitting still forever. Ewing's first seismic results on the seas showed a long gentle tapering slope. Field chose to announce to the world that science had disproved continental drift. Field, the promoter who brought together so many people, raised money to explore, and organized so much, had long opposed continental drift theory. He was quickly convinced that the initial seismic results had disproved Wegener's theory.

As the university newspaper would soon tell it, it would seem Professor Field, not Ewing, had been aboard the *Atlantis*, using seismic recorders. The paper reported that Field had been dropping dynamite into the ocean off the Jersey shore, recording the world's first marine seismic data. The exploded dynamite sent shock waves through the water, into the seafloor, past various layers of sediment, through the underlying basement rock, then echoed back to the *Atlantis* where it was recorded by electronic "ears" according to T.H. Wolf, a *Princetonian* reporter. Field told him that seismic would ultimately "enable geologists to map out the *terra incognita*" that covered two-thirds of the globe. This was the first use of seismic geophysics to try to determine seafloor structure and to establish whether continents moved. Field gave the newspaper a conclusion we now see as wrong.

---

i   Woods Hole Oceanographic Institute was founded in 1931 after a recommendation by the National Academy of Sciences that an independent non-profit organization be created to study the oceans. Its first funding was a 2.5 million dollar grant from the Rockefeller Foundation. Its independence attracted scientists such as Field and Ewing even while they remained associated with their universities. The Institute continues today with over 1,000 scientists and students working at its labs and on its ships.

Looking into the Earth

"A week's experimentation was able to uncover data of revolutionary importance. Professor Field found that the continental margin, which had long been thought of as a fundamental structural feature, is probably not. This discovery strikes a sharp blow at the Wegener theory of drifting continents."[292] The seismic data would later be interpreted quite differently.

From earlier work on fossil snails, Field had already decided against the idea of mobile continents. Although he had been at Princeton since 1923,[293] Field had not completed much original research. However, one of his few lines of study involved the vexing riddle of fossil distribution. To him, it meant a swarm of land bridges had once connected stationary continents. In a 1925 *New York Times* article, Field described the way his Princeton team, working with scientists from the Smithsonian, gathered fossils from both sides of the Atlantic. They found the creatures so similar that "the existence of the narrow strip of land connecting Europe and America is considered established." A red herring in the study was that fossils a short distance north – on both sides of the Atlantic – were different from those just to the south, but were identical to fossils on the distant, opposite sides of the ocean. In other words, Newfoundland fossils were identical to those of Wales, but different from those just to the south, in Maine. "This could only mean, according to Professor Field, that the marine life of the north and south never mixed, and that therefore they were wholly separated by a land barrier."[294] Field, in his *New York Times* interview, added, "The greatest difficulty is seeing how it would be mechanically possible for the continents to drift about on the Earth's crust, even allowing them millions of years in which to do it. The thousands of fossils that we gathered will be studied to show whether such movements of continents could have taken place." In his mind, they had not.

Maurice Ewing stayed with seismology. His first offshore experiment found sediments 3,800 metres thick near the coastline. This is a huge thickness and it meant that the sedimentary rocks could contain oil fields. After Ewing published his results about the unexpected thickness of the east coast offshore rocks, he tried to convince oil companies to pursue his discovery. They weren't interested. It would be years before technology and economics allowed oil collection from the eastern slope. Eventually fields like Canada's Hibernia, 320 kilometres from the east coast, would be operated from the world's largest oil platform. That field alone held almost a billion barrels of petroleum, and it proved that Ewing's idea was right. This and other discoveries were a huge boost for the sagging economies of Nova Scotia and Newfoundland. Meanwhile, Ewing had another disappointment. Not only had Field taken credit for the first marine survey, but Ewing's 1935 seismic study of the deep seafloor was attacked by his old partner and nemesis, Don Leet, who publicly accosted the whole operation. Leet said that Ewing's work had "an embarrassing absence of material" and "geologists will look in vain for acceptable data which may be classified as new facts about submarine geology."[295]

The Mountain Mystery

Ewing's work was not an embarrassment to science at all. In fact, it ultimately helped unlock the great mysteries of the Earth's wrinkled surface and showed the sea had a fundamental role in the creation of mountains. Until the late 1930s and the work done by Ewing and others, "the oceans had been a place where geologists could safely deposit many of their difficulties; almost nothing was known and almost anything could be assumed," said Edward Bullard.[296]

Exploring the seafloor with seismic was difficult work. And deadly. To get a vibration that would shake the seafloor (so seismic echoes could be recorded), the chief scientist tossed lit dynamite from the back of the ship. The scientist cut a stick in half, lit it, then threw it. John Hennion, an American ex-Marine and a geophysicist aboard the *Vema,* died when one stick exploded in his hands. He was buried at sea, somewhere south of Valparaiso, along the Chilean coast. On a different expedition, Maurice Ewing himself was washed off a schooner during a storm. He was trying to secure loose fuel drums on the deck when a wave swept him, his brother John, and two ship's mates into the ocean. The ship's first mate drowned by the time the boat was able to circle back for the men. Ewing was badly injured, taking a blow to the neck that left him partly paralysed, leaving him with a permanent limp. It did not stop his work, he continued on the seas and in the classroom for another twenty years. It became part of Ewing's legendary reputation as a man of single-minded determination. Edward Bullard once asked Maurice Ewing where he kept his ship. "I keep my ship at sea," said Ewing.[297] For forty years Ewing and his fellow scientists kept their metaphoric ships at sea, always probing, always listening.

15

The Revolution Begins

The Great Depression was ending, a new world war was beginning, and the first era of modern geophysics was drawing to a close. Scientists were not much closer to proving the origin of mountain ranges, but they had developed some impressive tools to explore the inside of the planet. The development of seismic recording – the inscription of echoes bounced inside the Earth and received by electronic ears – was a notable achievement. But equally important was the application of radioactive decay that finally settled the question of the planet's age. Geophysics would bound ahead even faster in response to the next big war. In 1939, Germany finished swallowing Czechoslovakia and then invaded Poland. England, honouring its alliance with Poland, declared war on Germany. The Allies – Poland, the UK, and France – were joined in days by Canada, Australia, New Zealand, India, China, and a dozen other countries. It would take two years before the United States joined the Allies, and by then the war was truly global.

Germany was powerful on the seas as well as on land. Japan had conquered half of the Pacific before attacking Pearl Harbor. The Allied navies were defeated in battle after battle. They desperately needed to know everything they could about the oceans and to apply everything they learned to gain some advantage in the war. The British and Americans needed to spot enemy submarines and to defend their own. Esoteric research such as the bathymetry and magnetic properties of the ocean floor was suddenly strategically important. Mapping the topography of the seafloor identified covert refuges for submarines; knowing seafloor magnetic properties helped expose submarines.

In addition, equipment failures at sea needed quick fixes. For example, sonar devices tended not to work well each day, shortly after lunch. It was as if the equipment took a daily siesta. Maurice Ewing, the geophysicist, seismologist, and oceanographer, was assigned the task of figuring out what was destroying sonar signals in the afternoon water. His team discovered that when water temperatures rose during the heat of the day, the returning blips of sound became disrupted. Ewing surmised warm-water zones bent the sonar sounds and created a dead shadow zone, an oceanic channel where sound waves were trapped.

The Mountain Mystery

This was not theoretical science at all, but something immediately useful and practical. It meant submarine captains could pilot their crafts at undetectable depths. Along with this discovery, Ewing's team also learned that ocean sound waves could be focused and broadcast huge distances horizontally while trapped within those sound channels. Whales had known about this for generations.

As whales descend past 500 metres, the cetaceans enter a depth where water temperature drops rather suddenly. As a result, sound waves sink, carried more readily in the lower, colder, denser water. But at slightly deeper points the ocean's enormous water pressure forces those sound waves back up. Between the depth where sound is bent down, and the deeper point where sound is bent up, lies a sweet spot where humpbacks sing their songs as if in a cathedral. At this point, their music can travel across an ocean. Ewing realized emergency communications in the zone would also travel far. Ultimately, he built a noise bomb that took advantage of sound channels and saved the lives of allied airmen.

The SOFAR Bomb (Sound Fixing and Ranging) was a two-kilogram sphere stuffed with TNT. If a crash were imminent, the navigator would look at a sea-chart and read the theoretical depth of the sound channel over his part of the ocean, somewhere below the impending crash site. The sound channel varies from place to place, depending on ocean temperature, currents, and salinity. As a plane began its crash, the navigator would calmly set the bomb's pressure sensor so it would make its explosion inside the ocean sound channel, neither too shallow nor too deep. The navigator would then open a window and drop the SOFAR bomb like a heavy grenade. Detonated at the right depth, the device sent distress signals 4,000 kilometres. Listening posts ashore triangulated the received signal and mapped a downed plane's position to within two kilometres. At the receiving hydrophones, the initial one-second sound burst expanded to 24 seconds, sounding like a kettledrum building to a sharp crescendo, then abruptly stopping.[298] It was a unique sound signature that ground and ship-based receivers recognized. They conveyed the location of the downed aircraft to recovery ships.

While seismologist Maurice Ewing was making noise bombs, geology professor Harry Hammond Hess[i] was commanding an assault transport ship in the Pacific. Harry Hess was born in New York City in 1906, a time when the city was bustling with new immigrants and phenomenal growth. The world's second largest city, it had nearly five million residents and was adding an incredible million people every five years.[299] Harry Hess's grandfather, Simon Hess, had started a construction company, which grew to include harbour dredging and dam construction. Harry's father, Julian, became a member of the New York Stock Exchange. Harry's mother came from Hammond, Indiana, where her father owned a liquor distillery. That Indiana town was the origin of Harry Hammond Hess's unusual middle name. From Manhattan's 22nd Ward, his parents moved to nearby Asbury, New Jersey, where Harry graduated high school with a speciality

---

i   Harry Hammond Hess (1906-1969), American geologist.

The Revolution Begins

in languages. He entered Yale at 17, tried electrical engineering, but switched to geology, then promptly failed his first mineralogy course. His professor advised him to change his major, convinced there would be no future for Hess in geology.[300] I have great sympathy for Hess's mineralogy fiasco. Except for the grace of my own mineralogy professor, I might have likewise been tossed from my studies. I am severely colour-blind, which I never disclosed to my instructor. Mineral identification depends on colour recognition and a good eye for shapes and patterns. I have none of that. The results of my lab exam were abysmal. My professor kindly looked for the few things I had accidentally identified correctly and (I suspect) nudged my mark up to a very slim pass. Coincidentally, my professor, Leslie Coleman, had been a Princeton student of the same Harry Hess who had failed his own introductory mineralogy exam. Despite Hess's lack of eye for mineral structure and thin-section identification, Harry Hess stuck with geology and became one of the most innovative scientists of the twentieth century.

Hess finished his geology degree. Then, in a brash move, he left America and spent two rough years in remote northern Rhodesia, working and living in the bush with a native African team. They were prospecting in what is now Zambia, a country rich in copper, nickel, and uranium deposits. After two years, Hess returned to school, enrolling in graduate studies at Princeton in 1929. He once said he would have preferred Harvard, but had noticed "No Smoking" signs in all the geology labs. Knowing he could never give up his cigarettes, he selected Princeton.[301]

Princeton, with its lax smoking regulations, was a good choice for other reasons. The geology program was solidly established. And Richard Field, the grandiose geologist who knew how to find money better than rocks, was established there. In three years, Hess had his doctorate. He studied a pale green rock called peridotite, an unusual igneous pluton made of olivine material originating deep below the crust, in the upper mantle. Those rocks – his doctoral

Harry Hess in camp in Zambia, 1929. This photo is courtesy Hess's son, Dr George Hess, who pointed out to me that the picture is an early 'selfie' – Harry is taking his own picture by pulling the string in his right hand.

197

thesis material – had a similar structure to the last minerals Hess would ever hold in his hands, forty years later, just before his early death from heart disease aggravated by a lifetime of smoking. The final rocks in Harry Hess's life were small chips of olivine which Neil Armstrong had brought for him from the Moon.[302] Hess had helped make Armstrong's trip possible – but we are jumping ahead in the Harry Hess story.

As a Princeton graduate student, in 1931, Hess assisted one of the first submarine explorations of the seafloor, working with Dutch scientist Felix Meinesz,[i] a geodesy[ii] expert who was developing equipment to measure seafloor gravity while on a pitching, rolling boat. This was a near-impossible feat, similar to trying to measure one's weight while swaying on a bathroom scale. Scales prefer quiet stability, the sort of calm a submarine offers when it sinks below the roiling seas. In the early 1930s, the submarine was the best vehicle to gather a reliable set of gravity data – but not every university had one. Through Field's formidable connections, Meinesz and Hess were outfitted with a US Navy sub to explore gravity anomalies in the West Indies. Six-foot seven-inch Meinesz worked cramped and hunched – submarines are built for much shorter men – as he measured the Earth's gravity field by timing the swing of a pendulum. It was the first of many treks to the Caribbean for the scientists. To help with paperwork and approvals, Field made sure the government ranked Harry Hess as a reserve Lieutenant officer in the Navy. Ultimately, Hess would rise to the rank of Admiral.

A few seasons after his first session of gravity work in the Caribbean, Hess was aboard the *Barracuda* with the Texan seismologist Maurice Ewing. Also on the research submarine that summer was Edward Bullard, a brilliant Cambridge physicist, later to be Sir Edward Bullard. (He preferred the name Teddy.) The team was again accompanied by Meinesz, who had still not perfected his gravimeter so it would ignore the bounce of waves. The researchers found unexpected gravity variations in the deepest part of the Caribbean, the arc-island trench just north of Puerto Rico. There the seafloor drops to 8,648 metres below sea level, the deepest water anywhere except at the Pacific's Mariana Trench. Hess and fellow traveller Meinesz decided the extreme low topography and the unusually weak gravity signal resulted from some active crustal deformation taking place at the bottom of the sea. It was an important insight.

Hess and Meinesz concluded, significantly, that the ocean basins are not static, but are shifting and changing dramatically. This idea ran against hundreds of years of scientific inquiry and intuition that had concluded the seafloors were permanent, immobile structures. Hess and Meinesz didn't immediately realize it, but they had discovered signs that the planet was eating its own ocean crust in the deep abyss. This discovery was just a few years after Wegener's death, and on the

---

i   Felix A. Vening Meinesz (1887-1966), Dutch geodesist and inventor.
ii   *Geodesy* is the study of the Earth's shape (roundness, or lack thereof) and gravity.

The Revolution Begins

Barracuda were Hess, Ewing, and Bullard, all 29 and 30 years old, three of the key scientists whose brilliant application of geophysics eventually vindicated Wegener's theory and advanced it from simple continental displacement to refined plate tectonics.

By gathering and examining reams of data from the Caribbean trench and its neighbouring islands, Hess soon knew more about the seafloor than anyone else on Earth. Within a few years, he would see similar island arc chains with deep trenches and active volcanoes while serving as a navy commander in the Pacific. Shortly after their last Caribbean expedition, the scientists were fighting World War II. Thanks to Field's meddling, Hess was already in the Naval Reserve. The morning Japan bombed Pearl Harbor, Hess reported for duty. His first assignment during the war kept him in New York City, using geophysics to uncover enemy submarine positions in the North Atlantic. The navy was so successful that within two years the German submarine threat was neutralized and Hess transferred to battle duty in the Pacific.

As commander of the attack transport *USS Cape Johnson*, Hess made four major combat landings – including Iwo Jima, the most fierce battle in the Pacific. Hundreds of the 1,500 troops he personally piloted ashore died, though the Americans eventually overpowered the small island's Japanese defenders. Ferrying troops and supplies around the sea was, of course, Hess's main assignment. But between his various engagements across the Pacific, Hess gathered oceanographic data for the navy. Like many transport ships, the *USS Cape Johnson* was outfitted with depth-sounding equipment so the ocean bottom could be mapped – with military applications in mind. The depth-device, the fathometer, helped ships avoid crashing into submerged reefs, rocks, and enemy traps during beach landings, but as the forces edged closer to Japan, vessels with fathometers collected data which the navy turned into the first crude Pacific seafloor maps. Such maps helped captains confirm their ship's position (often estimated by less reliable compasses and sextants) by noting features like submerged canyons and ridges. Hess and others mapped the ocean floor's topography, knowing it would help the war effort, but Hess knew it would also be useful data for his science, helping unravel the nature of the seafloor and its evolution. Hess was not disappointed.

By mid-1945, Hess had measured many of the Pacific's nether parts, including trenches ten kilometres deep among the canyons and crevasses that were nearly everywhere. He also discovered hundreds of submerged flat-topped volcanic mountains he named *guyots* in honour of Arnold Guyot, Princeton's first geology professor. The guyot seamounts were often in straight lines of decreasing height and increasing age, hidden under as much as two kilometres of water, and appearing to march toward deeper parts of the ocean. All this was significant to geologists who thought seafloors were broad and flat dirty sinks collecting mud from the continents. Quite the contrary, according to the Hess data. The seafloor was alive with unexpected geological activity and unexpected topography. Harry Hess and his colleagues would eventually use these unlikely observations to prove the floor of the ocean was in motion, creeping along in murky darkness.

The 18-Month Year

After the war, Harry Hess continued to champion oceanographic research. It would take fifteen years for him to synthesize all the data he and other navy captains had collected. Meanwhile the US Navy continued expanding its seafloor maps, uncovering hiding places for Cold War enemy submarines. Hess led other oceanography projects as well. He was a principal researcher during the International Geophysical Year, or IGY, which began in 1957 as a world-wide effort to understand the geophysics of the planet. From Princeton, Hess dispatched a dozen teams to three continents for the IGY. Hess himself sailed back into the Caribbean – Puerto Rico, Venezuela, the Dominican Republic – in yet another attempt to unravel the relationship between deep trenches, earthquakes, volcanoes, and mountain ranges.[303]

The International Geophysical Year began as a bit of conversation at a dinner party hosted by James van Allen,[i] the rocket scientist who eventually discovered the protective magnetic layer that shrouds the Earth from the worst of solar radiation. His April 1950 gathering was a chance for some American scientists to meet Sydney Chapman,[ii] a visiting British geophysicist. Chapman was in his early 60s at the time and had made his name as a mathematician specializing in the way random events interact, making him a pioneer in chaos theory. But he was more than a math wizard. Chapman was the first person to explain how the ozone layer is created, he predicted the magnetosphere that Van Allen later discovered, he was a noted geomagnetism expert, and coauthor of a two-volume tome that served as a geophysical reference and text for a generation. Chapman was an attraction for the ten scientists assembled to sip wine with him.

Talk drifted to politics. Physicist Lloyd Berkner[iii] complained that the Cold War interfered with his research of the atmosphere. Berkner needed data from all over the globe, including Siberia and other regions the American government said he was not allowed to visit. That evening he proposed the scientists attempt an international effort. The idea for the International Geophysical Year was born. By the time the night was over and the last bottle of wine had been opened, there was talk of international cooperation and world peace.

Over the next two years, Berkner, van Allen, and others worked on a proposal for the geophysical study. They pointed out science has no borders; nor should its investigators. The IGY was endorsed by the United Nations as an opportunity to

---

i   James Alfred van Allen (1914-2006), American geophysicist and rocket scientist.
ii   Sydney Chapman (1888-1970), English geophysicist and mathematician.
iii Lloyd Berkner (1905-1967), American physicist and engineer.

strengthen international ties among scientists and to encourage cooperation in other areas. The International Geophysical Year lasted 18 months – from July, 1957, to the end of 1958. The scientists selected mid-1957 as the start of their research because cyclical solar activity was expected to peak and several eclipses would occur during the period. A total of 67 countries signed up for the project, with China the only notable absentee. But the Soviets and Americans cooperated. Geophysicists laboured to create new tools for their investigations, establishing a remarkable 4,000 new research stations around the globe, all designed specifically for the International Geophysical Year's research goals.

Among other accomplishments, Russia launched *Sputnik* in October, 1957, as part of the IGY space study, and the Americans sent *Explorer I* aloft a few months later. Although racing to catch up, the United States built a much more sophisticated satellite. Van Allen, who had helped conceive the International Geophysical Year, loaded Geiger-counters into the American satellite to test Chapman's theory that a magnetic field enveloped the planet and trapped deadly radioactive particles. He found it; others named the field the Van Allen Radiation Belt. Lloyd Berkner obtained the atmospheric data he needed from Siberia. His research led to better radio communications. The study of the ocean crust was accelerated with oceanographic data from around the globe. International cooperation was achieved with a system for scientific exchange and development of three "World Data Centers" – the first large-scale computer-based data storage ever built. In preparation for arctic studies, the first permanent facilities were established in Antarctica. Although world peace did not result from the IGY, a level of detente was attained, including the Antarctic Treaty which banned weapons on that continent and declared it a scientific preserve. The treaty, ratified in 1961, grew directly out of the International Geophysical Year and was the Cold War's first arms limitation agreement.

An (Almost) Endless Mountain Range

The International Geophysical Year began in 1957, but Canadian geophysicist Tuzo Wilson argued in *Scientific American* that 1956 was the real beginning of the geophysical revolution. As a result of two key findings made that year, the dormant theory of continental drift was at last revived. A group of geophysicists demonstrated that magnetism in rocks, previously viewed as rather oddly random, were actually uniformly aligned if the locations of those rocks could be moved to reconstruct Wegener's Pangaea. Originally, while still a part of the supercontinent, magnetic alignment was as sure as a compass. But after Pangaea broke up and continental pieces drifted, the embedded magnetism of ancient rocks pointed askance. Also in 1956, Maurice Ewing and Bruce Heezen[i] realized the mid-Atlantic ridge stretched far, far beyond the mid-Atlantic. It weaved around the

---

i   Bruce Heezen (1924-1977), American oceanographer.

globe as a continuous 75,000-kilometre mountain range. The significance of this discovery wasn't immediately apparent. But soon it would indicate a dramatic new way of looking at the Earth and understanding continental drift.[304]

The mammoth mountain system growing from the ocean bottom was first inferred in 1850 by the American Matthew Maury, who noticed the ridge in bits and pieces on a variety of nautical maps. He attempted to connect the sparse data points, drawing a single long mountain range in the middle of the Atlantic. But few could believe the range was continuous from the north Atlantic to its southern limits. Although he published the world's first oceanography book, *The Physical Geography of the Sea,* and discovered the world's longest mountain range, few have heard of him. It did not help that Maury could only speculate – and not prove – the grandeur of his submerged mountain range. Nor did it help that he later served the Confederates, buying warships from the English and French to battle against the Union, and inventing an electrically triggered naval mine that sunk more Northern ships than all the sea battles fought during the war.[305] But it was his creation of oceanography as a science that should have earned him a more permanent memory than the placement of his name on a Virginia elementary school, a lane at an oceanography lab, and a research ship. His understanding of the oceans was prescient – one hundred years before it was confirmed, he wrote, "Could the waters of the Atlantic be drawn off, so as to expose to view this great sea-gash which separates continents, and extends from the Arctic to the Antarctic, it would present a scene the most rugged, grand, and imposing."[306]

Twenty years after Matthew Maury, the British science ship *Challenger* began its three years of oceanography. Charles Thomson, who had proposed the voyage, was instructed to use the ship to chart a path for a new transatlantic telegraph line. He discovered, in 1872, that there was no way around the mid-ocean mountains. Soundings showed that the ridge bisected the entire ocean. Thomson also found deep water temperatures east and west of the rise were different, also indicating the mid-ocean ridge was continuous and interfered with mixing of sea water. As a result, the first cable from Ireland to Newfoundland had to suffer the expense and risk of climbing up from the ocean bottom to cross a 2,000-metre-high submerged mid-ocean ridge, then plunge back 2,000 metres to the ocean basin on the other side. It would have been quite a shock if Thomson had learned the hidden ridge was ten times longer than even Maury imagined. It encircled the entire globe.

Although they missed the full extent of the biggest single feature on Earth, much was learned by the early bathymetrists. With cannon balls tied to hemp naval ropes and piano wire, over 20,000 ocean soundings were made by 1899. Prince Albert of Monaco chaired the first commission to create a world map of the ocean floors. It showed the pit of the Mariana Trench, the continental shelves, and some high points on the mid-Atlantic ridge. But much was missing. 20,000 soundings is a lot, but the oceans are vast. Spread out evenly (though they were not) this is an average of just one depth record for every 15,000 square kilometres

The Revolution Begins

of seafloor. It is the equivalent of trying to infer all the dales, mountains, and meadows of England with fewer than 20 elevation markers, randomly selected. One is quite likely to miss the Lake Country or Cheviot Hills entirely.

Beginning in the late 1940s, Maurice Ewing and Bruce Heezen tried to work together using seismic and sonar to map the seafloor. At first, they collaborated brilliantly. Ewing had discovered Heezen in Iowa, where Bruce Heezen was an undergraduate geology student. Ewing offered the young man a job alongside him aboard the *Atlantis*, collecting sea-depth profiles. Instead, when twenty-year-old Bruce Heezen arrived on the east coast, Ewing assigned him the role of chief scientist aboard a different ship. Heezen, not yet a senior at his college, was in charge of a research ship and its data collection. At the end of the summer, he was back in Iowa where he finished his degree, then headed to Columbia for graduate work directed by Ewing. But the two soon despised each other and their work was anything but cooperative. It was, however, successful. Their animosity probably grew out of rather self-assured egos. There were some similarities. Ewing grew up on a cattle ranch in west Texas; Heezen on an Iowa turkey farm. Ewing and Heezen both moved east, both fell into geophysics early, both were life-long outsiders, misfits, perhaps. Both played key roles proving Wegener's continental drift idea was right, yet both remained skeptical about the whole scheme. Heezen believed the Earth was expanding; Ewing wasn't sure what to believe and likely doubted continents move at all. Ewing was continually nervous that others would steal his thoughts, while Heezen was aggressive and given to throwing trash cans and kicking walls. So perhaps there was a bit of tension there. Ewing, the boss, tried to fire Heezen several times but Heezen was tenured and wasn't going anywhere. Ewing finally resorted to starving his colleague's budget. And then there was the brilliant Marie Tharp[i] complicating their dynamics.

The Rare Woman

The creation of the most fascinating maps of the oceans' basins involved the rare presence of a female geologist. In 1943, the University of Michigan was recruiting women into its geology department with enticing suggestions that oil patch jobs would be available afterwards. Tharp says ten women were accepted the year she started. She had already earned a music and English degree from Ohio State and in her free time had assisted her father, a soil science surveyor, using his scopes and transits. She was mature, smart, and already knew a bit of geology. As promised, after earning her Master's in two years, she quickly landed a petroleum exploration job. This put her in Tulsa, working for Stanolind Oil and Gas. During her three years as an oil prospector, she somehow also managed to earn a third diploma, a mathematics degree from the University of Tulsa. But she became bored with Tulsa and with her work and in 1948 she moved to New York

---

i   Marie Tharp (1920-2006), American geologist and cartographer.

The Mountain Mystery

City, scouting for a job that might use her diverse talents.

Marie Tharp was hired by Maurice Ewing, who was teaching at Columbia. He employed her because he wanted someone with both math and geology skills. He recognized that Tharp's extensive surveying, drafting, and mapping experience were vital to his department. Despite her qualifications and experience, Ewing assigned her as a technical assistant to his grad students, documenting their work and drawing their illustrations. Those graduate students included Bruce Heezen, with whom she would work for the next thirty years.

Marie Tharp, as it turns out, is the one who really turned oceanography on its head. She mapped the seafloors and made the key discovery that led others to prove Wegener's drift theory. Ewing directed the lab; Heezen collected the data; Tharp created the maps – and offered insights the others missed. Heezen was the consummate oceanographer. His credibility rose from his discovery of turbidites, huge undersea landslides that sweep millions of tonnes of rock and seafloor muck down the continental slopes. Usually triggered by earthquakes, turbidites race along the submerged continental shelf at horrendous speeds – up to 400 kilometres per hour. Carrying an enormous amount of debris, nothing in the path of a turbidite is safe. In 1929, an earthquake off Newfoundland forced a wide turbidite flow across the trans-Atlantic telephone and telegraph cables, ripping out 12 of them, clipping America's link to Europe. By understanding the cause, substance, and predictability of these undersea landslides, Heezen emerged a respected oceanographer, as unlikely as that may seem for a turkey farmer from Iowa. His reputation assured him of a steady supply of oceanographic data from a wide range of sources.

The hulking Heezen gave Tharp boatloads of data to add to her maps. The depth values were acquired in narrow straight lines, recorded by vessels crossing the Atlantic.

Rather than measuring ocean depths the old way, lowering and raising cannon balls suspended from wires, Ewing and geophysicist Lamar Worzel created

Bruce Heezen at sea, gathering depth soundings.

204

The Revolution Begins

the echo sounder the US Navy used during World War II. This new sonar system did a decent job of recording the depth to the sea basin by sending a repeating sound signal from the boat, down through the water, then measuring the time it took for the echo to return to a recorder on board. The system had a minor limitation. It required electricity which was often intermittent on the boats. If someone opened the refrigerator the moment an echo was recording, the circuit blew and the continuous data stream had a dead reading. Other than that, return times were captured on narrow spools of paper, day and night, as the ship sailed. Cruising atop 3,800 metres of sea water, the return trip for a blast of sound was almost five seconds. In shallow water, it would take less than one second. From return times, water depths were calculated, producing simple linear profiles.

Tharp used her impressive imagination and understanding of geology to extrapolate beyond the thin trails of data, broadening the tracks into robust three-dimensional maps. Although originally hired to assist all of Ewing's graduate researchers, Tharp soon found herself drafting exclusively for Heezen.

Working conditions for the mapping team were unusually comfortable. Ewing was a professor at Columbia University which had received a gift of land and a mansion from the Thomas Lamont family. Lamont had been a newspaper reporter, then an adviser to J.P. Morgan, and still later an investment banker. But money did not seem important to the family, which was worth about 25 million dollars when the senior Lamont died in 1948. The family was determined to give their fortune away. They contributed heavily to the restoration of England's Canterbury Cathedral after its Second World War destruction and they created Lamont Library at Harvard. In 1949, Thomas Lamont's widow gave Columbia University the family's weekend retreat – an estate overlooking the Hudson River. This became headquarters for the Lamont Geological Observatory, now the Lamont-Doherty Earth Observatory. Maurice Ewing was the Observatory's founding director. He moved Heezen, Tharp, and the entire oceanography and geophysical research team into the Lamonts' former weekend house. Marie Tharp was given a bedroom, remade into an office for her mapping. Heezen had the best spot, up on the second floor of the grand house where he took a huge suite. He remodelled its attached bathroom into a separate private study for himself.

Marie Tharp organized the profiles from Heezen's voyages, confirming their locations, removing the refrigerator blips, calculating time-to-depth conversions, placing everything in the correct order, west to east. By 1952, she had six complete, nearly parallel profiles across the North Atlantic. They were broadly similar. Shallow near North America and Europe, deepening to over 3,000 metres, and possessing a ragged mountain range near the centre of each profile – the mid-Atlantic ridge. But Tharp noticed something more.

The feature that was most similar from profile to profile was not the prominent ridge but instead a deep V-shaped valley exactly centred on the central mountain range. The jagged mountains, seen in Tharp's profiles, only roughly matched each other, but the valley was clearly the same one, nearly identical in size and shape across the various cross-sections. Almost immediately, she was reminded of Africa's rift valley. It was a profound insight – a rift valley forms where the Earth's crust is splitting and opening. If true, this would be both unexpected and puzzling. According to Tharp, when she presented the idea that the mid-ocean was widening, Heezen dismissed it and the talents of all female geologists in one clumsy swoop. "Girl talk," he said to her.[307] The snipe masked Heezen's real discomfort. He realized a rift valley would support the idea of continental drift, which he had long scorned. So he ignored her conclusion for a year.

While Tharp was arranging Heezen's profiles and proposing her rift valley theory, Bell Laboratories – the phone company's research division – approached Heezen for advice on positioning its new transatlantic cables. They needed to avoid another disaster like the 1929 landslide that had ripped out communications and cost millions to repair. Heezen knew the turbidite flow that had destroyed those cables began with an earthquake near the Grand Banks. A way to help Bell would be to map all areas showing seismicity and then suggest the least seismically active path. For this task, Bruce Heezen hired Howard Foster, a deaf student from the Boston School of Fine Arts. Foster posted thousands of earthquake signals on a new map.

Heezen now had two different sorts of data – water depth and seismic activity – and he required all the mapping to be drafted at the same million-to-one scale. A thousand kilometres of ocean became one metre of map. The widest stretch, New York to Spain, required almost 7 metres, or 22 feet, to display. The team's standardized scale made it easy for Tharp to blend the seismic and depth data and see if the earthquakes were centred on those steep rift valleys as she

Heezen and Tharp's physiographic map of the Atlantic. New York City and the continental shelf are upper left, followed by the continental slope, abyssal plains, and sea mounts to the lower right. This map, from a Bell Labs 1957 report written by Heezen, was the first of its kind published. *(Reprinted with permission of Alcatel-Lucent USA Inc.)*

The Revolution Begins

suspected. She used a drafting light-table and stacked the maps, one on the other. The earthquakes were indeed centred on the steep valley within the ocean ridge. To her, it was proof the valley was a seismically active spreading rift.

The stacked maps finally convinced Heezen. It took a bit longer before he discussed the rift valley with Ewing. But in Toronto, in 1956, the two presented the first of several papers about the rift discovery. A year later, at Princeton, Heezen again described the rift, its shape and earthquakes. Harry Hess, head of the geology department, congratulated Heezen, saying, "Young man, you have shaken the foundation of geology."[308] He, and Marie Tharp.

Tharp, Ewing, and Heezen collaborated on the 1959 book *The Floors of the Oceans* which was accompanied by their dramatic map of the North Atlantic seafloor. It was drawn in a single colour of ink. Marie Tharp plotted the depths and turned the sparse linear tracks into a three-dimensional map, extrapolating beyond the known to create a convincing diagram. But it was actually Heezen who picked up a pen and sketched a cartoon, or physiographic map, with shaded and cross-hatched patterns symbolizing the steep edge of the continental shelf and the bumps, hills, and mountain ranges rising from the abyssal floor. Heezen's sketch, which he wasn't quite pleased with as he was too impatient to be an artist, allowed the team to publish their findings. The navy, which had been financing the ships and their data acquisition, refused to let them release the actual sea

Marie Tharp, mapping the Atlantic Ocean at the Lamont Estate House, 1955.
(By permission of Debbie Bartolotta, Marie Tharp Maps LLC.)

depths. That was classified information. It might help Soviet submariners hide among cracks and crevices at the seafloor. So the military kept the numbers locked away. However, presenting the maps as a stylized physiographic work of art allowed Ewing, Heezen, and Tharp to release a copy of the first truly magnificent seafloor map ever drawn – without going to prison.

Their book and the presentations and papers about the rift and its huge size sparked public interest. As did a movie masterfully filmed by Jacques Cousteau. Cousteau had doubted the existence of Tharp's rift until he filmed it with a camera mounted on a seafloor sled dragged behind his retrofitted British navy minesweeper, the *Calypso*. Cousteau was already famously respected – his oceanography book *The Silent World* sold five million copies within a few years; his 1955 film based on the book won the Palme d'Or at the Cannes Film Festival. When he showed his new ridge and rift film at the first World Oceanic Congress in New York City, no one left the conference believing the ocean basin was flat. In the audience during that 1959 showing was Marie Tharp who had discovered the rift and had drawn it in ink from blips of data on rolls of paper. She said it was exhilarating to actually see the steep black cliffs of her imposing valley.

The popularity of Cousteau's book and film and a general interest in the structure of the seafloor brought attention to the team at Columbia University. *The New York Times* gave "Miss Marie Tharp" credit for noticing that the trench was the centre of large swarms of earthquakes in the North and South Atlantic. In a science article referring to a 'crack in the world' it does not really give her credit for discovering the trench itself. It was Ewing and Heezen who told the story of the discovery of the thirty-kilometre-wide, three-kilometre-deep crack in the seafloor, extending 75,000 kilometres around the globe. On each side of the deep valley they reported mountains rising 3,000 metres. Formidable heights, but nearly all the peaks are still below the deep water's surface. Ewing and Heezen both recognized the significance of the rift system and said so to the *Times*. Heezen declared the discovery "tended to weaken" the theory of continental drift, though it did quite the contrary. Even though Ewing had decided that continents likely do not move, he was neutral in his interview, explaining to the press that the earthquakes along the rift indicate the ocean crust is likely being "pulled apart."[309] Much later, Tharp, who lived until 2006, said she was neither bitter nor resentful of the sparse credit given her during those days, but was rather glad to have interesting work and to play a role in the remarkable discovery. Finding the rift valley and its huge earth-girdling length was something "you could only do once. You can't find anything bigger than that, at least on this planet," she said.[310]

Amid their success, the feud between Ewing and Heezen escalated from loud words and over-turned paper bins until Ewing, the director, cut off funding and marine data to Heezen. By then, Bruce Heezen was a tenured professor at Columbia and couldn't be fired. But Ewing fired Marie Tharp to hurt Heezen's research. However, Heezen had expected this and had already cultivated strong

The Revolution Begins

ties with other academies and with the US Navy. From these, he received funds and data so he paid Tharp through grants and continued to supply depth profiles to her. Tharp moved her work into her own home, using her own bedroom as her drafting space. By now she had completed the Atlantic and moved on to chart the Indian Ocean in preparation for the International Indian Ocean Expedition. Data arrived from the US Navy, various American oceanography groups, and scientific teams from the Soviet Union, Japan, the UK, and South Africa. At the same time, Heezen and Tharp's work attracted the attention of the National Geographic Society, which wanted a map for their coincidental 1964 magazine article on the Indian Ocean.

The National Geographic commissioned Austrian artist Heinrich Berann to paint Tharp's physiographic charts. He was discovered by the editor of the magazine when a letter from Berann's daughter arrived in Washington. The girl complained about the quality of the magazine's maps and suggested her father was a much better artist. So the editor visited Berann in Austria. He immediately agreed and hired Berann to paint all of Tharp's ocean basin maps. The Society printed these over the next ten years. Heinrich Berann's slate of colours and systematic brushstrokes brought the seafloor maps to life. They were hugely popular – the most memorable and enlightening series ever published, according to the National Geographic's members.

Tharp flew back and forth to Austria, collaborating with Berann on those monumental maps of the ocean basins. The one-to-one-million scale had to be cut back, which meant simplifying some of the features. The project also meant updating maps already completed as newer data arrived. Meanwhile, Heezen spent hundreds of hours in Navy research submarines, measuring and observing all he could about the ocean floor. By 1977, their opus was nearly complete. Heezen was carrying proofs of their final world summary map when he boarded the American nuclear-powered submarine *NR-1*, which held ten people and could stay submerged for 30 days. It was in the North Atlantic recovering a secret Phoenix missile that had sunk near Scotland because a US military plane had fallen from an aircraft carrier. The sub then moved over the submerged mountains of the Reyjanes Ridge southwest of Iceland where Heezen boarded to gather data. Charitably described by the *Des Moines Register* as a large man, Heezen died in the submarine, aged 53, of a heart attack.[311] His body was carried by helicopter from the sub's support ship to Iceland, then flown to the States. At the time, Heezen's thirty-year colleague, Marie Tharp, was in the eastern Atlantic, aboard the British research vessel *Discovery*. Their final world map of the ocean basins was published a few months later. Their work was over, the world's maps were complete. In his will, Bruce Heezen, who had never married, gave his house in Piermont, along the Hudson north of New York City, to Miss Marie Tharp.

The Tharp-Heezen map exposed the seafloor with the revealing clarity of a drained bathtub. But even Tharp's intuitive, conscientious, painstaking work could not eliminate the biggest problem her work faced. Accuracy was always limited to navy priorities. Ships with sounding gear never sailed the diverse paths needed for complete coverage of the globe's seafloors. There were huge gaps. It was up to Tharp to cull out bad data and to sketch something into the broad blank areas. She did this with flair and artistry. Her maps were stunning and inspirational, but not perfect. Ten years later, NASA and the American Defence Mapping Agency launched satellites with radar altimeters that traced the planet's geodesic shape with accuracy rivalling sea soundings. William Haxby at Lamont Observatory used a huge set of data from NASA's Seasat in 1983 to make the first truly objective seafloor map. It lacked the artistic beauty of Marie Tharp's life's work, but it was raw, complete, and void of thoughtful, though occasionally incorrect, subjective extrapolations. Haxby made the ultimate map – complete and correct.

The only catch was again military secrecy. The topography of the ocean floor revealed anomalies that affected the trajectory of the Navy's sea-launched Trident missiles. So Haxby's 1983 gravity-based map was classified. Scientists wanting the data pleaded for declassification of the Earth's shape. The government relented, doling out morsels. In 1987, maps of Antarctic waters were released – due to international treaties, nuclear warheads weren't allowed there, anyway. Five years later, all data south of the Tropic of Cancer was freed. Finally, yielding to pressure from the American Geophysical Union, all the world's data were released in 1995. We would like to think it was a scientists' pressure group that influenced the military's final decision, but the European Space Agency's ERS-1 satellite had already released identical data, making the American decision moot. Today, Marie Tharp's work is largely a piece of art, an inspiring historical artifact. Her artistic extrapolations have been supplanted by much more extensive and more finely detailed satellite data. However, her work excited a generation of oceanographers whose primary school education included the brilliant National Geographic maps hanging in the back of the classroom. And still does.

16

Poetry in Motion

Harry Hess, who had told Bruce Heezen that the rift in the mid-ocean ridge had "shaken the very foundation of geology," realized the deep fracture was a rift in the Atlantic Ocean, just as Marie Tharp had said. As a rift, the seafloor must be widening, which meant the continents were moving apart. To Hess, this meant the basics of Wegener's continental drift theory were right. However, when Tharp showed Heezen the rift, Bruce Heezen saw the crack as something entirely different. He thought the valley centred on the mid-ocean ridge indicated the Earth is growing. The Expanding Earth Theory predicted exactly such stretch marks. The planet was splitting apart as it grew. Heezen was not the first scientist to notice evidence for an enlarging planet. It made as much sense as the previous front-runner, the Contracting Earth Theory.

Contraction had been the leading theory of the Earth in 1900. But now it was time for the expansionists to present their case. John Joly[i], an Irish physicist best known for developing radiotherapy to treat cancer, thought mountain building was due to radioactive heating and expansion of sediments that had fallen into deep troughs near the edge of continents.[312] Joly was arguably the most diversely skilled scientist since da Vinci. At university, he took top prize in literature and eventually published two volumes of insightful and entertaining poetry. As a physicist, he invented the first true colour photography; as an essayist, he wrote "The Bright Colours of Alpine Flowers," contributing to evolution theory. His essays even embraced perpetual motion and the origin of the submarine. In 1903, he complemented Holmes' work with a new method to calculate the age of rocks, one based on radium decay. Ten years later, he and Ernest Rutherford computed that the Devonian period began 400 million years ago, a figure still recognized as correct. Joly suggested radioactive decay inside the Earth creates more heat than is lost at the surface. By his logic, the Earth continues to heat up and expand, creating cracks which fill with magma. Then the magma cools and solidifies, causing the planet to shrink a little. The system of expansion and contraction repeats indefinitely, according to Joly's theory of thermal cycles.[313]

---

i   John Joly (1857-1933), Irish physicist.

The Mountain Mystery

During the 1926 Continental Drift Conference in which Chamberlin and others debunked drift theory, John Joly presented radioactive heat and a form of convection as a mechanism which could move the Earth's surface, build mountain ranges and displace continents – but few were convinced.

Joly combined expansion with contraction, but others, including Heezen, were strictly expansionists. The most prominent advocate of an Expanding Earth hypothesis was Warren Carey[i], an Australian geologist. The Australians, like the South Africans, accommodated many independent-minded geologists supporting Wegener and other advocates of continental mobility. Carey rallied to the drift theory early. But he fully believed data showed the Earth growing larger daily, and without Joly's hesitant cycles of shrinking between growth spurts. Carey agreed with much of Wegener's theory as soon as he read *The Origin of Continents and Oceans*, in 1934. At the time, Carey, a farmer's son from New South Wales, was completing his graduate studies in Sydney. Continental displacement made sense to most Australian and South African geologists because it remedied issues they regularly encountered – the distribution of rocks, fossils, and landforms in the southern hemisphere. So it is not surprising Carey found value in the notion of a splintered supercontinent.

Warren Carey, Australia's Expansionist

One of Carey's great contributions to the continental drift debate explained why reconstructions of ancient Pangaea were seldom perfect. Although geologists and cartographers painstakingly drew maps of South America nestled tightly to Africa with Europe almost touching North America, there were always holes or gaps in the maps when the modern outlines of continents were moved and meshed together. The gaps in the reassembly of the supercontinent were pointed out as failures in the Pangaea theory. At the 1926 meeting hosted by petroleum geologists in New York, several scientists dismissed the whole hypothesis on the basis of poorly reconstructed gap-filled globes and maps, chiding Wegener for even suggesting the match-up of coastlines was possible. A few years later, Carey demonstrated the voids resulted because maps of the world miss the submerged parts of the continents. Much coastal property is under water. Each landmass has a swath of submerged continental shelf which would be visible only if the seas retreated a bit. With Carey's more accurately drawn continental outlines, building a supercontinent from stray parts is much more convincing. Giving the continents

---

i   Samuel Warren Carey (1911-2002), Australian geologist.

Poetry in Motion

their true boundaries, Carey correctly reconstructed Pangaea. He also proposed a solution to the other great weakness in Wegener's theory – he thought he had discovered a force that could move the continents.

Warren Carey proposed expansion as the power pushing the continents apart. Growth had broken the supercontinent. This resolved many of the issues confronting geologists when they encountered identical fossils separated by oceans, or when they observed layers of coal in the arctic. As early as 1956, Carey hosted a symposium on continental drift at the University of Tasmania where he taught. Growth of the planet was central to the discussions. And as late as 1976, he published *The Expanding Earth*, a lengthy discussion of the theory he clung to long after most of his colleagues had drifted away.

We have noted Charles Darwin himself had very briefly entertained a similar idea, based on formations he saw in South America. The Andes and the elevated plains and beaches of Patagonia brought this to his mind during his voyage on the *HMS Beagle.* He jotted the ideas into his notebook, but abandoned them quickly, back in 1835. There wasn't enough evidence. It seemed more likely to Darwin that the land had been forced upwards while the seas sank, leaving stranded beaches high above the ocean. For him, the Earth neither grew nor shrank. Seventy years after the *Beagle* voyage, Roberto Mantovani, the violinist-diplomat-geologist, published a paper on Earth expansion and continental drift. He assumed that a single continent had completely covered the surface of a smaller Earth. Thermal expansion, explained Mantovani, resulted in volcanoes which splintered the surface into continents. Those continental pieces – including the one you are resting upon at this moment – drifted away from each other as the Earth grew. This idea was difficult for his contemporaries to accept. A little expansion might roughen the world's smooth surface, but it would take quite a stretch to propel the continents far, far apart. Early proponents suggested that the world need only grow a tiny bit to make grand mountains and deep rifts. Just a tenth of one percent and mountains would jut skywards. But to split a supercontinent and send its parts half way around the globe would take much more dramatic growth. But there were those who claimed the world had done exactly that.

Dramatic expansion was proposed by László Egyed,[i] a creative Hungarian geophysicist. He published over 100 papers and articles, including the landmark textbook *A föld fizikája (Physics of the Solid Earth).* According to Egyed's calculations, the Earth's radius was increasing by an imperceptible millimetre, or less, each year. Imperceptible within the lifetime of a human, who would live on a planet that grew only a single hand's width in a hundred years. But the Earth is old. Since life appeared in the Cambrian, Egyed figured the planet widened 1,000 kilometres. The cause, said Egyed, was a change in the nature of the materials at the hot boundaries between the inner core, the outer core, and the mantle. Expansion, by his theory, occurred along those adjacent zones.

---

i László Egyed (1914-1970), Hungarian geophysicist.

The Mountain Mystery

A simple way to picture the Earth's various layers is to consider the planet as a chocolate-covered cherry. The centre is the solid fruit, surrounded by liquid cherry juice, enveloped in creamy fondant, all held together by a thin outer crust of chocolate. The proportions are wrong, but the order of the phases is correct: solid, liquid, gooey, then rigid. From the way seismic waves travel through our planet, scientists have inferred the central core is solid. Seismic also tells them the inner core is surrounded by a liquid outer core. The shallower mantle, where convection currents are thought to flow within its semi-solid plastic material, is the thickest layer. Encasing it all is our comfortable solid surface.

Earth as a chocolate-covered cherry

In physics parlance, a phase transition occurs when a material transforms between solid, liquid, or gas states. The Earth has phase boundaries at its solid-liquid-solid interfaces. The Budapest professor explained there was a slow, steady, irreversible transition of the material of the inner core into outer core material, and, closer to surface, outer core into mantle.[314] At each phase change, the Earth's inner stuff becomes less dense, so the planet expands. Less than a millimetre a year is not dramatic and Egyed seems correct about phase changes. They may really be happening, with inner core morphing into outer core, outer core into mantle. If Egyed is right, the planet's rotation must slow so angular momentum is conserved. The Hungarian geophysicist proposed a geomagnetic test of this, but the results are not yet clear. So far, geophysicists are **not in general agreement** that our planet has expanded in the way Egyed suggested. But his physics are valid and his estimates are cautious. Unlike a contemporary of Egyed, Ott Christoph Hilgenberg,[i] who invented a new physics to show the Earth has doubled in size since its formation.

At 17, Hilgenberg was drafted into the German army at the start of World War I. He suffered permanent hearing loss in battle and was wounded by a shot through his knee. Regardless, he was stuck in combat until the war ended, then he began university in Berlin. With a mechanical engineering degree and strong physics background, he was hired by a geophysics company looking for oil in the United States, twenty years after Anthony Lucas drilled the first Texas salt dome. Hilgenberg was frustrated by the ambiguity of geophysical signals coming from the subsurface – the data simply weren't explicit enough to consistently trust. Today, a hundred years later, honest geophysicists complain about the same thing.

A fundamental problem with geophysics – then and now – is ambiguity. The

---
i   Ott Christoph Hilgenberg (1896-1976), German engineer, scientist, and humanist.

scientist uses tools to gather faint signals either from a huge treasure buried deeply in the Earth or a tiny treasure buried at a shallow depth. It is often frustrating work distinguishing the two, although geophysicists have recently become better at sorting them out. In the end, exploration remains an informed gamble. After four years in America, Hilgenberg was back in Berlin, at the Technische Hochschule where he had earlier studied electro-mechanical engineering. His goal was to make geophysics data more reliable and less ambiguous. During his work, he also became interested in Wegener's theory.

Like Carey, Hilgenberg saw expansion as the power source that parted continents from their original Pangaean positions. If the Earth had doubled in size, the continents would have moved to their present locations. To anyone with interest, Hilgenberg showed a series of globes of increasing diameter from a half-size ancient Earth with continents jammed together, to a current model of the planet. Doubling the size of the Earth is not an intuitive solution. Difficult enough to believe continents could move; nearly impossible to picture the entire planet expanding to twice its size.

Hilgenberg, who devised his hypothesis around 1929, held on to it for life. He dedicated his 1933 book, *Vom wachsenden Erdball* (*Of the Expanding Globe*), to Wegener, who had died in Greenland three years before the book was published. But Wegener had earlier distanced himself from the expansion idea. Hilgenberg persisted. He was wholly convinced Earth expansion resolved the geology issues around

Ott Hilgenberg

fossil distribution and palaeoclimates. Fluent in Russian, Hilgenberg was heavily influenced by Soviet scientists, many of whom also favoured the expanding Earth notion. He tightened his hypothesis by amalgamating theories from a variety of sources. But his mechanism of planetary growth was entirely different from Egyed, who considered expansion a result of the physical changes in the structure of inner-earth materials. Hilgenberg created an entirely new physics to explain himself.

According to Hilgenberg, all the mass in the universe is expanding. Every atomic particle is absorbing "energy-mass" from empty space, which he said wasn't empty, but instead contains "energy density." An object expands by "feeding itself on space."[315] The concept of dark matter had not yet been developed and in some ways Hilgenberg may have been presaging that notion. His ideas about absorption of energy-mass and some of his theories about gravitational waves were presented in his *Solution to the Mystery of Gravitation* in 1929. His proposals were loosely supported by some experimental evidence, but rejected by nearly all physicists. His theories ran counter to quantum physics and were contrary to the theory of relativity which was proving more and more convincing. During the early 1930s, Hilgenberg was academically isolated.

Nevertheless, as a brilliant and innovative physicist, he was preparing to chair his school's department of mechanical engineering. But first he was told to become a member of the NSDAP – the Nazi Party. He refused, and found himself assigned to working in the university library. Near the end of the Second World War, Allied air raids were a daily occurrence and Russian snipers hid in the city. Hilgenberg, over fifty years old, nearly deaf, wounded in the first war and now shot through the arm in the second, set about saving the university's library. He drove a truck, raced around the city, and hid the university's science books in Berlin cellars to protect the collection from war damage. Immediately after Germany's surrender, he reversed the process, arguing the books out of the Russian sector and securing release of 10,000 documents from the British so he could repopulate his library. Hilgenberg was also involved in relief efforts, which resulted in recognition for humanitarian work from the new West German government. Believing the west was rewarding his charity for political reasons, he rejected the honours they tried to bestow. For the next five years, until 1950, his home was in Berlin's west sector while his job was in the Russian-controlled east. He risked being trapped in East Germany, so finally left his work and his library.

Hilgenberg continued to develop mathematical proofs of his expanding-earth theory – merging geology, palaeoenvironment, and plate tectonics. He expanded the concept of his novel physics of energy-into-mass, explaining the system by invoking black holes, redefining Cartesian aether-space, and developing a new theory of gravity. He was editing his last attempt to unite cosmology and earth-expansion when he died, age 80, in 1976. Little has developed in the last thirty years to support his theses and physicists continue to reject most of his expanding matter theory, although there may be some eventual redemption amid newly theorized properties of dark matter. At best, Hilgenberg gave tangential support to the Tasmanian geology professor, Warren Carey, who continued to promote Earth expansion into the 1980s.

Warren Carey, like Hilgenberg, never gave up his fundamental hypothesis of planetary inflation and promoted it with congenial vigour. Late in his life, he continued to attend conferences and held court with a dwindling number of like-minded expansionists. His support was weakened when satellites indicated the Earth is not likely expanding, although expansionists have pointed out that the rate is so small our instruments may not yet detect the few millimetres growth that have occurred in the twenty years since this sort of measuring technology has been developed. But there are other problems with simple expansion as the explanation for the planet's morphology. The most troubling is the fact that the oldest seafloor rocks are only 180 million years old, so expansion would have been recent. There is also evidence that Pangaea was preceded by earlier continents that had also clustered before breaking apart. There is still much to laud among the ideas of Egyed, Hilgenberg, Carey, and others. It is still possible that some degree of expansion based on phase transitions (or Hilgenberg physics) have played a role in the way the seafloors open and spread and continents travel across the planet. Expansionists do not necessarily reject the entire idea of plate tectonics.

A current and tireless advocate of the earth-expansion model is Giancarlo Scalera of the National Geophysics and Vulcanology Institute in Rome. He cites similarities between Australian and South American features that are best explained by reducing the Earth to half-size. At that point, which he indicates may have been in the Triassic, 220 million years ago, every continent would be in communion and there would have been little or no ocean on the globe. And there are other expansionists. In 1991, Jan Koziar published an expanding-earth model he described as "an orange peel effect,"[316] invoking memories of Suess's shrinking apple planet. In 2001, a PhD was granted on the thesis of the earth-expansion hypothesis to James Maxlow of Perth's Curtin University.[317] When I read Maxlow's thesis, the overwhelming data collected to support his theory did not thoroughly convince me, but certainly alerted me to the realization that information, observation, and evidence can be interpreted in a variety of interesting and possibly valid ways. As Bruce Heezen thought, the mid-ocean ridges and rifts may indicate expansion. However, geophysicists began to see the rifts as evidence of a spreading seafloor, but not an expanding planet. The former Pacific navy captain Harry Hess became the most significant advocate of seafloor spreading and was in a position to draw attention to the idea.

Geo-poetry

The rift in the great mid-ocean mountain chain was discovered by Marie Tharp in 1952 and announced to the world in 1956. That led to Harry Hess's most important contribution to science. In 1959, he had an idea that tied together all his earlier investigations, an idea which is considered the most important part of the most important concept in earth science. Hess realized that the Earth's crust was creeping away from the long, volcanically active ocean rift. This singular idea is the heart of plate tectonics and is valid today, fifty years later. Hess built on the ideas of many others, but he especially pointed out the work of Bruce Heezen. Harry Hess also added to Arthur Holmes' theories, which had included a convection scheme for the mantle and an early hypothesis similar to seafloor spreading. But it was Tharp's discovery of the rift itself that gave Hess the confidence to promote his conjecture. In 1960, Hess informally presented his unpublished theory in a manuscript that became widely xeroxed and circulated.

Two years later, Harry Hess published "History of Ocean Basins," within an

The Mountain Mystery

obscure collection called *Petrologic Studies*. He anxiously described his work as *geopoetry*, apparently in an attempt to diffuse some of the criticism he knew it would attract.

Hess hypothesized that the seafloor widens at the great global rift system's volcanic fissure where hot magma emerges. Extreme pressure along the rift forces new ocean crust away from the ridge, then moves the crust across half the ocean. At the end of its enormous conveyor belt, it sinks at subduction zones, disappearing in the mantle. The basaltic ocean crust doesn't pile up or accumulate. Hess's spreading, subducting, mobile seafloor explains why so little sediment accumulates on ocean floors. And it explains why ocean crust is relatively young. Hess calculated the oldest ocean crust has survived 300 million years, but continental crust is more than ten times older. Within 300 million years, the new seafloor has journeyed from rift to trench, then descended into the mantle, never to be seen again. Meanwhile, continental material mostly stays aloft. By this insight, Hess explained the puzzling age difference between marine fossils found on the ocean floor (actually less than 200 million years old) and marine fossils lifted to mountain slopes (sometimes over 500 million years old). Any fossils not safely thrust upon the non-recycling continents remain in the seafloor mud and are eventually destroyed at subduction zones.

"History of Ocean Basins" had critics, of course. But the paper was well-received by the new generation of scientists. For the first time, a scientific position on mobile continents was attracting more supporters than detractors. Hess's history of the ocean basins was the most important contribution to the development of modern plate tectonics – his paper was cited more frequently than any other geophysics research paper, geopoetic or not. For a dozen years, it was

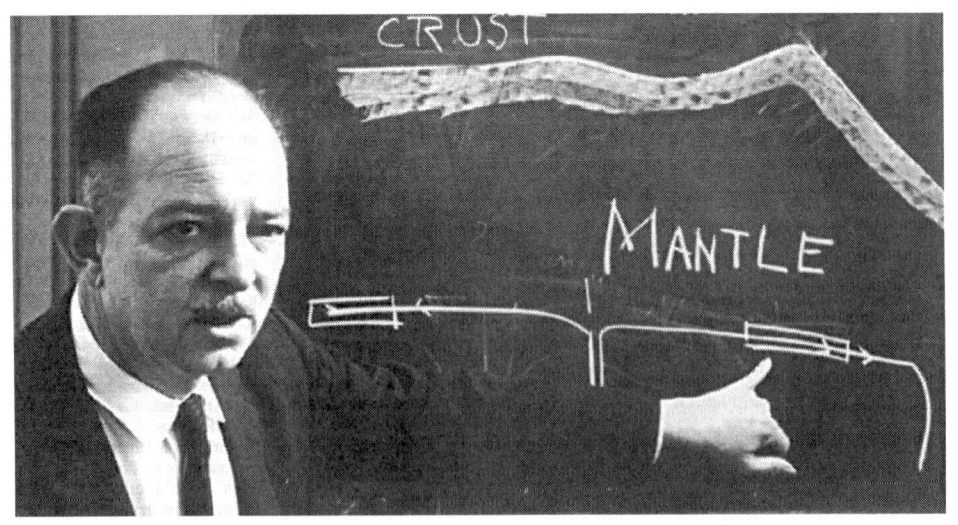

Harry Hess describing crust upwelling and subduction.

the source material other geophysicists used when they wrote their own papers. Hess stated that the evidence for mobile continents was strong, however, in presenting his paper, he cautioned "It is hardly likely that all of the numerous assumptions are correct."[318] But the most important ones were. The oceans spread; the continents move. In just a few words, here is how he summarized his interpretation of the facts scientists had amassed over the preceding century:

> "The Mid-Atlantic Ridge is truly median because each side of the convecting cell is moving away from the crust at the same velocity, ca. 1 cm/yr.[i] A more acceptable mechanism is derived for continental drift whereby continents ride passively on convecting mantle instead of having to plow through oceanic crust." [319]

Hess knew Wegener's displacement theory provided the best solution to all the unsolved geologic mysteries. Nothing had been discovered that made the basic points of Wegener's theory incompatible with new geophysical data, oceanic maps, or samples of fossils and rocks. But a key difference with the Hess proposal was that Wegener had believed the continents somehow pushed through the seafloor, which critics rightly dismissed as physically impossible. Hess proposed that the continents were sitting on a huge conveyor belt that moved the ocean basins and landmasses alike on the same assembly line. In 1962, the scientific community was still warming to the Hess hypothesis, but within a few years, thermal and magnetic studies of the seafloor confirmed it.

Not only did the new geophysical findings explain the mid-ocean rifts, dissolve the problems of misplaced fossils, coal seams, and glacial debris, and clarify the youthful age of the ocean crust, but the discoveries also supported the growing understanding of the differences between continent and seafloor. The oceanic bottoms and the continental tops of our planet are different in almost every way you can imagine. Hess pointed out that the visible part of the continents, on average just of a mere 300 metres above sea level, make up only 29 percent of the Earth's surface. This small portion of the planet barely has its head above water. If the oceans were somehow drained to reveal the continental shelves, we would see a rapid drop beyond the shelf edge where the average ocean depth quickly plunges to 4,300 metres.

The elevation difference between ocean and land is due to rock type and density. The rock of the seafloor is basalt, made of minerals like peridotite and serpentine which are dense, heavy, and rich in magnesium and iron. Continental rock is of an entirely different fabric. It is lighter than seafloor rock, which is why

---

[i] The Hess estimate of seafloor spreading (one centimetre per year in each direction) is quite close to current measurements. The spreading rate is fairly constant at any single locale, but the speed of spreading varies around the globe. For example, Atlantic spread is about 2.5 cm/year while the Nazca Plate off Chile is rushing away from the Pacific Plate at 15 cm/year.

the average continental crust is perched an average of 5 kilometres above the average ocean-floor crust. Our continental rock is largely in the granite family, enriched with potassium, silica, and aluminum. The continents are not ocean rock lifted into mega-islands. This knowledge was confirmed only recently, in the middle of the twentieth century. Hess put an end to speculation that ocean crust was disgraced, fallen, submerged continental crust serving as a timeless basin collecting the dirty decay of continental erosion.

The Velikovsky Affair

Part of the success of the curious and creative Harry Hess was that he did not dismiss novel ideas without giving them a hearing. The best example of this was his long association with a strange man who had written some peculiar books. Immanuel Velikovsky's[i] theories about the Earth and its ancient history were speculative thought-experiments which were seldom grounded on data or even scientific principles, but heavy on Old Testament religion. During the 1950s and 1960s, his imaginative books sold by the millions. Thousands of supporters vigorously argued his theories in churches and campuses across North America.

Velikovsky's ideas were such a strange porridge of pseudo-science that it is odd someone like Harry Hess would befriend the man and defend his right to lecture at Princeton. Perhaps it was Hess's own situation – defending speculative continental drift – that made him somewhat sympathetic and professionally respectful towards Velikovsky. Or, more likely, it was simply that Hess felt everyone was entitled to an audience. At any rate, the *New York Times* called it the "Velikovsky Affair" in their headline when they linked the two men.[320] Dr. Harry Hess was the head of the geology department at Princeton at the time Velikovsky spoke to a standing-room-only crowd there in September, 1966. Hess defended the controversial scholar. It was a brave stand – the mere mention of Velikovsky's name sometimes made scientists spit.

Immanuel Velikovsky was trained in medicine and psychiatry but gained fame for his best-selling *Worlds in Collision* which offered an alternative view of the evolution of the solar system. His book claims Venus was once a comet ejected from Jupiter and the new comet Venus swept near the Earth in 50-year intervals, disrupting the planet's spin and orbit, leading to some of the dramatic events portrayed in the Bible – including the plagues of Egypt and the parting of the Red Sea. At one point Venus stalled near the Earth, giving Joshua the extra hours of sunlight the Bible said he needed to slaughter the Amorites of Gibeon,[ii] following

---

i   Immanuel Velikovsky (1895-1979), born in Belarus, raised in Russia, lived in Israel and the USA, a psychiatrist and author.
ii   From Joshua 10:12-14, "The sun stopped in the midst of heaven and did not hurry to set for about a whole day. There has been no day like it before or since, when the Lord heeded the voice of a man, for the Lord fought for Israel."

Poetry in Motion

God's orders to kill every Canaanite on Earth. Not only did Venus help Joshua in his battle, but the planet carried pestilences of flies that originated on Jupiter. Those flies and other Jovian creatures were the source of Earth's petroleum deposits. Decaying organic bodies of creatures from Jupiter, said Velikovsky, were carried to Earth on the comet tail of Venus. He also argued persuasively that Saturn had once gone through a nova state, ejecting a huge quantity of water into space, which found its way to Earth and caused Noah's flood. Velikovsky also proved (at least to his readers) that the planet Mercury was involved in the collapse of the Tower Of Babel, and Jupiter in the destruction of Sodom and Gomorrah. *Worlds in Collision* was enormously popular in the 1960s. It appealed to many who wanted to believe the literal word of the Bible, yet also needed to support their belief with some type of tangible proof. Although Velikovsky had no training in physics or astronomy, he persuasively explained Old Testament miracles with an imaginative flair that seduced millions of followers.

Geologists dismissed these and other charming Velikovsky hypotheses, inadvertently strengthening his standing among those who recognized the dismissive tones as a conspiracy among mainstream scientists to hide the true nature of the solar system. Velikovsky offered a non-mathematical solace with his entertaining religious-scientific synthesis. It did not take much depth of thought to be swept away by his tales – in fact, that was perhaps a prerequisite. Readers came to the defence of the doctor whom they saw as a persecuted scientific martyr attacked by the establishment. When geologists objected, Velikovsky's books sold better.

Velikovsky grew wealthy from *Worlds in Collision*, *Ages in Chaos*, and *Earth in Upheaval*, which were printed in condensed form in the *Reader's Digest* and in large book runs by Doubleday. But he was no charlatan. In 1945, before his fame and success began, he recorded his theories in a court affidavit, to prove that he alone invented his many peculiar theories, thus preventing others from stealing credit for what he sincerely believed and promoted. He was intellectually

Joshua Commands the Sun, by Dore, 1885

221

brilliant. Velikovsky had a respectable career as a medical doctor and psychiatrist. Among his many accomplishments, he and Albert Einstein co-edited two volumes of Jewish contributions to science and he helped establish Jerusalem's Hebrew University. But his amalgamation of Bible and science had little basis in fact.

Although Velikovsky was rejected by scientists, Harry Hess, head of Princeton's geological sciences, publicly came to the defense of Velikovsky's right to an audience. In September 1966, when Velikovsky came to present his ideas at Princeton's new Woodrow Wilson School of Public and International Affairs, Hess came to his support "in the interest of fair play." Hess told *The New York Times* that he was impressed by Dr. Velikovsky's sincerity and by his phenomenal memory. "He quotes whole paragraphs of Newton," said Hess and "spends much of his time in the libraries" of Princeton. Velikovsky was often found there, sifting through documents, searching for support for his theories.[321] Meanwhile, Velikovsky, in his memoir *Stargazers and Gravediggers*, claimed that Harry Hess knew his book *Earth in Upheaval* by heart.[322] Velikovsky outlived his friend Hess by a decade. Velikovsky wrote, "In Hess's passing, I lost the only member of the scientific elite who demanded a fair treatment for me and my work."[323]

Harry Hess died while working, collapsing at a meeting of the Space Science Board of which he had been chairman for years. It was within a month of the arrival of the first moon rocks. Hess had been one of ten men in America scheduled to perform tests on those first lunar samples. He made a few preliminary examinations and he hoped further studies would reveal evidence about the Moon's formation. Hess, the navy admiral, scientist, and geopoet is buried at Arlington. He lived just long enough to see his ideas confirmed and generally accepted as the principle idea of our planet's physical evolution.

Smokers and Worms

New evidence of Harry Hess's seafloor spreading came from a variety of places. Geophysicists measured the temperature of the seafloor at the mid-ocean ridges. They discovered it is hot nearly everywhere along the narrow rift. Not necessarily with scorching volcanoes, but the rift zone showed a definite, unexpectedly high thermal anomaly. In some places, cracks in the newly forming crust allow cold seawater to seep down to where the rocks are truly searing. Geysers erupt, kilometres below sea level. We now know ocean rifts are dotted with sea vents spewing boiling water which, under pressure, is several hundred degrees hotter than neighbouring seawater. These geysers lift minerals, including sulfides that seem to be a favourite snack for some odd types of bacteria. The bacteria, in turn, are eaten by a variety of predators. An entire unique ecology has developed in the blackness of the deep ocean bottom.

The vents, geysers, and smokers were discovered in 1977. They create an amazing seascape. While most of the ocean is unnervingly quiet at 2,100 metres, the sea smokers are a haven for bizarre wildlife. An unmatched ecology exists there, totally disconnected from the sun's energy. At such a depth, the Earth's core itself is animating a realm entirely independent of the rest of us, providing the energy which sustains life. Among its bizarre creatures, the Pacific has giant tube worms, some as tall as an adult human. These live in superheated hydrosulphuric-laden water bubbling from the boiling sea geysers. Rather than eating the bacteria thriving near the vents, about half of a tube worm's weight is composed of the friendly microbes. In their symbiotic friendship, tube worms suck passing particles of hydrogen sulfide, carbon dioxide, and minerals from the hot vents and pass these as food to the bacteria adhering to their bodies. Lacking a digestive tract, the worm feeds the bacteria, then ingests the bacteria's excreted organic material. The tube worm turns this into more tube, thus attracting more bacteria. For us, there should be comfort in knowing that this distant and distinctly alien environment exists. We have no need to fear the total annihilation of Earth's life. After being pelted by an asteroid, or suffering a cataclysmic volcanic event, our planet's crust may melt, the atmosphere may evaporate, surface life may vanish. But two kilometres below the sea's surface, tube worms will flourish undisturbed near the planet's mid-ocean rifts. Life on Earth will continue.

Before geophysicists found the rift system snaking its way around the globe, there was no reason to suspect anything hot existed in the middle of the frigid dark oceans. Quite the contrary. The ocean depths were supposed to be intolerably cold. It was expected that the deep seafloor was stark and lifeless. When the *Challenger* completed her journey in 1876, it was estimated that over 90% of the seafloor had a temperature of 5° Celsius or cooler. But in 1962, a heat-flow profile was conducted across the Atlantic. Although sea vents weren't discovered until fifteen years later, the heat anomaly at the rifts was definite and surprising. Measurements were made at each of 14 stations from Martinique in the Caribbean to the Canary Islands near Europe. Heat-flow emitted from the ocean crust, in calories per square centimetre, varied by a factor of twenty, with the greatest heat radiating from a belt 200 kilometres wide, centred on the Mid-Atlantic Ridge.[324] This was the residual heat of new ocean crust hardening as lava cooled from 1200°C, its crystallizing temperature, in contact with frigid seawater.

No one in 1962 could measure the gradual widening of the seafloor. Seafloor spreading remained a probable, but unproven, theory. However, to complement the thermal evidence, Maurice Ewing, director of Lamont Observatory, gathered another set of supporting data. He sent drilling ships towards the ridges. Starting at the Sierra Leone Rise, a high spot in the seafloor, research vessels drilled a series of test wells in straight lines leading away from the rise. Cores recovered after poking into the Atlantic seafloor contained tiny fossils entombed in the muck just above the basalt crust. The fossils were youngest at the ridge where crust was new and significantly older in cores drilled farther away from the rifted centre. This agreed with the idea that newer crust formed at the ridge, became covered in sediments and fossils, then drifted away. Fossils brushed from silt just above the new crust were 50 million years younger than fossils a thousand kilometres away from the ridge. This meant the seafloor conveyor belt was moving at 2 centimetres a year, carrying a record of the evolution of Earth's creatures.[325] With these microfossils, heat recordings, and Harry Hess's geo-poetic description that tied the clues together, it was becoming more difficult to dismiss the notion that the planet's crust is moving. But skeptical scientists wanted more. They demanded something ironclad and irrefutable, something set in stone. They soon had it.

17

Written in Stone

Harry Hess predicted that the mid-ocean rifts were the birthplace of the planet's ocean basins. His 1962 paper showed how the fissures ooze mantle material, pushing the ocean apart. His argument was strong, it explained the youngish age of oceanic crust, it accounted for the varying ages of those tiny fossils found in Ewing's cores, and it tied in the heat variation across the seafloor. But it was entirely circumstantial – neither Hess nor anyone else had irrefutable proof of seafloor spreading.

The key evidence that ocean basins widen came from palaeomagnetism, something akin to a rock's memory of the planet's dynamic geological history. This memory is embedded in ordinary igneous rock, mostly the grey basalts that Reginald Daly once declared the most important material on the planet. As Daly predicted, basalt revealed the final truth about the Earth, its evolution, and the origin of mountains. Remnant magnetism of ancient rock proved ocean crust was spreading. It was irrefutable evidence. By interpreting the crust's magnetic finger prints, the continents' former locations within Pangaea could be reconstructed. Discovering this technique was a brilliant piece of science.

Scientists had long known that when melted iron cools, its magnetism aligns in the direction of our planet's magnetic field. So too, with the iron-rich rocks of the ocean crust. However, around 1906, two French scientists were startled to discover remnant magnetism in rock samples is occasionally oriented exactly opposite to the expected direction. As if the North Pole had temporarily visited the southern hemisphere. Independently, Bernard Brunhes[i] and Pierre David measured natural magnetism at various sites in the French countryside. Brunhes, in particular, is credited with working out a test that measured remnant magnetism in the field. At first called *fossilized magnetism*, a name that invokes a once lively vibrant condition now captured as a relic, today the phenomenon is known more cumbersomely as *palaeomagnetism*. The scientists' magnetic data were found in clays from central France that had been baked much earlier when a hot layer of lava flowed over the earthy strata.[326]

---

i   Bernard Brunhes (1867-1910), French geophysicist.

The Mountain Mystery

Clay, of course, is the raw material of one of our greatest inventions – the brick. Sun-dried or kiln-fired, often with straw added to prevent cracking, clay bricks are the artificial stones that housed civilizations from Ancient Persia through Victorian England. The French researchers found natural brick was created at the contact between clays buried under a Miocene[i] volcanic flow. At a particular level, the brick material had the expected northerly magnetic orientation while an adjacent older layer in the same sequence, baked by lava, had the opposite polarity. For over twenty years, Brunhes's results were a geological befuddlement. At first, it was assumed an undiscovered property in the clay created the anomaly. The actual cause, a complete reversal in the Earth's magnetic field, is more stunning.

Magnetic polarity reversals were conjectured in 1929. Motonori Matuyama,[ii] after studying a series of solidified basaltic lava flows from China and in his native Japan, offered the startling speculation that the world's magnetic field had reversed. Matuyama had found a series of igneous rocks where a normal polarity was consistently measured in all samples up to 780,000 years of age, but his older samples had an exact opposite magnetic orientation. Since the mineralogy hadn't changed in those samples, he guessed that the Earth's magnetic field had flipped. As it turned out, he eventually found that with even older rocks, a group beyond two and a half million years, remnant magnetism was back to normal again.

For decades, most geologists were unconvinced that old rocks had captured remnants of global magnetism. They understood how a compass needle retains the planet's magnetic orientation when forged, but there were issues with the raw rocks of nature doing the same thing. Most geologists were certain the data were flawed. In fact, the majority of rocks hold no clear magnetic signature. Or it is faint. So faint that the pounding of the geologist's hammer while knocking a sample from a formation can knock the magnetism right out of the rock. At other times, stone samples acquire new magnetism while simply sitting on a laboratory shelf. Most scientists felt the apparent polarity reversal was from poorly handled samples, or perhaps some undiscovered rock property.

Worldwide only a few hundred samples had been collected. More data were needed. Allan Cox[iii], a Berkeley graduate student, travelled north to Idaho's Snake River basalt fields and took careful measurements of remnant magnetism in the thick beds of rock he found there. He used the radioactive decay of potassium into argon to accurately determine rock ages for the youngish lava flow he was inspecting and he built a convincing case supporting Matuyama's theory. Allan Cox was just one of several scientists, including Brunhes in France, Runcorn in England, Hospers in Iceland, and Irving in Australia, who corroborated reversals at specific time intervals. Consistent results from widely varying locations and

---
i   The Miocene Epoch began 20 million years ago and lasted 15 million years.
ii Motonori Matuyama (1884-1958), Japanese geophysicist.
iii Allan Cox (1923-1987), American geophysicist.

different rock types confirmed the Matuyama hypothesis that the Earth's magnetic poles had flipped polarity.

Matuyama was honoured by having the first magnetically-reversed era named for him, the *Matuyama reversed chron*. Prior to the Matuyama era, the planet's field was normal for the million-year magnetic era called the *Gauss chron*, named after the great German physicist, Carl Friedrich Gauss, who put numbers to magnetic field strengths and fluctuations. The man credited with discovering the Earth is a huge magnet is remembered in the even older *Gilbert chron*. We are living in what we have chosen to call a normal polarity period, the *Brunhes chron*, named for the scientist who first noticed opposite polarities in baked clays in France a hundred years ago. Researchers have created a catalogue of age-related reversals stretching back tens of millions of years. At first a bewildering curiosity, magnetism eventually played a key role in confirming the spreading-seafloor theory. But it was complicated science, clouded by more than magnetic reversals.

Hundreds of years ago, sailors noticed the magnetic north pole moves around quite a lot, and sometimes rather quickly. Scientists also recognized the unpredictable habits of the Earth's magnetic field. But the cause of that wandering is still not clearly understood. Our two north poles – the magnetic north pole and the map's geographic, or rotational, north pole – rarely coincide. When they do, it is a brief encounter. There is no lingering. The magnetic pole doesn't slow, or stop, but glides past its geographic counterpart. Observations of the wandering have been noted at least since the time the scientists in Paris started measuring the declination, or variance, back in the 1500s, as we have already seen. Last century, the magnetic north pole moved north ten degrees – a distance of over 1,100 kilometres. In 1900, compass needles pointed towards 70.5° North; in 2001, the Canadian Geological Survey found magnetic north was near Ellesmere Island at 81° North. In 2012, the magnetic north pole left Canada, slipping even farther north and west, over Arctic waters of disputed nationality – waters claimed by Russia, but usually considered by most others international.

In 1539, a map called the Carta Marina, drawn by Olaus Magnus, portrayed an *Insula*

Magnus's imaginary Insula Magneta on his busy 1539 map

227

The Mountain Mystery

Magnetum, or *Island of Magnets*, as the cause of compass attraction. Magnus placed the mythical magnetic island off the European Arctic coast, north of Russia's Murmansk and close to the spot Norway, Finland, and Russia meet on today's maps. Many early European navigators thought their compass needles were attracted to a magnetic mountain on that island. Some even claimed to have seen it, and called it *Rupes Nigra,* or *Black Rock*, a 16-kilometre-wide island of black magnetic stone.

The famous sixteenth-century mapmaker Gerardus Mercator, copying a monk's older text dating from the 1300s, quoted, "under the Pole lies a bare Rock in the midst of the Sea. Its circumference is almost 33 French miles, and it is all of Magnetic Stone."[327] More sophisticated navigators dismissed the legend of the island that had never been seen by reliable eyes, believing their compasses were attracted towards the north star, Polaris. On clear nights, sailors preferred using the star as their guide and ignored their compass. By a stellar coincidence, at the time Europeans began exploring the seas, Polaris was perched almost perfectly over the geographic North Pole. It is still close, but won't always be so usefully aligned. The Earth has a precessional wobble, which means the rotational pole gradually points to different parts of the northern sky. In a few thousand years, the North Star will be a north-by-northwest star and not quite so effective as a night-time guide.

A serious search for the magnetic north pole was conducted between 1818 and 1826 when Edward Sabine, William Parry, and John Franklin engaged in separate expeditions to find a route through the Arctic Islands, the coveted Northwest Passage, that would take merchant vessels from Europe to Asia without travelling around the bottom of either South America or Africa. The explorers found only ice and islands, they could not cross Canada's Arctic with their large ships. Nor did they discover their consolation prize, the magnetic north pole, which they also sought. Of the three explorers, only Sabine[i] was a noted scientist – an astronomer, ornithologist, geophysicist, and president of Britain's Royal Society, the world's oldest organization of scientists.

Soon, however, other Europeans set foot upon magnetic north. On June 1, 1831, James Ross[ii] and his men located the magnetic pole on the western edge of Canada's Boothia Peninsula, at about 70 degrees north latitude, 2,200 kilometres south of the geographical north pole. After walking in the Arctic for five days they reached Cape Adelaide Regina where Ross measured a downward magnetic dip of 89.983° – as close to 90 degrees the measuring tools carried to the arctic in the 1830s could achieve. A perfect 90° marks the spot where imaginary lines drawn to match the magnetic field go straight into the Earth. At the equator, the same imaginary lines are parallel to the surface, and measure a dip of zero degrees. At the magnetic north pole, the needle of the compass doesn't spin in circles trying to

---

[i] General Sir Edward Sabine (1788-1883), British scientist and explorer.
[ii] Sir James Clark Ross (1800-1862), British arctic explorer.

find north. Instead, if it could, it would point straight downwards, 90° into the Earth. In the southern hemisphere, the magnetic field emerges from the Earth off the coast of Antarctica, about 2,800 kilometres north of geographic south and just 2,600 kilometres from Tasmania. The magnetic south pole is presently closer to Australia than to the geographic south pole. In that hemisphere, the north end of a free-range compass needle points straight up into the sky. For Captain Ross in Canada's arctic in 1831, the needle was pointing as close to straight down as his equipment could determine. He was standing at the spot where the magnetic field entered the planet. During a 24-hour period, the pole wandered around a circle with a 26-kilometre diameter while Ross and his men chased after it. He finally planted the British flag, taking possession of the magnetic north pole in the name of King William IV.[328] But magnetic north soon wandered off, leaving Ross and the Union Jack behind.

The next European explorer to encounter magnetic north was the Norwegian, Roald Amundsen[i], who located it while crossing Canada's dangerous Northwest Passage, a journey that took three years in a small craft that hugged islands, waited for thaws, and floated over passages as shallow as a single metre. His vessel was a refitted fishing boat that once caught herring in its nets. She was called the *Gjøa*, and at a place he called Gjøahavn, Amundsen operated a magnetic observatory from November 1903 to May 1905. During his long series of observations, the magnetic north pole wobbled east and west while roaming about 240 kilometres north and south.

The third encounter of the pole had to wait until August, 1947, when Canadian scientists Paul Serson and Jack Clark caught up with it at Allen Lake, on Prince of Wales Island. Their pilot landed a Canso amphibious military aircraft on the shore of Allen Lake, having flown 325 miles from a Canadian Air Force Base on Victoria Island.[329] They found that the pole danced around a 40-kilometre radius on magnetically quiet days, and 80 kilometres on magnetically nervous days. It was clear to them

Roald Amundsen, arctic explorer, 1909

---
i   Roald Engelbregt Gravning Amundsen (1872-1928), Norwegian polar explorer.

that the magnetic north pole was a temporary area, not a point, even moment by moment. The variation in location was not due to solar magnetism or any other disturbing celestial events as some had guessed. The meandering of the magnetic pole position was due to activity deep inside the Earth. One side-effect of the constantly dynamic position of magnetic north involves surveying. For hundreds of years, some plots of land were measured and staked based on compass notions of the location of north. Before it was apparent that variations in the magnetic field could confuse orienteering, fences between neighbours were sometimes skewed. Later, more accurate measurements led to rare but acrimonious reconciliations between written deeds and ancient fence lines.

North becomes South

One of the great abiding geophysical mysteries is the Earth's magnetic field reversal. North becomes south, we are certain of that fact. But we are not certain of the cause. Nor do we know the damage that will occur to the planet's creatures during the shift. Animals that use magnetism for navigation will undoubtedly be confused. Loggerhead sea turtles, for example, are guided by the planet's magnetic field on their annual 8,000-kilometre migration. Species as varied as bats and salmon use the force on their long-distance journeys. Further, the magnetic field protects the Earth from solar radiation and gamma rays. It captures and deflects dangerous particles that would otherwise bombard our planet. During a reversal, the protective field weakens, and may disappear temporarily before re-engaging as its polar opposite. How this will affect our health is unknown, but an increase in cancer and genetic alterations is anticipated. Some scientists have tried relating past absences of protection with accelerated evolution from genetic mutations caused by cosmic ray damage. Their success at making this link is still debated. Complete reversals usually occur less than a million years apart. The next reversal is due momentarily. We should also consider that the magnetic field weakens and disappears more frequently than it actually reverses. There is no reason to believe our magnetic field always changes direction when it dies and restarts. It may simply regroup and strengthen with the same polarity. This could mean its complete absence is rather frequent.

To appreciate how all this relates to the ocean floor and the origin of mountains, we need to briefly visit an old French pottery collection. Among the 50,000 pieces of ceramic held by the Musée National de Ceramique de Sèvres along the Seine in Paris are samples Emile and Odette Thellier tested for magnetism. Beginning in the 1930s, the French couple measured remnant magnetization in the elegant ceramics made by Sèvres. Then they examined ancient earthware recovered by archaeologists. Through their tests the Thelliers discovered a technique to measure the Earth's magnetic field as it existed when clay was fired to create ancient cups and saucers.

Written in Stone

The husband and wife team studied 200 primitive oven sites and determined magnetic intensity for the past 2,500 years. The Thelliers found intensity peaked around the year 850 A.D., about the time the Vikings began to conquer the north – likely aided in their navigation by compasses which were especially responsive under the spell of the enhanced magnetic field. The field has since dropped to half that peak,[330] decreasing nearly ten percent in just the last century. The couple inferred that our field strength is waning by comparing modern ceramics to historic clay bricks, old ovens, and associated shards of pottery scattered around ancient baking sites.

Magnetic earthenware

The Thelliers later expanded their use of archaeomagnetism and surmised that the magnetic field completely disappeared 41,000 years ago, then returned with the same normal polarity it previously had held.[i] It is likely such magnetic waning and waxing without polarity change is much more frequent than the documented million year reversals.[331] Geophysicists expect the present waning to continue; our field may soon disappear altogether. Humanity has no memory of the last collapse, but we know it is definitely coming. The next temporary extreme weakening of the magnetic field will not be a good day, particularly if one considers the havoc that may be wreaked upon electronic equipment. On the other hand, the Earth's creatures have survived scores of previous lapses and reversals – and such events may have accelerated the random genetic mutations that helped species adapt and survive in their changing environments.

We have already noted that natural magnets, or lodestones, sometimes form when lightning briefly melts rocks containing magnetite, and that ancient craftsmen in China heated iron needles in ovens to make compasses. When cooled, those rocks and compass needles retained the Earth's magnetic orientation. Materials such as iron, steel, or any ferrous lava (which is most of it) become demagnetized at about one thousand degrees. At that temperature, the Curie point, minerals jiggle randomly. In its agitation, orderliness that once contributed to magnetism is lost. The effects of the Curie point are sometimes visible to the trained eye. At the Curie point, iron shines with a glow that backyard sword-makers recognize as best suited for tempering. They increase the heat slightly past that point – watching the colour carefully – then slowly cool

---

i   Recently scientists proposed that a 250-year long mini-reversal may have actually occurred.

their hardened weapon. Within such swords as well as lightning-struck outcrops, fresh lava spills, and fragments of prehistoric French pottery, the direction of the Earth's magnetism is captured at the moment of cooling when ferrous minerals realign in a permanent palaeomagnetic record.

Not long after the French scientists learned to use bits of ancient pottery to infer the environment of the kilns that fired them, other scientists put the procedure to more general use. They realized remnant magnetism found in rocks of the same age, but from different locations, once aligned with ancient magnetic north. Extending imaginary lines of magnetic orientation away from such rock and rotating them until an intersection occurs can reveal palaeo-north.

But it gets complicated. As we now realize, the crust is in motion. Discovering the ancient position of the magnetic pole becomes somewhat like mapping the location of a radio tower while snapping photos from a moving car. The tower stands motionless while the camera is rotated to get a clean picture. If we know the camera position and the angle it is pointing, we can later reconstruct the tower's location from our series of snapshots. In principle, this is used by geophysicists to pinpoint an old magnetic pole. Unfortunately for the scientists, their task is not so simple. At irregular moments, imagine the tower suddenly disappearing and being replaced by an equally dramatic hole in the ground. Then the tower is back, the hole is gone. Geophysicists recognize these events as reversals. We can add another wrinkle to the task. Let's say the tower can move around a bit. Not a lot, it stays in the general direction of north, but nevertheless dances around unpredictably. Now you have photographs (rocks with embedded magnetic signatures) taken from a moving vehicle (the drifting tectonic plate the rock sits on) photographed towards a tower (the north pole) that may be randomly replaced by a hole (magnetic reversals) while moving a bit (magnetic wandering). All of this is even more challenging on days with heavy fog and a faulty camera. Geophysicists needed rather good math skills to develop this particular film. But with enough patience, the reversals, the wandering, and the various pole positions were sorted. One of the men who did this was a dyslexic failure at grammar school. But Sir Edward Bullard,[i] whom we first met sailing the Caribbean with Harry Hess and Maurice Ewing, was determined, innovative, and brilliant.

The Magnetic Generator

In Britain, Teddy Bullard began his career as a theoretical physicist before unravelling magnetism, geodetic contortions, and crustal heat flow. His family could have given him a wide range of advantages, but they did their best to wreck the young man instead. Bullard's great-grandfather owned *The Goat*, a popular Norwich pub, and started a brewery, making the family rather wealthy. Bullard's grandfather was thrice elected mayor and was Member of Parliament for Norwich

---

i   Sir Edward "Teddy" Crisp Bullard (1907-1980), British geophysicist.

Written in Stone

– until he was expelled for bribery. Bullard's father was dyslexic and performed miserably at school. Edward Bullard himself was similarly afflicted, claiming his inability to spell was hereditary. Bullard's maternal grandfather was a lawyer whose client list included the Japanese Imperial Navy. Sir Frank Crisp also inked the contract for cutting the largest diamond ever found, the 3,100-carat Star of Africa, or Cullinan Diamond. Crisp was also an amateur scientist, an officer in the Royal Microscopist Society, an enthusiastic gardener, and was rumoured to be insane. He lived on an estate where he erected a replica of the Matterhorn featuring a cave populated with garden gnomes. At the gate to Friar Park, the retired solicitor erected a statue of a monk holding a frying pan punctured by a pair of holes. He called the statue "The Two Holy Friars." Young Teddy Bullard, the future geophysicist, enjoyed spending a month each year at his eccentric Grandfather Crisp's 120-room house near Henley, until Sir Crisp died bankrupt in 1919, when Bullard was twelve. Years later, ex-Beatle George Harrison bought the holy friar estate and turned it into a comfortable home for his final years.

For his education, Sir Edward Bullard was first placed in a girls' school. At age 9, his parents transferred him to a stodgy grammar school which made him so miserable he considered suicide. Anxious because of his grandfather's mental state, the family alerted a psychiatrist who recommended that the child, by then 11, be sent away to a boarding school. Teddy refused, but his parents sent him anyway. There, at age 12, he was tested and placed next to last in a classroom of eight-year-olds. He survived the experience and was rewarded with promotion to a strict secondary school with equally privileged classmates where he once again performed poorly. But a physics teacher with a doctorate arrived and took a serious interest in young Bullard's future. He set the young man loose in the library with problems to solve and lectured him privately two or three times a week. It worked. Previously, Teddy expected he would be an apprentice at the family brewery; instead he applied to Cambridge and became a theoretical physicist.[332] Against such disadvantages of birth, Edward Bullard became one of his generation's most brilliant and hard-working geophysicists.

At Cambridge, Bullard studied physics under Lord Patrick Blackett, his doctoral adviser for a quantum mechanics thesis. Then Bullard worked with Nobel laureate physicist Ernest Rutherford, also at Cambridge. But by 1931, the Great Depression dried up funding for pure research – advancing the science of quantum physics was not deemed essential while millions of unemployed roamed England's streets. Bullard's boss, Rutherford, reluctantly sent Bullard away to teach applied geodesy – glorified surveying techniques that might help farmers and navigators. Through geodesy, Bullard became a gravity expert. He understood the physics behind surveys and maps and he determined the Earth's gravity distortions and nuances of the planet's non-spherical shape. Geodesists calculate innumerable perturbations, all of which affect the orbits of satellites and the drafting of maps. Before the Second World War, Edward Bullard also researched

The Mountain Mystery

Earth magnetism and heat transfer. During the war, he found ways to use geophysics in Britain's fight to survive. He devised a method to demagnetize British ships so they could avoid German detection and he used magnetic disturbances to locate enemy mines. After the war, Bullard headed the University of Toronto physics department, then went to La Jolla, California, to work for Scripps Institution of Oceanography. At Scripps, Bullard designed a tool to measure heat as it flowed from the ocean's floor. But eventually Bullard returned to England to unravel the mystery of the source of the Earth's magnetic field.

Also adapting to the post-war science environment was another brilliant physicist, the co-inventor of the cosmic ray cloud chamber, Bullard's doctoral adviser, Lord Patrick Blackett.[i] Blackett had long been at the leading edge of nuclear physics. Understanding the fundamental nature of Nature appealed to him, but he regretted that his wartime work helped create atomic weapons.[333]

The same year Blackett won the Nobel Prize for Physics, 1948, he published *Fear, War, and the Bomb*. Lord Blackett was vilified because his book challenged the build-up of nuclear arsenals and warned of the environmental consequences of escalation. In the eyes of British and American politicians, Lord Blackett was a traitor, even though he understood the politics and power of nuclear fusion better than the elected politicians. Blackett left nuclear physics and channelled his energy into geophysics, applying all he knew to deciphering the manners of the planet. First he examined magnetism.

Lord Blackett tried to explain the Earth's magnetic field as a simple expression of a rotating

Sir Edward Bullard, being presented the Albatross Award by the American Miscellaneous Society, a group of geophysicists organized to recognize the "lighter side of science" and dedicated to the motto *Illegitimi non Carborundum*.

(Photo used by permission of Scripps Institution of Oceanography, UC San Diego.)

---

i   Lord Patrick Maynard Stuart Blackett, Baron Blackett (1897-1974), British physicist.

mass. Any mass. He suspected that a spinning object creates a magnetic field. The more massive the object and the faster the spin, the stronger the field. His idea that a spinning object generates a magnetic field developed from his observation that the Earth and Sun have magnetic fields proportional to their angular momentum, that is, their combination of mass and spinning velocity. If correct, the 17 kilograms of pure gold he borrowed from the Royal Mint would make a magnetic field when spun on a lathe. It was very significant that it did not.[334] After three years of intermittent, but determined experimentation, Blackett was finally convinced that his idea was wrong. Separate evidence came from his colleague, Keith Runcorn,[i] who lowered super-sensitive magnetometers down mine shafts. There was no anomalous change in magnetism, although Blackett predicted there would be, based on changes of angular momentum with depth. The spinning lump of gold and the mine shaft experiment showed Lord Blackett's theory of magnetism was wrong. With a positive answer, there may have been a second Nobel Prize for Blackett, but there is no such reward for proving one's own fundamental concept is false.

Lord Patrick Blackett, 1950

A better explanation of the source of the Earth's magnetic field came from Edward Bullard. About the same time Bullard was explaining the generation of the Earth's magnetic field, he was recognized for his war-time service and his management of England's National Physical Laboratory. Queen Elizabeth made him a knight. Henceforth to be addressed as Sir Edward, he continued to insist on Teddy as his proper title. Knighthood was bestowed, not particularly for science, but for management and service to his country. This amused Bullard as he realized he was often perceived as forgetful (he was preoccupied with his science puzzles) and a bit reckless (he was naturally impatient and impulsive).

In geophysics, Bullard's greatest proposal was his idea that slowly swirling convection currents inside the Earth's liquid outer core are the planet's magnetic generator. Bullard's dynamo theory stated that an electrically conductive material such as iron, in motion as a fluid molten rock, provided with kinetic energy from the planet's rotation and from radioactive heat, produces the global magnetic field.

---
i S. Keith Runcorn (1922-1995), British geophysicist.

Nothing about this model is clean – the iron is not pure, the heat source is not uniform, the convection current is not steady. These messy irregularities likely contribute to the faltering and reversing of our magnetic field. Bullard had discovered the haphazard generation of our magnetic field. He and others would soon learn how it proves the continents are in motion.

Zebra Crossing

During the Cold War, submarine detection remained as important as it had been throughout World War II. Both sides knew their enemies' submarines were spying off their shores and could even launch atomic bombs at their cities. To uncover every possible oceanic anomaly, the US Navy infused millions into San Diego's Scripps Institution of Oceanography and New York's Lamont Geological Observatory for geophysical and oceanographic research. Scholars at Scripps had been studying the oceans since 1903, when San Diego was a frontier town of only 20,000 people. Lamont was a 1949 east coast newcomer. Its founder and first director was Maurice Ewing, the Texan seismologist. These and other labs amassed huge quantities of data. Ocean depths, heat levels of the sea floor, sea core samples with fossils, gravity and magnetism were all there. Unfortunately, much was classified by the sponsor, the Department of Defense. Scientists might look at the data, but not talk about it. Without permission to lecture or publish conclusions, scientists mostly didn't bother to look. But secrecy rules were slowly eased, or creatively by-passed, and scientists were gradually using military data to tackle planetary geophysics problems. The navy owned the most wide-ranging set of data. This included information about seafloor magnetism, which unexpectedly furnished the most convincing evidence of continental drift.

The key scientist who gathered the raw data for this part of the story was Ron Mason[i]. He had just finished his physics degree at Imperial College, Arthur Holmes' *alma mater*. Mason says, "I was looking for an alternative to spending my working life in a laboratory when I discovered geophysics."[335] Other physicists had entered the field for a similar reason. Pure physics is often a mental – almost philosophical – exercise, exploring unseen bits of the universe through sketches and equations on a chalkboard or squinting into a computer screen. Geophysics offers the possibility of experiencing science from an experimental perspective – a chance to squeeze rocks, jab thermometers into lava, shake the ground with seismic thumpers.

Fortuitously, Mason took a one-year sabbatical at Caltech, starting in 1951, which led to a stint aboard a military research vessel plying the Pacific. His sabbatical year grew into eleven, spent with La Jolla marine scientists, recording magnetic data from the ocean floor. Scripps oceanographers were already measuring gravity and collecting depth soundings. Mason convinced the group to

---

i   Ronald George Mason (1916-2009), British-American geophysicist.

add a magnetometer to the electronics aboard their research ship, the *Pioneer*. The first results weren't especially good because of magnetic interference from the ship's generator and various instruments, so Mason persuaded the military to tow his device far behind their vessel. A ship's crew is reluctant to drag anything – it can be dangerous. The vessel must be slowed cautiously each time the tool is reeled aboard. If the ship stalls or stops urgently, the cable's forward momentum may cause it to wrap around the power shaft. Dragging the magnetometer was not immediately approved. Mason and his fellow geophysicists had to show that the extra equipment would not add more than two hours a week to the ship's time at sea. The scientists also had to agree that the *Pioneer's* captain could cut the cable and abandon the magnetometer if he felt the safety of his ship was being compromised.[336] Consent was reluctantly granted; Mason was appointed ship's operator of the tool tugged behind the *Pioneer*. The equipment survived months at sea and was never lost to the murky ocean floor.

It took nearly two years for the *Pioneer* to loop back and forth from the west coast, 850 kilometres seaward, then back to shore, again and again. North to south the massive marine survey spread 2,000 kilometres, from southern California to offshore British Columbia. Acquisition was a monotonous exercise, but it yielded a magnetic survey as large as all of Mexico. Then the data had to be edited – an even less interesting task. Mason and his colleagues struggled to identify and remove specious anomalies caused by local geology and depth variation, cleaning smudges from their window over the seafloor. "We did this by overlaying our map on the map of the Earth's magnetic field published by the Hydrographic Office, and

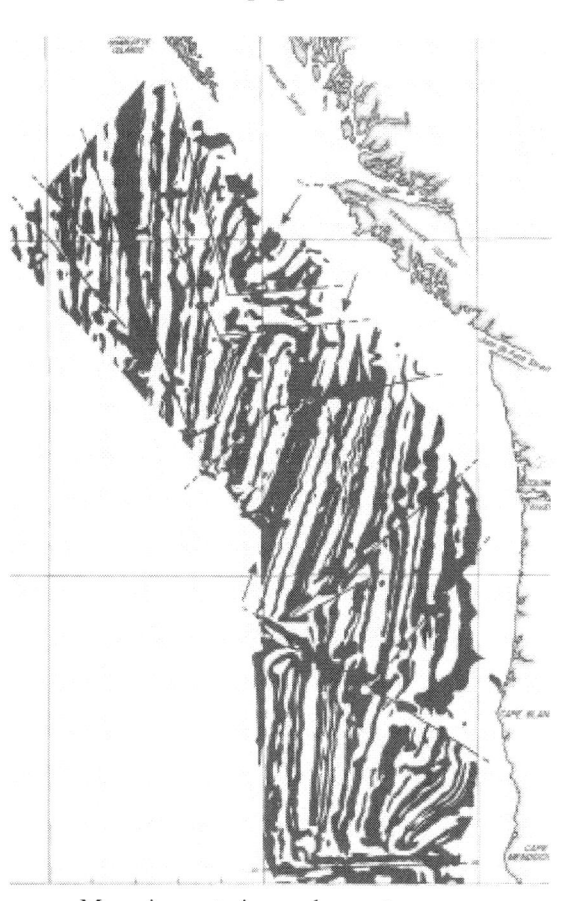

Mason's mysterious zebra-pattern map
(Reprinted by permission of the Geological Society of America.)

237

subtracting the one from the other graphically. This rather tedious procedure greatly simplified the original map."[337] In the end, it was indeed a simple map, but the result was unexpected blotches of black and white stripes. It took several years before anyone could explain why a map of the magnetism of the ocean floor should look like the hide of a zebra.

One reason it took so long to find a cause for the striping was because the entire collection was a classified military secret. The *Pioneer* expedition's bathymetry, gravity, and magnetic data were not to be published. The Navy worried the data might reveal places to hide submarines to America's enemies.

Mason convinced his bosses that if he subtracted the raw magnetic data from the previously published maps of background magnetism, and removed absolute values and any hint of seafloor topography, the residual could have its secrecy clearance lifted.[338] Mason published his map with the curious stripes. He had spent years gathering the data and removing regional trends and background noise. But he and others who examined the result couldn't explain the pattern. He thought the zebra stripes were due to differences in inherent rock properties, perhaps various rocks were more magnetic than others.[339] His 1958 paper sat in the scientific literature where seasoned geophysicists might explain the odd stripes. For four years, no one did.

Then two geophysicists, two decades apart in age and with quite different experiences and educations, tackled the alternating magnetic seafloor independently. Neither knew what the other was doing. The result of their efforts to publish their discoveries left as bitter a disappointment as there has ever been in the history of scientific discovery. Both men, by all accounts, were likeable, reputable, and extremely intelligent. Frederick Vine[i], in England, had his paper published; Lawrence Morley[ii], in Canada, saw his own identical hypothesis rejected. By the same journal, at almost the same time.

Lawrence Morley thought he'd work in insurance. His career would be in actuarial statistics, a method of applying maths to the calculation of insurance premiums. So Morley enrolled in mathematics and physics at the University of Toronto. His first summer job was horrendously boring. It landed him at an insurance company, armed with a mechanical calculator, computing policy dividends in a stuffy office. He said the work was so dull that he returned to university in the fall as a geology-physics student. It was an unusual choice in 1939. So unusual that only he and one other student were enrolled in the rare dual discipline; and there had only been one graduate of the program in six years.[340] The university had no actual geophysics program. The degree was earned by taking half the lectures in the Department of Physics and the other half in the Geology Department. Morley wrote, "Geologists thought geophysical exploration was something akin to water divining, while the physicists thought geology was

---

i   Frederick Vine (1939), British geophysicist.
ii Lawrence Whitaker Morley (1920-2013), Canadian geophysicist.

an inexact science involving the gathering of numerous rock samples and categorizing them, rather akin to stamp collecting."[341] Morley stayed with the combination water-witching-stamp-collecting program and eventually became Canada's second geophysics graduate. Even that wasn't a full four-year degree as his education was interrupted by military service.

During the Second World War, Morley served as an officer in the Canadian Navy, engaged in the Battle of the Atlantic.[342] He spent five years in the service, using radar as a kind of remote detection, something akin to the geophysics he had been studying. Later in his career, he would be engaged in a wide range of geophysical sensing systems, including magnetotellurics, radiation, seismic, remnant magnetism, and airborne electromagnetism. But his first post-war job was with a small geophysical company prospecting for magnetic iron ore in the north. His magnetometer – the best available at the time – was not much better than the tool Sir William Gilbert, who discovered the Earth is a magnet, had used 300 years earlier. Morley measured the intensity of magnetism by simply watching a needle balance horizontally on a knife edge. If he trudged across a magnetic vein embedded within the granite at his feet, the needle deflected further downwards. "This was high-tech at the time. It would take a whole day to collect a mile-long magnetic profile of data. This tedious work, combined with unbelievable clouds of mosquitoes in the Canadian Shield, led me to think that perhaps I had made a mistake in dropping actuarial science."[343] Hiking all day across brutal terrain was the only way to capture a few magnetic data points in 1946. But a year later, his data collection would speed up a hundred-fold.

When Morley heard about the potential use of airplanes towing magnetometers on long cables, he vowed never to go back into the bush with a hand-held device. The airborne tool had been perfected by Gulf Research and Development Corporation just before the Second World War. The oil company planned to use it to outline potential hydrocarbon basins, but loaned the equipment to the Navy during the war as a search tool against enemy submarines. The war was over when Morley, by then working in the Canadian Shield and nursing mosquito bites, heard a talk about the airborne magnetometer. He realized he'd rather fly in a recycled navy warplane than spend five minutes trying to get a point reading.

Gulf was headquartered in Pittsburgh; Morley was in Ontario. Getting across the Canadian-American border was a problem. According to historian Henry Frankel, Morley said, "I couldn't get a job unless I had a visa, and I couldn't get a visa unless I had a job. I actually sneaked across the border."[344] Morley presented himself to Gulf, but they wouldn't hire him as a researcher without a PhD. Instead, they referred him to a small independent contractor. Morley spent 1947 and 1948 as their party chief, flying an aeromagnetic survey in Venezuela and Colombia over the Llanos Basin, east of the Andes. It was the world's first commercial aeromagnetic survey and Morley proved that a huge tract of land could be surveyed from the air by peering through thick inaccessible jungle and cloudy rain forest to assess mineral deposits and oil basins. Morley returned to Canada, determined to use aeromagnetic surveys to explore the vast Canadian Shield.

Back in Ontario, he realized that learning theoretical geophysics would be useful. Because of the war, his BSc had been truncated. He now wanted to understand his science more thoroughly. So Morley returned to the University of Toronto where Tuzo Wilson[i] became his PhD supervisor. Tuzo Wilson was already legendary in both geology and geophysics. Canada's first geophysicist, Wilson would later play a pivotal role in plate tectonics theory by solving a thorny problem with crustal motion along mid-oceanic ridges and for proposing that the Hawaiian islands came to life from a hotspot under the moving Pacific plate. Wilson, in 1949, introduced Morley to the idea of continental drift. Although the great Wilson had not yet made up his own mind about the theory, Lawrence Morley became intrigued with it. He was even more captivated when, as part of his preparation for graduate research, he encountered a new paper on palaeomagnetism. Its author suggested that the magnetism of ancient rocks might offer evidence for continental drift.[ii] If one could plot old magnetic pole positions,

Lawrence Morley, left, with the Geological Survey of Canada
survey plane and its pilot, 1952
(Photo used by permission of the Geological Survey of Canada)

---

i John Tuzo Wilson (1908-1993), Canadian geophysicist.
ii The paper that deeply influenced Morley's thesis choice was John Graham's 1949 monograph, "The Stability and Significance of Magnetism in Sedimentary Rocks."

one could track the ancient movement of the continents. This was fifteen years before plate tectonics would enter mainstream geology and it was still a fringe science with many more detractors than supporters. Nevertheless, for his doctoral thesis, Morley pursued the remnant magnetism of the Precambrian Shield's rocks. But the polar drift data he sought eluded him – geophysical equipment in 1950 was simply not accurate enough. Background noise and primitive equipment overwhelmed his efforts. "I could not get a constant direction. I couldn't clean the samples enough to get a consistent direction," he said.[345] Morley's effort to prove continental drift through remnant magnetism failed; nevertheless, his research was sound and he earned his doctorate.

Morley says he was excited about that line of research, but as would happen several times in his career, he was side-tracked by the need to earn a living. In 1952, he became the first geophysicist to work for the Geological Survey of Canada. Most of his time was now spent planning and supervising government aeromagnetic surveys to encourage mineral exploration. Data collecting was contracted at a cost of $30 million dollars and took 17 years to complete. "The benefits of this survey to the mining industry in Canada have never been calculated, but they must be more than several billion dollars and are still going strong," he wrote in his reminiscent article *The Zebra Pattern*.[346] It was this government data that helped inspire diamond prospecting and led to ore discoveries from Newfoundland to the Yukon.

As an expert in magnetic properties of rocks, Lawrence Morley was drawn to Ron Mason's zebra-pattern magnetism map of the Pacific Ocean. Morley, working at the Geological Survey of Canada, was the first to correctly interpret the alternating stripes as direct evidence of seafloor spreading. He called it his "Eureka Moment." The stripes, he believed, were because new seafloor was being created by magma at oceanic rifts, and as it cooled into solid rock, the magnetic polarity of the moment was captured. Hence, the stripes. Morley was right. He tried to get his paper published in *Nature*, the prestigious British science journal, in February 1963. He penned a simple, non-analytic interpretation of Ron Mason's zebra-striped map, but the journal and the geophysical community were not ready to entertain musings that linked magnetic anomalies to the spreading seafloor.

Here is part of what Lawrence Morley wrote in his failed submission:

> "If one accepts, in principle, the concept of mantle convection currents rising under the ocean ridges, traveling horizontally under the ocean floor, and sinking at ocean troughs, one cannot escape the argument that the upwelling rock under the ocean ridge, as it rises above the Curie Point geotherm, must become magnetized in the direction of the Earth's field prevailing at the time. . . it stands to reason that a linear magnetic anomaly pattern of the type observed would result."[347]

Morley had perfectly captured the solution. Independently, and just a few months later, the British team of Frederick Vine and Drummond Matthews[i] came to the same conclusion. They were successfully published, in *Nature*, in September, 1963. It was unfortunate timing for Morley – and for Vine and Matthews, who likely didn't know about the Canadian work. But the entire episode points to a fundamental problem with peer-reviewed publication. As Morley himself noted, "I knew that when a scientific paper is submitted to a journal, the editors choose reviewers who are experts on the topic being discussed. But the very expertise that makes them appropriate reviewers also generates a conflict of interest: they have a vested interest in the outcome of the debate."[348] Morley's reviewer, who remains anonymous, may have felt seafloor magnetic zebra stripes prove nothing, and may have been staunchly opposed to the concept of continental drift. Morley was told that *Nature* did not have room to print his short paper. Morley quickly dispatched it to another research journal, which also rejected it. There, the anonymous reviewer scolded that although Morley's idea was interesting, it was best discussed over martinis, rather than published in the *Journal of Geophysical Research*.[349]

Perhaps the real problem was that Morley's explanation of the zebra pattern was qualitative, not quantitative. The zebra map's raw data were still classified, so Morley did not have the actual numerical values in hand, nor did he have other marine data that might have corroborated the Pacific magnetic set. At Cambridge, scientists had been gathering similar, albeit unclassified, data from the Indian Ocean and North Atlantic long before Frederick Vine became a student there, so the Matthews and Vine article had the university's considerable quantitative material to draw upon for their analysis.

"Magnetic Anomalies over Ocean Ridges" by Matthews and Vine was submitted to *Nature* (the same journal which had rejected Morley) in early July and was printed in September 1963. The Cambridge paper had numerical data which Morley's lacked, but the interpretation and results were the same. When it became apparent Lawrence Morley had previously tried to publish the same conclusion, he was also belatedly credited in the Morley-Vine-Matthews Hypothesis.

Magnetic Fruit from a Vine

Frederick Vine says a page in a school book first aroused his curiosity about continental drift. He was in Chiswick, a west London suburb with wide streets, trendy cafes and popular theatres. Chiswick – *Cheese Farm*, in Old English – began as rolling farmland and meadows. For professionals living there, it is an easy commute from large Georgian and Victorian homes into London's centre. Frederick Vine attended nearby Latymer Upper School, an expensive prep school

---

i   Drummond Matthews (1931-1997), British marine geologist.

for elite, academically gifted boys. In April 1955 Vine was 15 and beginning to study for his final exams. He said he opened his geography text book, "probably for the first time" and saw a diagram of South America snuggling Africa. The book described a splintered supercontinent, "but geologists had no idea if there was any truth in this hypothesis. I was struck at once both by the boldness of the idea that seemingly stable continents might have drifted across the face of the Earth in the past, and by the fact that we do not know whether this had occurred."[350] Within ten years, Vine was in the midst of the proof.

Frederick Vine placed well in his exams. He qualified for a spot at St John's College, part of Cambridge University, and decided to become a high school physics teacher. St John's is a small school, with just a few hundred undergraduates and perhaps 400 graduate students, and is focused on educating the brightest. Now 500 years old, St John's had originally been granted a charter approved by King Henry VIII, the Pope, and the local Bishop. The latter two were involved to allow conversion of the three-hundred-year-old Hospital of St John into a school. Traditions abound on the old campus. Fellows of the College are the only people in England, outside the Royal Family, allowed to catch, cook, and eat the beautiful wild mute swan. (Traps were used until recently to collect the dinner birds.) Ghosts in the dorms and silver service three-course daily meals add to St John's ambiance. Quaint customs didn't prevent nine of the college's alumni from earning Nobel Prizes.

At Cambridge, Frederick Vine's work with the magnetism of rocks was accidentally assigned to him. By 1962, he had finished his undergraduate studies, was no longer considering a teaching career, and had begun to work on a PhD in marine geophysics at Cambridge. "At any one time," says Vine, "there was always at least one graduate student working on each of the main geophysical techniques – gravity, heat flow, refraction seismics, magnetics, and so on. There was a vacancy in the magnetics area."[351] One can assume that if refraction seismic had been short one student, Vine would have landed there and the chance discovery of magnetism in seafloor spreading would have fallen to some other lucky young scientist. As it was, Cambridge physicists had built a number of magnetometers and had collected mounds of data which were piling up in boxes, uninterpreted. Someone had to look at it.

In preparation for his geomagnetism assignment, Vine consumed the literature and studied the theory. Data which Cambridge scientists had acquired over the Indian Ocean showed alternating normal and reversed remnant magnetism in the rocks of the seafloor, linear patterns similar to the more spatial zebra patterns from the Pacific. Coincidentally in May 1962, Vine gave a short talk about Hess's seafloor spreading and mantle convection ideas. His supervisor, Drummond Matthews, asked if Vine thought the polarity flips might be related. Vine couldn't say. The data were complicated and confusing – earlier geologists had expected some correspondence between topography and magnetism and had spent considerable time and effort without finding any. The anomalies had no relationship to local elevation or rock thickness. Plans were made to build complicated models from clay mixed with iron filings to see if the striping could be mechanically reproduced. Matthews and Vine considered duplicating the known topography and applying various magnetic fields with a scale model, then measuring the remnant field. In the end, they decided to attempt computer modelling.[352]

Vine's data included several profiles that crossed an ocean ridge. Each had a positive polarity at the ridge, negative signatures equally spaced on either side of the ridge, then switched back to positive again. Today it seems a simple step of logic, a game of connect-the-dots. The seafloor was spreading, the rock was acquiring the polarity of the Earth's field at the moment the new igneous rocks solidified. But the data had background noise, there were varying spreading rates, and even Vine and Matthews' Cambridge colleagues had little sympathy for a spreading seafloor solution. Few of them were able to connect any dots in 1962.

Vine thinks he developed the answer in perhaps February or March, 1963. He and Matthews submitted to *Nature* their solution to the magnetic polarity reversals of the seafloor, complete with analysis of data from the Indian Ocean and the North Atlantic. Vine and Matthews were working with linear strips of data – the survey ships passed just once across spreading seafloors, but the data still showed a regular switch from positive to negative polarity, flipping back and forth about 20 times on the sinusoidal curves they published. The authors attributed this to seafloor spreading and reversing magnetic poles. In *Nature*, they wrote:

> "...the whole of the oceanic crust is comparatively young, probably not older than 150 million years, and the thermo-remanent component of its magnetization is therefore either essentially normal or reversed with respect to the present field of the Earth. Thus, if spreading of the ocean floor occurs, blocks of alternating normal and reversely magnetized material would drift away from the centre of the ridge and parallel to the crest of it."[353]

Thus Vine and Matthews published the same hypothesis as had already been suggested by Morley. It was likely a huge surprise to them that Morley had tried to publish the same idea – in the same journal – a few months earlier. Vine later wrote, "Lawrence Morley, of the Canadian Geological Survey, penned a letter to *Nature* proposing exactly the same idea. He was unable, however, to draw on a survey of a known ridge crest and had to make the case with reference to the linear anomalies mapped in the northeast Pacific, which were not obviously related to a mid-ocean ridge. Morley's paper was rejected by *Nature*, and subsequently by the *Journal of Geophysical Research*, for being too radical and speculative."[354] Much later, Vine even more clearly recognized the priority claims of Morley. In 2007, Vine received the Geological Society's Prestwich Medal and described his plate tectonics work saying, "Four years later I learned that Lawrence Morley, a Canadian geophysicist, had had the idea some months before I did, and had tried to get it published . . . without success."[355]

The magnetism reversal idea of Morley, Matthews, and Vine was incredibly significant to the acceptance of the spreading seafloor hypothesis and that, in turn, is the underpinning of the entire modern plate tectonics theory. One would expect trumpets heralding a new age in earth science, but as Vine put it, "Initial reaction to the paper was, to say the least, muted."[356] Either muted or critical. Even in his own Cambridge department, his colleague John Sclater recalled senior scientists there did not even invite Vine and Matthews to present their findings at an upcoming conference, "It was considered highly speculative and worthy only of barroom gossip."[357] Sclater added that after the seafloor spreading hypothesis was published, it was a topic of humorous conversation in the lunch room. "Most of us still believed that the continents were fixed, and we were all surprised that *Nature* published what we considered idle speculation." After the paper appeared, Frederick Vine spent the next year writing his PhD thesis, applying for a temporary job at Princeton, getting married, and running a scouting troop.[358] Very little attention was given to the stripes theory for months.

There was certainly a strong element of luck in Frederick Vine's success. In 1962, he was an undergraduate who heard about the idea of seafloor spreading for the first time when Harry Hess was keynote speaker at the university's geological conference. A year later, Vine was co-author of a paper published in the prestigious journal *Nature*. In the autumn of 2003, there was a nostalgic revisit of the paper's 40th anniversary of publication, receiving considerable notice in the various geology journals. Vine paid tribute to his good fortune in a peculiar way: "Undoubtedly, my greatest debt of gratitude must go to pillow lavas, whose fine-grained, single-domain titanomagnetites preserve a particularly strong and stable remanent magnetization. Without them, this would have been a very different story."[359] Indeed it would have been – you will recall that Lawrence Morley selected a much more stubborn continental igneous rock for his own research and it almost sunk his dissertation.

Vine once said, "The contribution that I was able to make to this subject, during the 1963-1966 period, was a classic example of 'being in the right place at the right time'. I was lucky. I think, I could claim that, to some extent, I maneuvered myself into an area that struck me as being fertile ground for a possible breakthrough."[360] He continued to create his good fortune by working with one of the most progressive geologists of the twentieth century, John Tuzo Wilson. Together they wrote a 1965 paper that focused on Mason's zebra-pattern survey and further developed evidence for seafloor spreading and plate tectonics. That paper pointed out that the magnetic seafloor was symmetrical and had a constant spreading rate – new crust was emerging from the Earth's hot inner realms and spreading at a rate of 3 centimetres (about one inch) each year.[361] The same sequence of magnetic anomalies was being found all over the world. Maurice Ewing, director of the Lamont Observatory made sure that every survey he ordered into the Atlantic included a magnetometer. The world's oceanic data collection became enormous and it all confirmed the seafloor spreading hypothesis. Once the science was understood, it was almost as easy as rewinding a tape player to reconstruct Pangaea from its dispersed continental pieces. The proof to Wegener's theory had been found, captured in cooling basalt on the ocean's floor, written in stone for future generations to decipher. As it turned out, Wegener had been right. Almost. There were still some rough edges to trim.

18

Rough Edges

Harry Hess, with his geopoetry, was largely correct about the ocean floor. One of the few things Hess had wrong was his idea that the ocean's crust changed to serpentine when magma contacted cold sea water. It was Robert Dietz[i] who recognized the extruded rock was basalt, straight from the mantle with no major modifications. It was also Dietz who coined the term *seafloor spreading*, the descriptive name for the whole process of the widening ocean. Dietz was a marine geophysicist with the US Coast and Geodetic Survey when, in 1953, over lunch with another scientist, he speculated about the Emperor chain of seamounts extending northwest from the Hawaiian chain. He told Robert Fisher that something might be carrying the old volcanic mountains northward, like a conveyor belt.[362] This was certainly prescient of the moving seafloor idea. Eight years later, he expanded this table talk in his paper, "Continent and Ocean Basin Evolution by Seafloor Spreading," better known by its subtitle, "Commotion in the Ocean."[363] Among his commotions were these thoughts:

> "Seafloor spreading is envisioned as the fundamental process creating continents and ocean basins. Accordingly, the sea floor moves out in opposite directions from the mid-ocean rises. The gap is filled by new strips of sea floor created from the ultrabasic mantle. . . Thermal convection cells in the mantle provide the fundamental driving force and the mid-ocean rises mark their divergence while the continents tend to lie over the convergences. . . The principle novelty of this concept is that no fixed layer separates the sea floor from the convection process; rather the ocean bottom is the exposed and outcropping limb of this convection. Although perhaps alarming at first thought, seafloor spreading is an orderly, evolutionary and actualistic process consonant with geologic history."[364]

It sounds like something Harry Hess might have written. In fact, the similarity between the ideas of Dietz and Hess led to a rather unfortunate controversy, says

---

i   Robert Sinclair Dietz (1914-1995), American geophysicist and oceanographer.

Sebastian Bell, who arrived as an Oxford student at Princeton just as plate tectonics was being sorted. Bell was assigned to Harry Hess for his graduate studies. When they met, Bell was "somewhat taken aback." There was Harry Hess, surrounded by disordered books, some on the floor, his shirt partly open, and wearing tan canvass trousers, the blue jeans of the 1950s. At Oxford, Bell's adviser always wore a three-piece suit and tie – even on field trips. The next day, Bell bicycled past his professor on the street and Hess waved and said, "Hi there, Sebastian." British professors at the time would neither wave nor be so informal as to "Hi" a student. Bell also recalls Hess taught the new grads about seafloor spreading with "chalkboard, grubby slides and the ever-present cigarette."[365]

Bell recalls that Dietz published his short ocean commotion paper in *Nature* in 1961, after discussions with Hess, and before Harry Hess had published his own version. Many saw it as Hess's idea. Allies of Hess wanted to make trouble, but Hess responded, "Don't bother, it's probably all wrong anyway."[366] Although it looked as if Dietz ran away with Hess's work and beat him to publication, Harry Hess was not particularly irritated with Robert Dietz. The two were actually good friends. Others were not so forgiving. When geophysicist Allan Cox collected ground-breaking plate tectonics papers in a landmark 1973 anthology, *Plate Tectonics and Geomagnetic Reversals*, Robert Dietz's paper was conspicuously missing. Dietz himself later clarified the issue of priority by remarking that Hess deserved recognition for the seafloor spreading theory. Hess was credited, he said, because of "priority and for fully and elegantly laying down the basic premises. I have done little more than introduce the term *sea-floor spreading*."[367] Harry Hess created the idea, distributed pre-publication copies of his work in 1960, and discussed it with Dietz (and many others). Dietz published a version of the theory in 1961, but Hess offered his more refined geopoetry in 1962. This clarifies the controversy, but in 1968, geologist Arthur Meyerhoff suggested the ideas of Hess and Dietz were both a rehash of the old theories of Arthur Holmes.[368]

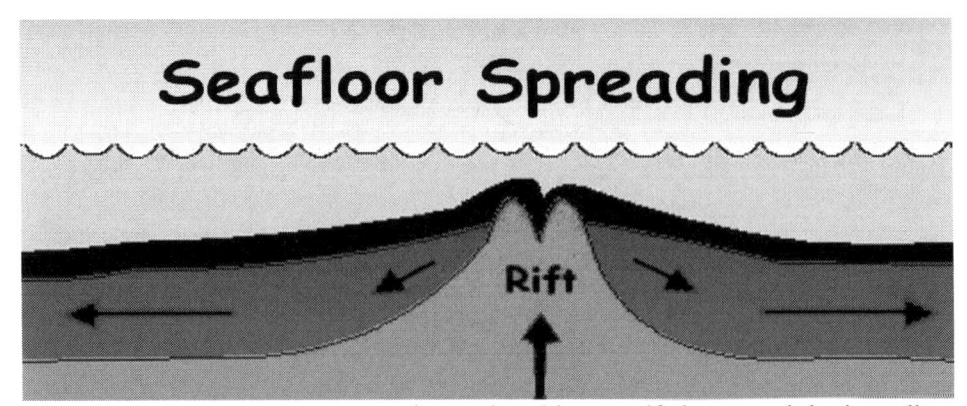

Ocean crust is generated as magma rises at the mid-ocean rift then spreads horizontally. Dietz called this activity *seafloor spreading*. Drawing based on Holmes, Hess, and Dietz.

Rough Edges

Meyerhoff had a point. As early as 1931, Arthur Holmes described continental drift much the same way Harry Hess would later. Holmes had sketched a rising convection current which parted the ocean floor. In his update to *Principles of Geology*, the most successful earth sciences textbook of the 1940s, Holmes ends with a chapter, "Continental Drift," which includes the subsection "Search for a Mechanism." He includes a diagram of a rifting continent, spreading seafloor, and crust descending into an ocean trench. Holmes's 1945 diagram is virtually indistinguishable from anything that might have been published by Hess and his contemporaries twenty years later. Holmes wrote:

> "To sum up: during large-scale convective circulation, the basaltic layer becomes a kind of endless travelling belt on the top of which a continent can be carried along, until it comes to rest when its advancing front reaches the place where the belt turns downwards and disappears into the Earth."[369]

With that simple description, Holmes had essentially written the entire plate tectonics scheme. The difference between Holmes and Hess largely arose from new evidence discovered in the intervening years. Hess was not a revolutionary, nor did plate tectonics suddenly originate with him, brilliant though he was. Hess was able to expand the simmering work of Holmes and synthesize a more comprehensive story for continental drift. Within the decade bracketing Harry Hess's "History of Ocean Basins," new geophysical clues rolled out like new cars on an assembly line: the great rift amidst the oceanic ridges; heat flow near those rifts; the magnetic zebra stripes of the seafloor; variously aged microfossils in ocean cores. More oddly-distributed continental fossils were also discovered – in the early 1960s palaeobotanist Edna Plumstead[i] analyzed enough relic bits of plants to show that Antarctica, South Africa, South America, India, and Australia all had sedimentary rocks of the same age with the same fossils.[370] Thus fossils were yet again confirming the existence of an ancient supercontinent.

Supporting evidence for tectonics was everywhere. I had dinner with a pioneer of continental drift, Alice Payne, who grew up in the arctic helping her father, a successful gold prospector.[371] It was a natural choice for her to include radiometric analysis of samples from the Canadian Arctic, Greenland, and northern Scotland when she wrote "A Line of Evidence Supporting Continental Drift" as part of her graduate research in 1964. Her paper predated acceptance of tectonics, so it was a daring title. But her evidence indeed supported drift. Payne showed that various intrusions, hundreds of kilometres apart, line up on a great circle – if the ancient rocks are relocated to their palaeo-positions. She told me that some of her colleagues quickly distanced themselves from her paper and anything dealing with mobile continents. But her conclusions were right and strengthened the theory. And even more evidence of plate tectonics was being discovered.

---

i   Edna Pauline Janisch Plumstead (1903-1989), South African palaeobotanist.

The Mountain Mystery

Tuzo

Tuzo Wilson mentored Lawrence Morley in Toronto and co-authored a seafloor paper with Frederick Vine. Geologists remember him for tirelessly promoting science, for boundless curiosity and energy, and for creatively solving three problems the critics of plate tectonics spotted. As presented in 1962, the idea of continents in motion implied Hawaii and Yellowstone shouldn't exist – they are remote from the energy sources of spreading ocean rifts and colliding continents. Nor could geologists make sense of swarms of awkwardly positioned faults near rift zones – they weren't within the style of typical faults. Finally, there was evidence that some mountain-building had occurred during continental collisions that predated Pangaea. These three issues suggested flaws in drift theory. Wilson solved these problems without complicated mathematics. He used something he described as visual imagery. Hawaii's islands, a new style of seafloor fault, and sequential supercontinents that grouped, dispersed, then regrouped were imaged in his creative mind. His visions rebuffed most critics of mobility theory. His ideas were at first outlandish, but geologists listened to Tuzo. He had a long-established reputation for getting things right.

John Tuzo Wilson's middle name honoured his mother's family, French Huguenots who settled in Virginia three hundred years earlier, then resettled in Canada. The name is also remembered in the Canadian Rockies – Mount Tuzo is named for Wilson's mother, Henrietta Tuzo, the first mountaineer to climb that massive peak. Tuzo Wilson's father was an aeronautical engineer who arrived in Canada from Scotland and played a leading role in developing Canada's civil aviation – in part by choosing sites for both the Montreal and Toronto airports.

Tuzo had a precocious interest in geology. As a 15-year-old high school student, he worked summers for the Geological Survey of Canada. He loved geology and excelled in math and physics. It is not surprising that he earned the first geophysics degree awarded in Canada, at Trinity College in Toronto. He received a Rhodes scholarship to St John's at Cambridge, then finished his PhD at Princeton, in 1936. Wilson's doctoral adviser at Princeton was Harry Hess, the early champion of rifts and seafloor spreading. After Princeton, Wilson returned to the Geological Survey, mapping Canada's enormous frontier, the Northwest Territories. With the start of war in 1939, Tuzo Wilson enlisted in the Canadian Army, finishing his service with the rank of colonel. After the war, he led Exercise Musk Ox, a 5,000-kilometre vehicle excursion through the Arctic – it remains the longest arctic automotive trip ever undertaken. Upon discharge, Colonel Wilson became Professor Wilson at the University of Toronto. He stayed there for almost thirty years.

If it involved geology, Wilson had probably studied it. His interests were so wide-ranging that colleagues called him the cyclone scientist. He was internationally regarded for his work on glaciers, which led him to draft the first glacial map of Canada. As part of that study, he searched for ice patches on the arctic islands and became the second Canadian to fly over the North Pole. His experience in flight, mapping, and the Arctic merged into pioneering work in aerial photography. His photos were a key element in creating Canada's geological maps. A view from above was an essential step in that tedious project.

Even today, mapping the arctic's geology is challenging. Geologists hike through the tundra, gather rock samples, make notes and sketches, then merge these with thermal, radar, and Landsat images captured by satellite, or colour photographs from aircraft cruising above the frozen north. On the arctic islands of the Sverdrup Basin, or farther east on Baffin Island, the intrepid geologists may be draped in gloves, overalls, and netting. They appear as tundra astronauts to the billions of black flies and mosquitoes which would otherwise suck blood until the unprotected become anaemic or worse. In winter, geologists wear other armour – cumbersome layers protecting against temperatures dropping to minus forty. Winter has the advantage of being bug-free and the tundra, bogs, and marsh are frozen and more passable. Although winter might be preferred, these field geologists are often graduate students and professors. Winter duties of teaching labs, attending seminars, and analyzing field data usually force these peripatetic scientists northwards during the buggy, marshy, sloppy summer months.

In the field, suffering blistered feet and carrying tents and food while watching for bears and wolves, the indomitable geologist drags heavy sacks of samples destined for laboratories where the stones will be sliced or crushed, then subjected to chemical and microscopic examinations. Set among such challenges, the hunter-gatherer must map the location and orientation of each sample. Before nudging a lump loose, the angle of the stratigraphic layer is noted – the pitch in or out of the ground. The geologist then tries to guess if the outcrop among the sedges and berry bushes continues unseen under nearby scrub and marsh.

In the days when Tuzo Wilson was creating the first techniques to incorporate ground samples with photographs taken from propeller planes at a thousand feet, the geologists on the tundra also took a few photographs. These were grainy grey images that might jar memories and help reconstruct the field layout. But photos were few and expensive and often featured geologists bathing in icy puddles or steering clear of wolves. And the traipsing geologists had a problem similar to the men in the crafts working overhead during short intervals of clear skies between frequent unexpected storms: they often didn't know where they were.

In Wilson's earlier days, it was nearly impossible to ascertain a location in the tundra. No GPS satellites circled overhead. No radio signals triangulated positions. Close to the North Pole, compasses were unreliable – the magnetic field is nearly useless for navigation where flux lines crash perpendicularly into the Earth and change positions each minute. Errors of 50 kilometres were common, even on seismic-recording research ships in the 1960s.

Early Arctic geology maps were based on careful counting of strides taken towards lakes, or islands on lakes, or sharp bends in rivers – features large enough to spot on aerial photographs. Those images were organized beside the drawings of earlier mapmakers who had charted the area, perhaps with sextants and telescopes from boats dodging polar sea ice in late summer. Finally, the grainy photos from land and air, the bags of rock chips, and the clutches of hand-written notes were incorporated into rudimentary geological surface maps. A pale grey colour on a surface photo where a geologist saw a wide quartz vein in a granitic layer disappearing into the ground might join similar parallel patches scattered upon the same image. These might be extended by aerial photographs and the geologist's imagination. By plotting such evidence – a few stones here, a bit of structure there, photos of shorelines, scribbled notes – geological maps slowly emerged that described the natural history of one of the most remote parts of the planet.

Creating the techniques that shaped these maps were among Wilson's early contributions to geology. But there were other dimensions to his life. He loved the mountains and scrambled up their slopes as often as time allowed. In 1935, during field studies associated with his Princeton doctoral thesis, he was the first person to climb the formidable Mount Hague in the Bear Tooth Range, just north of Yellowstone. And there was more. Wilson wrote two books about China, recounting his journeys there, including a trans-Siberian train trip after a conference in Moscow which was part of the International Geophysical Year. According to historian Martin McNicholl, Tuzo's Cold War-era books helped open relations between China and the West.[372]

Wilson enjoyed performing. In the 1970s, he created and hosted a 12-episode television series – *The Planet of Man* – in which he explained plate tectonics, fault systems, and mountain growth. These shows opened with Tuzo cruising lakes in northern Canada on his Chinese junk, a 12-metre craft that he said was well-suited for approaching isolated outcrops. Another venue for public education came late in life – he spent the eleven years up to his 76th birthday as director of the Ontario Science Centre. Because he had made such outstanding contributions to support plate tectonics, the Ontario Science Centre built a three-metre-tall memorial to commemorate Tuzo Wilson. The sculpture includes a spike jabbed into the ground. It indicates the distance the science centre has drifted away from Europe since Wilson's birth. A remarkable monument that celebrates a man who spent most of his life skeptical of Wegener's continental drift theory.

For years, if Wilson had any opinion about continental drift, it was a guarded one. Those who knew him said he didn't find much value in the idea until about 1960. But after he contemplated the new evidence, Wilson decided the idea of moving continents fit the data and he became an unwavering advocate. By the time continental drift was becoming plate tectonics, Tuzo Wilson was already past 50. At an age most scientists find change and innovation difficult, he made the first of his three greatest contributions to science.

Rough Edges

A New Source for Volcanoes

As a result of a quick but fortuitous stop in Hawaii, Wilson theorized something called *plumes*. It was Wilson's idea that hot plumes, or streams, of magma rise through the crust from deep within the mantle, forming volcanic mountains at the surface. He said he saw the scheme as an image, or rather a movie, unfolding in his mind. He created the unnerving scene of someone lying face up in a shallow stream, blowing air upwards, bubbles breaking at the surface, then swept away by the current.[373] This was how he saw the Hawaiian islands forming in his vision. He was 55 at the time and attempted to publish "A Possible Origin of the Hawaiian Islands." The scientists who reviewed it at the leading American geophysical journal found his idea so radical they rejected it.[374] Tuzo quickly turned to the *Canadian Journal of Physics,* which published his paper in 1963 – probably because they didn't realize it was such a controversial subject.

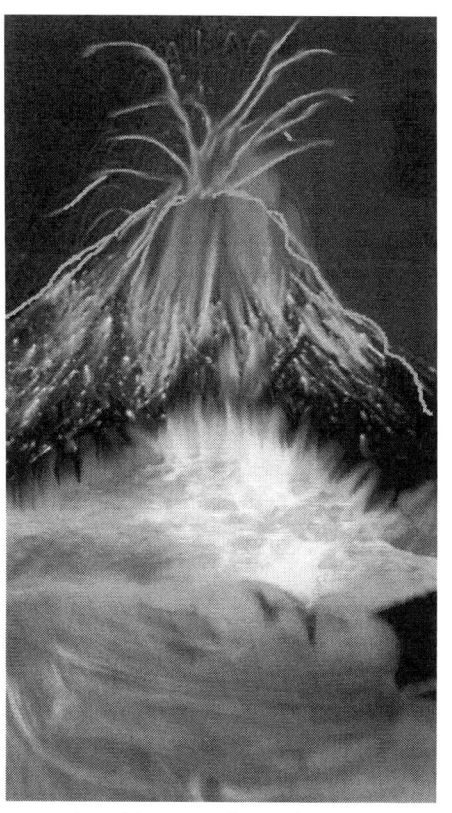

A goddess creating volcanoes.

Tuzo Wilson's paper explained that the Hawaiian islands formed in a chain because the Pacific tectonic plate slowly slid across a fixed hotspot while a plume of magma rose directly from deep within the molten mantle.[375] We don't know exactly why a hotspot is hot. Some hypotheses flip the cause on end – maybe there is nothing anomalously hot, perhaps instead there is a rip in the fabric of the crust allowing lava to seep through while the world drifts past. Others are not sure plumes are involved at all, suggesting the material that builds the Hawaiian islands and their kin is sourced close to the surface. The jury is out; however, most geophysicists studying this phenomenon agree with Wilson's original thesis: a hot mantle plume is erupting from the core-mantle boundary, expelling lava to the surface as it moves past. But no one has yet explained the reason hotspots stay fixed for tens of millions of years while tectonic plates move. Similar streams of hot mantle rising in narrow ribbons are likely the cause of volcanoes in Réunion, Yellowstone, Galápagos, Tahiti, Iceland, and at least forty other places scattered about the globe. With his near-mystic eerie vision, Tuzo Wilson created a whole new source for volcanoes.

The Mountain Mystery

The volcanoes of Hawaii are the most studied in the world. Kilauea and her cousins are approachable, splatter predictably, and their lava can almost be fetched in a tin bucket. These mild-mannered fountains of molten rock are so well-studied because Hawaiian volcanoes are located in Hawaii. Few scientists reject a grant to study geology, oceanography, geophysics, volcanoes, mountain building, and surfing on the islands. From Captain Cook to Major Dutton and Professors Daly to Tuzo, hundreds have trekked to the islands to discover at least one more fact. But none came away with as hefty an insight as Tuzo Wilson.

Hawaii was one of the last places to experience the delicate touch of humans. The Hawaiian volcanoes were discovered when settlers from western Pacific Islands arrived a bit over a thousand years ago. The Polynesians arrived in waves, sometimes hundreds of years apart, with successive groups generally displacing the indigenous. But a long-lasting legend about the birth of the island chain survived all the invasions. Their stories include the fire goddess Pele, poking up from the ocean bottom. She surfaced first in the northwest, creating Nihoa, Niihau, Kauai and Oahu, then Molokai and Maui, and finally Hawaii, her current work of art. In the Pele legend, it is clear that the islands to the northwest were created first and are the oldest. They certainly show the most age – erosion has slayed the mountains and cut deep valleys. The Big Island of Hawaii, youngest of the creations, is still the simmering home of Pele. In legend, the body of the goddess of fire is Mount Kilauea, the world's most active land feature. Since her most recent eruption, which began in 1983, Kilauea has destroyed homes and buried farmland under lava piled to the height of a ten-storey building.[376] But the current destruction and Tuzo Wilson's discovery of the birth and evolution of the islands changes nothing of the ancient narrative.

Iapetus begot Atlas

In Hawaii, Tuzo Wilson imagined Pele under the water, blowing bubbles, creating mountains. Wilson then turned his attention from Pacific to Atlantic. His next revelation startled some geologists who were reluctantly agreeing that the planet once had a supercontinent. Wilson rewrote geological history to include a whole series of prior supercontinents which he imagined had formed, then shattered into pieces, reformed, then broke again. His 1966 paper "Did the Atlantic Close and Then Re-Open?"[377] was less a question than a statement of another of his visions. Among other evidence, Wilson used the similarities of Cambrian and Ordovician trilobites found in Scotland and Newfoundland to show that long before Pangaea existed, parts of North America and Europe had already been in communion. Then they were separated by an ocean, and again united, this time as Pangaea. The Atlantic Ocean precursor, hundreds of millions of years older than the current ocean, is called Iapetus, named for the Greek mythical gentleman who was father of Atlas, from whom arises the Atlantic's name.

Rough Edges

It seems Scotland and Newfoundland were once joined, nurtured identical trilobites, but then Iapetus, the son of Uranus and Gaia, came between them. The ocean of Iapetus wrung dry after the god begat Prometheus, Epimethus, and Atlas – who in turn sired the entire human race. As Iapetus Ocean drained, the Earth's landmasses collided in the Caledonian orogeny,[i] forming the supercontinent we call Pangaea. At that time, a quarter of the Earth was Pangaea, the rest was water. But then Atlas squeezed back into his father's old sea bed – creating a new rift between the western and eastern hemispheres, the chasm we call the Atlantic.

Wilson offered this model (without the Greek gods and titans) to describe cycles of opening and closing ocean basins, explaining the action in terms of plate tectonics. Just when geologists began accepting Pangaea had once existed and splintered, Wilson gave them several sequential supercontinents to consider. He said fossils and structures indicated the oceans had opened and closed repeatedly over the course of Earth history. According to Wilson, we are living on the latest of a long series of fractured supercontinents. The entire scheme, with some modifications, is the *Wilson Cycle* – a cycle that has repeated at least four times. Some geologists suspect seven cycles have occurred and have named their series of speculative landmasses. Ur was the planet's first supercontinent. Then Vaalbara, followed by Kenorland. Each of the seven survived dozens or even hundreds of millions of years, then broke up, then reassembled again, at a rate of about half a billion years per cycle. Again and again continents merged into supercontinents, then rifted, oceans filling the voids. The present Atlantic Ocean, having ruptured Pangaea, is perhaps now at its widest. This suggests the Americas and the eastern hemisphere are planning another reunion.

The Atlantic closure may have already begun. This brings us back to the devastating unexplained Lisbon earthquake of 1755. It killed thousands and wrecked Portugal's capital, but according to plate tectonics, it should have never happened. The Iberian Atlantic coast is considered a quiet passive margin, calmly receding from the Americas as the Atlantic Ocean widens, in harmony with our current expanding phase of the Wilson Cycle. So geophysicists had been puzzling over the violent earthquake. But in 2013, they began to see things differently. João Duarte, of Monash University in Melbourne, says his team has mapped the ocean floor near Portugal and found the region is beginning to fracture. "What we have detected is the very beginnings of an active margin — it's like an embryonic subduction zone," Duarte reports in *Geology*.[378] It is still a hypothesis and will take years of study to prove. However, we may be witnessing the beginning of a big squeeze which will eventually once again drain the Atlantic and refresh the eroding Appalachian mountains. Because it will take millions of years to close the Atlantic, many of us will not be around to witness this coming compressional stage of the Wilson Cycle.

---

[i] The Caledonian orogeny began around 500 million years ago. Caledonia is Scotland's Latin name; *orogeny* means the genesis, or creation, of mountains.

The Mountain Mystery

Abnormal Faults

Wilson seemed to add insight to plate tectonics as casually as he might lure trout from a mountain stream. After hotspots and ocean cycles, the next puzzle he tackled dealt with the way massive slabs of crust slide past each other when plates are in motion. For this, he invented *transform faults*. The term *fault* arose in the late eighteenth century when geologists first noticed weaknesses in rock layers. They were faulty; they broke. The word now indicates a sheet of rock has sheared and one of the pieces has moved past its partner. Until Wilson, geologists recognized three main ways rock sheets shuffle after splitting.

When we think of a fault, we normally picture the sort that tosses one side up while the other slides down. A *normal* fault occurs when the angle is steep and the part known as the hanging wall slides down. (Geologists describe the two sides of a fault as the foot wall, which you might walk upon, and the hanging wall, which was sometimes occupied by intruders snooping around a gold mine's fault zone.) A steep-angle normal fault is a simple scheme, but occasionally the results are spectacular. Normal faulting has formed Wyoming's Teton Mountains, producing a gorgeous elevator-like uplift of ten kilometres which continues to rise a metre every thousand years.

A *thrust* fault is a sort of low-angle reverse fault – a normal fault with the hanging wall sliding up while the foot wall drops. This is typical in the Rockies where west meets east and west ends up on top. When Suess and others were promoting a shrinking planet, contracting as it cooled, mountains were imagined as mostly vertically projected rock piles, forced aloft as steep normal faults. But Suess was also aware of the great compressional thrust faults of his native Alps that piled rock layer upon rock layer. Thrust faults helped kill Suess's contraction theory because no one could explain how one rock strata might ride over another for a hundred kilometres if the Earth were simply cooling and shrinking. Accounting for long thrust faults is much easier when plate tectonics are recognized as causing crumpling collisions.

A third great class of faults befuddled geological explanation for a long time – chunks of crust sail past each other, rubbing shoulders, moving along, neither rising nor falling. The most famous *strike-slip fault* is in California where the San Andreas causes farmers to veer right as they harrow between their dextrally[i] displaced lemon trees. Those types of faults represent the rending of crust as tectonic plates amble along. A special sub-class of the strike-slip fault was

---

i   Dextral faults are displaced to the right. Rows of trees across such a fault-line appear rightward after a dextral displacement. It doesn't matter if you are on the west or the east side of the fault, as you can test yourself with a simple thought experiment: Picture a line of trees, cut by a strike-slip fault. As you gaze at the once continuous row, you see the most distant have slid right. Now carefully walk across the fault, go into the displaced grove and look back to where you were. The trees now on the other side also appear displaced dextrally. Fault systems displaced leftward are sinistral faults, taking their name from the belief that left-handedness is evil, or *sinister*.

imagined by Wilson when he considered how the seafloor rifted and its pieces diverged.

Tuzo Wilson was puzzled by an idea he learned as a geology student in Toronto. It bemused him that strike-slip faults, such as California's San Andreas, eventually die out. Logically, faults can not continue forever, but then again, what causes rocks to slide 50 kilometres in one location, yet not at all further along the same ripped zone? What happens to the rock between the place it moves and the place movement has stopped? Do rock layers pile up? Do they simply shatter, disintegrate, disappear? No one could answer Tuzo's questions. The problem came back to him again when he considered the rifted seafloor where faults allow pieces of crustal plate to slide past other pieces. He imagined short zigzags that accommodate seafloor spreading. These *transform* faults would neither create nor destroy crust; they would simply allow slabs of rock to slide past each other at different rates of motion. If transform faults existed, they would produce movement opposite from what geologists expect and should be detectable on seismic recording equipment. Wilson's new way of understanding the faults along the ocean floor explained plate motion on a tectonic scale without destroying lithosphere. These conservative plate boundaries are important to the mechanics of plate tectonics.[379]

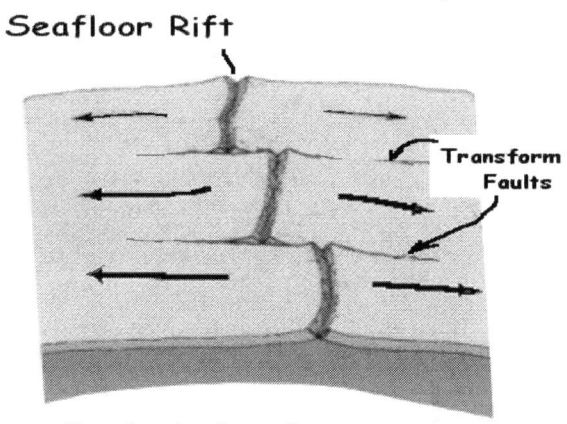

Transforming the seafloor

As the seafloor splits, newly formed lava on either side moves in the same direction, away from the rift, though some pieces move faster and farther than others. Transform faults allow this activity, letting some slabs amble past others. But they are not like the aggressive strike-slip faults found along the San Andreas. Instead, these are comparatively mild good-natured neighbours, moving in the same direction, but at different paces.

Wilson's idea that transform faults accommodate crustal motion was just another of his visions – a theory buried deep below the sea's waves. For a while, no one could confirm their existence. Confirmation would require detailed seismic data recorded in a wide range of remote places, particularly on mid-ocean islands. No one was funding such expensive esoteric research, but in an unlikely chain of events, a nuclear test-ban treaty provided the data geophysicists needed. The new data showed that Tuzo Wilson was right once again.

The Mountain Mystery

A Seismic Shift Towards World Peace

Jack Oliver[i] was still finishing his doctorate when he noticed some odd patterns appear on his seismograms. He was working at Maurice Ewing's Lamont Observatory, just outside New York City. He had never seen seismic wiggles such as his new samples. Several things seemed immediately peculiar about the wiggly shapes on his strips of paper. The seismic activity arriving in New York seemed to be coming from the deserts of Nevada, a place where strong earthquakes were rare. The second peculiarity was that the seismic wave form, the wiggle's shape, was quite different from recordings of normal earthquakes. A seismogram from an earthquake starts with strong sideways shear waves, accompanied by weaker pressure waves. Pressure waves are as you might expect – forward pressure along the ground, while shear waves result from side-to-side shaking. During an earthquake, rocks break along a fracture zone, shearing raggedly. That's why the arriving shear wave is strong and shows side-to-side motion. But Oliver's records began with a sudden sharp forward-moving pressure wave instead. The third strange thing about the seismic record was that the signal seemed to radiate from a single point, not a fault zone. Jack Oliver deduced that the single-point was a nuclear explosion in the Nevada desert. Oliver was inadvertently spying on a top-secret test of American atomic weaponry, 4,000 kilometres away.

Jack Oliver admitted he was a bit nervous when he realized what he had discovered. He had the practical common sense of a midwesterner and suspected the army might not be pleased when they learned that a civilian was looking over its shoulder. Oliver was born into a small heartland community, Massillon, Ohio, a steel town just west of the tire factories of Canton. He was a talented football player, a member of a high school team coached by Paul Brown, later the first coach with a professional football team named after him – the Cleveland Browns. Paul Brown's help and Jack Oliver's talent earned him a sports scholarship. University was interrupted in his second year by World War Two, which put Oliver in the South Pacific. He was thirty years old when he finally finished his geophysics doctorate at Columbia, and accidentally uncovered a way to monitor secret atomic weapons tests.

Rather than being arrested for eavesdropping on the military, Jack Oliver was soon fêted as a celebrity scientist – the world's expert on detecting nuclear explosions. He showed that seismic waves from an atomic bomb could be recorded by any backyard tinkerer, anywhere on the planet. Secretive nuclear tests could no longer be hidden from scrutiny. Oliver was invited to the White House. Eisenhower asked him to advise on the first draft of the Nuclear Test Ban Treaty. In the late 1950s, Jack Oliver was a delegate to negotiations in Geneva. With his discovery, American scientists could monitor Russian activities (and, of course, vice-versa) making nuclear tests verifiable and a non-proliferation treaty possible.

---

i   John "Jack" Ertle Oliver (1923-2011), American geophysicist.

Rough Edges

Geophysicists triangulated blast locations and estimated weaponry power. But to do this well, seismic detectors had to be installed around the world. Small nuclear tests, such as the recent renegade explosions in North Korea, have a force equal to a mild 4.5 magnitude earthquake. Their significance might be missed without a world-wide array of seismometers. The government gave the geophysicists millions of dollars to build a detection system. Arrays of seismic recorders were established everywhere, increasing the accuracy of the snooping. Since these data were not classified, geophysicists were soon gleaning a wealth of new information – and finding all sorts of interesting anomalies.

In addition to hearing the seismic roar of nuclear bombs, the new network of seismometers picked up earthquake activity near mid-ocean rifts. Lynn Sykes[i], also at Lamont, examined ten earthquakes from the mid-ocean ridges. The data showed him that Tuzo Wilson's transform faults were real – the blocks located at the ridges were moving along transform lines, sliding away from the widening centre in the pattern of Wilson's prediction. Thus another of Wilson's visions was proven, further strengthening plate tectonics. With Jack Oliver and Bryan Isacks,[ii] Lynn Sykes also examined deep earthquakes along the ocean basins' edges, in offshore Chile and in the deepest Pacific trenches.

Some of the extensive grid of Cold War seismic detectors were in the South Pacific. The region exhibits the most earthquake activity of any place in the world, so Jack Oliver and his former graduate student, Bryan Isacks, installed extra recording equipment on Fiji and Tonga. Isacks said he and Jack Oliver consistently followed the idea that "if you really want to understand a natural phenomenon, figure out where in the world that phenomenon is most clearly and most vigorously manifested, and go there to study it."[380] They could not have selected a better location.

The islands and nearby deep trench had a long and violent history of explosive volcanoes and destructive earthquakes and had been studied by a long string of scientists. A hundred years earlier, the *Challenger* had discovered that this part of the Pacific reaches an almost unfathomable depth – nearly 11 kilometres. In the 1930s, Felix Meinesz arrived at the trench with his gravimeters and found a strange gravity anomaly. He measured extremely low readings, indicating rocks were either missing or the ones present were peculiarly light-weight. During World War II, Harry Hess bounced sonar signals off the steeply dipping trench.

During the International Geophysical Year, the king of Fiji asked geophysicists to set up seismometers and to listen to the rumblings under his restless island. Recently, employing modern GPS equipment, scientists have learned that the spot Oliver and Isacks selected for their 1960s seismic study has the fastest plate movement on the planet – a rate of 24 centimetres per year,[381] ten times faster than the global average of less than 3 centimetres. Movement of ocean crust at the

---

i Lynn R. Sykes (1937), American geophysicist.
ii Bryan Isacks (1936), American geophysicist.

Tonga Trench makes it the planet's most energetic zone of seismicity. So, Isacks and Oliver were brilliant in their selection of an area where their object of interest was "most clearly and vigorously manifested." As Jack Oliver noted, in order to make an important discovery in science, be at the right place at the right time.[382]

Luck is involved. Oliver writes in his book *Shocks and Rocks*, plate tectonics was proven because "serendipity prevailed."[383] Serendipity, discovery by accident. Repeatedly, scientists have made accidental discoveries, often by gleaning data originally gathered for military use. Geophysicists scanned it and extracted tidbits that revealed nuances of the Earth's physical properties. The military wanted to find enemy submarines, the data pointed towards zebra-striped magnetic anomalies; the military had problems communicating at sea, the whale's echo channel was discovered; the military wanted to launch satellites, geodesy uncovered gravity anomalies; the military wanted to spy on nuclear tests, the seismic data proved Wilson's transform faults. And on it went – GPS, side-angle radar, communications satellites – all originally military creations that unlocked scientific discoveries. Jack Oliver pointed out that the world-wide grid of seismic stations was not built for him to find slabs of ocean crust descending into the Tonga Trench, but scientists used the data for that purpose.

To investigate the deep-seated earthquakes revealed by military data, Bryan Isacks spent 15 months on various tropical islands, tinkering with seismometers and recording seismic stirrings within the mantle. With his strategically located gadgets, Isacks found himself eavesdropping on the death groans of a gigantic slab of ocean crust sinking far below the geological structures that had created Tonga and Fiji. The seismic activity analyzed by Isacks, Oliver, and Sykes was originating at least 600 kilometres below the Pacific's calm surface. These scientists knew earthquakes result from the violent shattering of brittle slabs of rock. Material six hundred kilometres below surface should be soft and pliable – bendable, not breakable, not able to generate earthquakes. Enormous heat and pressure make such rock malleable – the material softly deforms, it should not snap or crack. Nevertheless, they recorded deep earthquakes. The researchers concluded that a thick slab of cold ocean floor was being thrust deeply into the planet's interior. But this was not a simple and direct conclusion. The idea contravened accepted wisdom – everything in the mantle should flow like cold molasses, not break like glass.

Although Arthur Holmes and Harry Hess had both written about subduction zones, it was not until Isacks, Oliver, and Sykes published "Seismology and the New Global Tectonics" in the *Journal of Geophysical Research,* that subduction zones were proven. Their paper was the first to document how the Earth recycles crust in ocean trenches. The paper was a true collaborative effort, as they indicate with their second footnote: "order of authors determined by lot." All three deserved equal credit for their landmark study. For such an important paper, it is an easy read and a good review of the development of plate tectonics up to late summer, 1968. Their article acknowledges Wegener's continental drift, Hess's seafloor spreading, and Wilson's transform faults. After that preamble is a nod to mid-oceanic ridges, which are explained as growth areas of the Earth's crust. They show that the spreading seafloor is revealed by seismicity, earthquake swarms, and the "young ages measured by radioactive and palaeontological dating and the general absence of sediment."[384] Thus the authors acknowledged the birthplace of the planet's rocky crust. Logically, if the crust is spreading from mid-ocean rifts, and not surviving to an extremely old age, it follows there must be places where the (comparatively) youthful rock is being destroyed. Their seismic proof of the crust descending into deep trenches was the fundamental contribution of their paper. Their seismic observation of the digestion of ocean crust was as convincingly clear as an X-ray image of a python swallowing a gerbil.

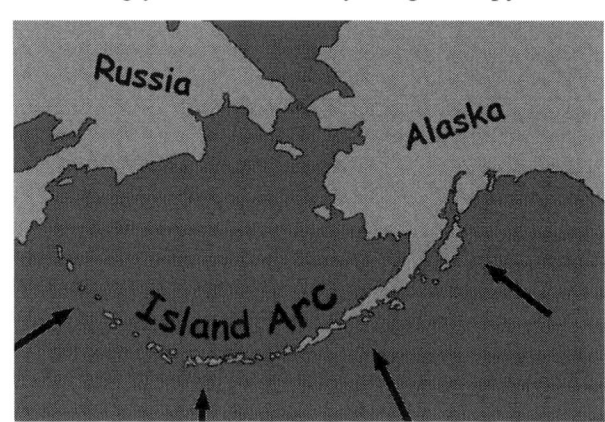

Prominent Island Arc: Here the Pacific Plate from the south is subducting under the North American Plate.

The authors point out that "almost anyone who glances casually at a map of the world is intrigued by the organized patterns of the island arcs." If you are one of those who missed casually glancing at a world map, you have missed some of the planet's great arcs – Alaska's Aleutian Islands and the nearby similarly curvacious Russian Kurils, for example. Once you know what to search for, you find these dramatic features all over the globe. Among the most notable are the arc-island systems in the South Pacific (the Philippines is a giant example), the Antilles of the South Caribbean, and the Greek arc in the South Aegean Sea. These island arcs have unusual volcanic and earthquake activity including deep murky oceanic trenches – the graveyards of the seafloor.

Although gravity studies in the 1930s by Vening Meinsz and Harry Hess suggested descending vertical convection at island arcs, it was the trio of Oliver, Isacks, and Sykes that concluded island arcs are indeed zones of destruction, not quite vertical, but rather like downward escalators carrying ocean crust into a deep dark recycling depot. As proof, Isacks captured those seismic earthquake signals – noises not dulled by soft, pliable rock but instead the crisp sounds of rigid surface crust cracking within the abyssal trench. The low-frequency seismic rumblings recorded on Fiji were the cacophonous crushing of ocean crust within the Earth's mantle.

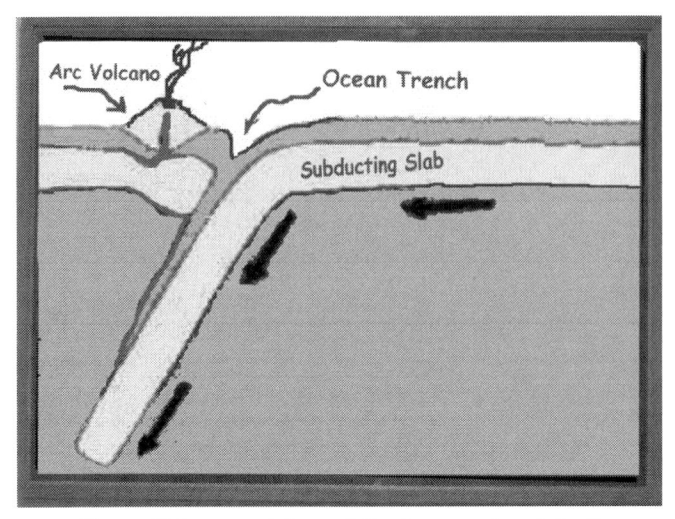

"the well-known picture of the down-going slab..."

Until the work of Isacks, Oliver, and Sykes in the South Pacific, the theory of tectonics was floating on thin crustal plates. Underlying dungeons were not much considered. Oliver himself said, "As far as I know, no one before us had thought in terms of such a large-scale thrusting phenomenon that moved a 100-kilometer-thick slab of lithosphere from near the surface to depths of at least 720 kilometers, or had even brought the lithosphere-asthenosphere structure into the picture."[385]

Oliver summed up the moment when he and Bryan Isacks first realized they had proof that trenches were swallowing ocean crust. "In retrospect, the final interpretation of the data seems obvious, but we pondered the data for months," said Oliver. Then, one day, he says, the team compared their records from Fiji to data from the Caribbean and "almost immediately, the well-known picture of the down-going slab beneath island arcs appeared on the blackboard."[386]

By 1968, plate tectonics was widely accepted. It had taken centuries to reach this point and to realize the origin of mountain ranges was a clash of continents, a tectonic train wreck driven by the ocean's floor. As proof piled up, even some of the most wary scientists conceded the idea that the continents are mobile. However, as Tuzo Wilson, speaking on his Canadian science show, *The Planet of Man*, said in 1975, "a lot of people weren't happy about it, because they weren't brought up with it – it's like asking a middle-aged man to change his religion, and they don't really like it. They would have been delighted to see something happen that would destroy it all, and go back to fixed continents!"[387]

19

Moving Back

Proof of continental mobility was almost irrefutable, but not everyone had boarded the tectonic bus. By the mid-1960s, fifty years ago, it seemed clear that the solid Earth is made of dozens of huge pieces, wandering around, bumping into each other. But stubborn doubters still rejected the idea. Listen to the words of Gordon MacDonald,[i] a geologist intimate with the discoveries leading to plate tectonic theory. He was a wondrously smart scientist who would later become President Nixon's science adviser for the environment.[388]

> "In all science there is a strong 'herd instinct.' Members of the herd find congeniality in interacting with other members who hold the same view of the world. They may argue vigorously about details, but they maintain solidarity when challenged or criticized by those outside their comfortable herd. If individual scientists stray too far from accepted dogma of the day, that of the herd, they are gently (or not so gently) ostracized. The herd instinct is strengthened enormously if the paymasters are members of the herd. Strays do not get funded and their work, sometimes highly innovative, is neglected as the herd rumbles along. When leaders of the herd decide to strike out in a new direction, the herd often follows. Before the 1950s, the North American herd of geologists found it comforting and amusing to ridicule those foreign geologists who advocated continental drift. In the early 1960s, Harry Hess, Tuzo Wilson, and Bob Dietz, all respected leaders of the North American geology herd, decided to shift directions and the herd soon followed."[389]

It was a bit frivolous of MacDonald to claim Hess, Wilson, and Dietz had "decided to shift directions." The data had something to do with it; no conspiracy was involved. Despite his enduring opposition to the plate tectonics model, MacDonald was recognized by his colleagues as an intelligent scientist and was encouraged to participate in their discussions and conferences. MacDonald was

---

i Gordon James Fraser MacDonald (1929-2002), American geophysicist, policy wonk, and environmental scientist.

born in Mexico where his Scottish father had settled and met his American-born wife. Gordon MacDonald was brilliant and was readily accepted at Harvard. But he said Harvard was at first a disastrous experience. He blamed his professors, whom he said were extremely poor teachers. Quite unusual for a geophysicist, MacDonald had only half a year of formal physics training. He said the teaching was so poor he refused to take any more classes. Nevertheless, MacDonald completed his geology degree and began a long career in Earth and atmospheric sciences by first studying twisted and deformed rock outcrops in New England, contributing to what we know about the way heat and pressure transform rock. He became a professor at Massachusetts Institute of Technology and an adviser to Nixon and the CIA. He consistently and, it seems, gleefully, ended up on the contrary side of the tectonics debate. His unyielding skepticism regarding plate tectonics was reasoned, thoughtful, vocal, and wrong.

At the White House, he worked for Nixon as the president's chief adviser on all things environmental. It was largely because of MacDonald that the United States has an Environmental Protection Agency. Two weeks after Nixon became president, MacDonald was invited to a photo-op staged at a California beach which had just been spoiled by a massive spill from a Union Oil Company oil well. They walked along the shore, Nixon in the centre, Union Oil's boss to the left, MacDonald on the right. The beach had been raked clean and the oil executive pointed out to Nixon that no damage had been done. This infuriated MacDonald who knew the oil was just below the manicured surface. He said so and impulsively kicked a clump of sand which, according to MacDonald, planted a large oil stain on the American president's trousers. In an awkward spot. That was the beginning of his long relationship with the president. The scientist spent over four years at the White House, advocating for clean air and water, voicing early concerns about global warming, finally resigning when Watergate was revealed.[390]

MacDonald later served Reagan and Bush, continuing to warn about the effects of increasing carbon dioxide, global warming, and rising seas. He said he got Reagan's support when he pointed out that the president's recurring bouts of skin cancer were due to the changing climate. Although MacDonald argued that climate change is part of Earth's normal cycle and atmospheric carbon dioxide had reached similarly high levels in the distant past, he was concerned the current greenhouse effect's rising seas would destroy many of the developing nations of the world. In a 1994 interview, MacDonald said, "You have something like 800 different languages in the South Pacific and many dozens of distinct cultures and you just wipe that out. We should be concerned about preserving the diversity of humans and cultures as well as biodiversity. How is a poor country like Bangladesh going to deal with increased tidal surges and storms? And on and on and on. And so what we're doing is transferring wealth from the poor in the future to the rich of today. And I think that is just morally unjustified in every way,

Moving Back

taking what could be a livelihood for future generations and using it up today."[391]

MacDonald had no doubt global warming was happening. As early as 1990, he was on the cover of *People* magazine, standing on the steps of the Capitol building, his feet where water will be when the ice caps fully melt. He fought misinformation when he could. Many in his camp (including Reagan) claimed that more carbon enters the atmosphere through volcanoes than anthropogenic means, but he and the United States Department of Energy disagreed. Volcanologists tallied up the global volcanic CO_2 emissions while nations around the globe determined how much carbon is released by human activity through burning fossil fuels. Globally, volcanoes release about 200 million tonnes of carbon dioxide each year. This seems like a huge amount, but in 2002, when MacDonald died, CO_2 from global fossil fuel emissions was over a hundred times greater, tipping the scale at 26.8 billion tonnes.[392]

It was a good move for MacDonald to leave geology and advocate for the environment. His opposition to plate tectonics was at first quite reasonable – his ideas were based on the work of a powerfully influential British geologist who reviled plate tectonics theory. Sir Harold Jeffreys[i] was most notably a mathematician specializing in probability theory. But in geophysics, he is remembered for his obstinate rejection of continental mobility based on his idea of the strength of the Earth's crust. In his opinion, continental drift was "out of the question."[393] The crust, he insisted, is too strong to either break or move. Sir Harold Jeffreys was once considered among the most formidable of geophysicists, solving problems related to the structure of the planet, especially deep within its shell. For decades, he convinced nearly every geologist he met that the continents do not move. Jeffreys refused to read contrary papers or listen to the arguments of anyone else. At age 91, in a 1982 piece for a prestigious geophysical journal, Jeffreys wrote that subduction of ocean plates was as likely as "cutting butter with a knife also made of butter."[394] He lived nearly 98 years, passing away in 1989, twenty years after plate tectonics was rather universally accepted. By then he was one of only a handful of scientists on the planet who still could not imagine continents adrift.

MacDonald agreed with the inflexible assessment of Jeffreys and cited him as the core of MacDonald's own long-standing stubborn rejection of the theory.[395] Along with Jeffreys, MacDonald thought continental roots extend 500 kilometres below the surface, making it impossible for them to budge. If they could move, he figured the result would be a deadly mix of mantle materials that would stir continents with ocean crust, putting an end to both. If continental granites blended with oceanic basalts, the result would be a smooth flat crust buried under kilometres of ocean water. It is only because they are thoroughly differentiated that the lighter continents are buoyed by isostasy, keeping us above water. But MacDonald also felt mantle convection wasn't powerful enough to move

---

[i] Sir Harold Jeffreys (1891-1989), British mathematician and geophysicist.

continents and the mantle itself would never flow like a fluid. He was wrong about all of this. But as he'd said, he had never been schooled in physics.

Nearly all the evidence and observations were making MacDonald's objections meaningless. For this, he complained that commentators dismissed him as "a troglodyte who was slowing the convergence of thought"[396] leading to general acceptance of plate tectonics. In the mid-60s there was still resistance to drift theory, but at a 1967 American Geophysical Union meeting, everything changed, and MacDonald knew it. It was a tectonic shift, if you will. Suddenly a vast majority of geophysicists agreed that the continents move. With powerful evidence from a range of overlapping sciences, paper after paper explained why plate tectonics made sense. Almost no one was left at the conference to defend stationary continents. At the close of the meeting, Edward Bullard read a summary statement favouring the new global tectonics model. MacDonald had agreed to present the rebuttal, but he was called away at the last moment and did not address the gathering. So Bullard gamely produced what he thought MacDonald would have said – and he reminded the audience that these were not his own words.

MacDonald never conceded he was wrong. Instead, he claimed that he moved away from the controversy when he found more valuable things to do with his time, such as serve as Nixon's science liaison. As late as 2001, in his contribution to *Plate Tectonics: An Insider's History*, historian Naomi Oreskes's great collection of scientific reminiscences, he dismissed the whole of drifting continents by writing that "neither continental drift nor plate tectonics has had much influence on the health of society. For example, earthquake prediction was impossible before the acceptance of plate tectonics and has remained so afterwards."[397] With this, he has finally indicated his irritation with the theory, suggesting it had become an excuse for geologists to become sloppy and lazy in their work: "Rather than working with six continents, the geologist now has 11 plates and can suggest more if the geologic evidence points that way. . . The lack of any discipline required by geophysical observation places few, if any, limits on the creativity of geologists in interpreting the past. Long ago, the geologic herd overcame the physicist herd in the battle about the age of the Earth. They now have a comfortable confidence that they have found truth in plate tectonics, even if there are a few troublesome details yet to be dealt with."[398]

MacDonald's dismissive tone towards unlimited creativity makes him sound unfairly stuffy. His words do sound like hisses from a troglodyte, but he is right in the liberating power of plate tectonics. It allows many degrees of freedom, in both the philosophical and mathematical sense. Geophysicist Tanya Atwater[i] obliges MacDonald's prudish observation with an excited freshness in describing how geophysicists Dan McKenzie and Robert Parker presented the theory to her at a pub in Del Mar, north of San Diego. "Plate tectonics really set us free and flying,"

---

i   Tanya Atwater (1942), American geophysicist.

Moving Back

she wrote. At the pub, McKenzie scribbled a simplified plates scheme on a napkin. Atwater dared him to show how plate tectonics could deal with something as complicated as the San Andreas fault, which was central to the doctoral thesis she was writing. "Easy!" said McKenzie and he quickly sketched in three more plates to make the fault work.[399] For MacDonald, this attitude was too much artistic licence. However, for him and a few others, the most troublesome detail in plate tectonics remained the strength of the crust and the rigidity of the plates themselves.

Nature is flowing with examples that disprove MacDonald and his fellow naysaying geologists. Some materials stretch and bend, then crack abruptly: frozen honey, tar, rock layers amid mountains. Wegener himself had pointed out that the ocean's crust might act as dried tar does under different stresses: brittle when hit with a hammer, but under pressure and with time, able to flow. Geologists were familiar with bent and twisted rock layers, some rolled over upon themselves as recumbent anticlines, curling like the first yellow bricks on a long road.

Physicists rank fluidity in terms of viscosity. The ideal fluid (found only in the ideal supercold laboratory) has a viscosity of zero. It offers no resistance to flow. At a degree or two above absolute zero, liquid helium may flow up, out of its container. Stirring it is like stirring air – it offers no resistance. Cold water, with a viscosity of about 1 (milli-Pascals per second) is stickier than warm water, which has a viscosity approaching that of money – trickling through fingers rather easily. Blood, incidentally, really is thicker than water, with a viscosity of 3. Room temperature honey, even more viscous, has a value approaching 10. Geologists have found that on this scale, the Earth's mantle is somewhat more sticky than water, blood, or honey. About a million quadrillion times stickier – it has a viscosity rated as the number 10, followed by 24 zeroes. Nevertheless, with enough time and enough pressure, even the mantle flows.

An explanation of how the rigid, yet broken, crustal pieces move above the stubbornly viscous mantle was included in the landmark 1968 paper by Jack Oliver, Bryan Isacks, and Lynn Sykes. They suggested that subduction of crustal slabs helped draw the spreading seafloor apart. Surface crust cooled, sank, forced the convection cycle to continue, kept the continents in their controversial motion. As an undergraduate, Lynn Sykes was told "good scientists did not work on foolish ideas like continental drift."[400] He may have heard this from Gordon MacDonald himself, who taught at Massachusetts Institute of Technology when Sykes studied there. Sykes left MIT and began his graduate work at Columbia in 1960. He says he thought his adviser, Jack Oliver, also did not believe in continental drift at the time. Other key scientists who had trouble with drifting continents included Bruce Heezen, who never gave up the expanding Earth theory, and his equally cantankerous boss, Maurice Ewing, who whispered to Edward Bullard at a small gathering in 1966, "You don't believe all this rubbish, do you Teddy?"[401] Ewing, ubiquitous at Lamont, was still working 18-hour days, controlling, micro-managing, domineering. He could have prevented scientists at his facility from developing plate tectonics. But his colleagues pointed out that Ewing had hired many pro-drift scientists and did not begrudge his charges when they published plate tectonics papers. Reversing a lifetime of quiet opposition, in 1967, Ewing decided they might actually be right.

Theories were merging and corroborating, building a new grand model of the Earth. The Hess and Dietz spreading-seafloor theory was confirmed by the Morley-Vine-Matthews reversed magnetism evidence, which itself suggested to Wilson that transform faults had to compensate for irregularities in seafloor motion, while Sykes discovered the data that Wilson's hypotheses predicted. The idea of mobile continents was finding independent confirmation everywhere.

Drifting into Tectonics

By the late 1960s, continental drift, refined as plate tectonics, had replaced contracting/expanding and the various stationary theories of Earth evolution. The idea began to be called by its new name, *plate tectonics*, displacing the quirky phrase *continental drift*, which had been cast to create the unlikely image of continents adrift on the seven seas. *Plate tectonics* was more mature – and more accurate. *Tectonics,* from Late Latin, *tectonicus*, is a term the Romans borrowed from Greek, *tektonicos,* which meant building, especially carpentry. An apt expression for the notion of hammering together the topography of the planet.

Since 1899, the word *tectonics* has been applied in the geological sense to events that transform the Earth's surface. The new expression, plate tectonics, reflects the notion that solid plates of crust are involved as rigid units. Geophysicists signalled they had discarded the idea of continents drifting across oceans – or plowing through them – when they adopted this new phrase.

I spent several tiring hours reviewing journal and newspaper articles, looking for the first use of the phrase *plate tectonics*. Although British geophysicist Dan Mackenzie suggested that it first appeared in a paper[i] he co-wrote in 1969,[402] McKenzie quickly adds that the term by then was "widely known."[403] Actually, according to Anthony Brook of London's History of Geology Group, scientists had used *raft* tectonics, *continental* tectonics, and *global* tectonics before *plates* stuck as the popular metaphor.

A full year before Mackenzie, in "Seismology and the New Global Tectonics," by Isacks, Oliver, and Sykes, the word *plate* appears a remarkable 49 times, and the expression *plates of lithosphere* 18 times. Simply on the basis of density of use, that paper is the first to *almost* publish the term *plate tectonics*. The authors narrowly miss using the phrase when they refer to the *plate model of tectonics*.

---

[i] In October 1969, Jason Morgan and Dan McKenzie's paper "Evolution of triple junctions" appeared in *Nature* and included the line ". . .areas of the Earth's surface move as rigid spherical caps, and for this reason it is often called *plate tectonics*."

Perhaps an earlier draft expressed it as the *model of plate tectonics*, but was unfortunately rubbed out in an edit.

Anthony Brook, of the British history group, tells us that Jane Dore, an information specialist at Worthing Reference Library, West Sussex, unearthed an article by James Schopf which describes mountain-building in Antarctica.[404] Schopf's paper, published in *Science*, predates McKenzie's claim and states that if a geologist "considers continental drift in the light of *plate tectonics*, displacement of the Ellsworth Mountains can readily be explained."[405] As far as it can be determined, this was the first published use of the now common expression. Outside scientific literature, the phrase was tentatively legitimized by *The New York Times* in a January 1970 newspaper article "Theory is Upheld on Earth Plates" in which the report wraps uncertain quotation marks around *plate tectonics*: "Research on the crustal plates, known as "plate tectonics," is a refinement of the earlier study of ocean-floor spreading, which itself grew out of the study of how the continents move."[406]

Still not Convinced

The mystery of the mountains seems rather simple now. The Earth's surface is composed of jostling rigid plates, driven by internal convection currents which thrust those crustal slabs into each other, building mountains and levelling cities. But nothing is ever so simple. Although it has been 50 years since the theory of plate tectonics was generally adopted by Earth scientists, there have been some very bright geologists who continue to find faults in this clever synthesis. Plate tectonics is a good model for describing the planet's landforms, but not everyone is convinced.

In 1972, the father and son team of Howard[i] and Arthur Meyerhoff[ii] took a principled stand against plate tectonics, producing vigorous pleas alerting fellow Earth scientists to the pitfalls of the theory. The elder Meyerhoff was a science professor at the prestigious women's school, Smith College, when he first came to notice as a contrarian – as far back as 1936 he disagreed with a study, published in *Science* by environmentalist Paul Sears, which blamed the Depression-era dust storms on farming practices that stripped off the nation's topsoil. For uncertain reasons, Meyerhoff chose to counter, "It is unfortunate that Dr. Sears has tried to inject the important question of soil preservation into a situation where it has absolutely no application... I find Dr. Sears's viewpoint that soil preservation will solve the problems connected with [floods and dust storms] much too elementary."[407] Meyerhoff has since been proven clearly wrong – farming practices that wreck topsoil do result in dust storms and floods.

After his brief skirmish with environmental scientists, not much in the field of

---

i   Howard A. Meyerhoff (1899-1982), American geologist.
ii   Arthur Augustus Meyerhoff (1928-1994), American geologist.

geology is heard from Howard Meyerhoff for years. He spent a decade helping various companies battle striking workers. After subduing coal miners and then forcing an end to a strike at a textile factory, Meyerhoff was appointed a director and later chairman of the board of the same company. From there, he somehow became executive secretary of the powerful American Association for the Advancement of Science, which publishes the prestigious journal *Science*. Meyerhoff worked as its editor for a time. By 1963, Howard Meyerhoff had returned to geology research, after a 25-year absence.[408] This also heralded the teaming of Howard Meyerhoff with his son, Arthur, as the pair assembled data that contradicted the growing influence of plate tectonics.

The Meyerhoffs expertly assembled obscure facts that disputed tectonic motion and they released reasonable commentaries voicing opposition to the nascent hypothesis. They were usually correct in pointing out weak and contradictory aspects of the theory. Australia's Warren Carey repeated some of the Meyerhoffs' work to complement his expansion model, quoting from their unyielding papers – citing, for example, that India has always been part of Asia. They disagreed with the tectonics idea that the subcontinent had drifted into place from somewhere near Madagascar and knocked the Himalayas into existence by striking Asia. The Meyerhoffs asserted their opposition was based on "geological fact, which nothing can change."[409] But new information can change "geological fact" as we have seen repeatedly. Almost invariably, a scientist who claims to hold facts that "nothing can change," is proven wrong. The facts are always changing, always open to investigation and confirmation. Today, geologists believe India was transported atop a north-bound convection current. For them, this best fits the geological facts – as understood at the moment. The Meyerhoffs protested much about mobile continents, but didn't provide an alternative to explain earth history as convincingly as the mobile continent model. This resulted in slow ostracization from the earth sciences community.

Stopping the theory: as impossible as stopping the plates.
(Photo courtesy Lisa Padilla)

Nevertheless, they had a few supporters. Their 1972 paper, "The new global tectonics: Major inconsistencies,"[410] was readily published by the American Association of Petroleum Geologists, an organization that encouraged vigorous

Moving Back

opposition to plate tectonics well into the 1980s – and an organization for which Arthur Meyerhoff served as publications manager. You may recall it was this group that funded the bitter 1926 New York City conference organized to end Wegenerian chatter. The younger Meyerhoff's opposition to plate tectonics was usually seen by friendly colleagues as a devil's advocate role, but his opposition was sincere and stubborn. However, Arthur Meyerhoff realized the role of naysayer wasn't enough. He needed to add something positive to the discussion. By 1988, he proposed a creative alternative earth-model: an interconnected near-surface world-wide plumbing system that conveyed melted igneous rocks. This plumbing system, he suggested, was being misinterpreted as plate tectonics.

Meyerhoff's new idea attempted to explain some Earth phenomena in terms of mantle flow, not as convection currents, but as fluid motion within a series of channels just below the Earth's crust. Although briefly popular during the 1980s and early 1990s, his alternative hypothesis did not attract lasting converts. Interestingly, according to his daughter, who edited his posthumously published book, *Surge Tectonics*, one scientist whom Arthur Meyerhoff admired greatly was Sir Harold Jeffreys,[411] who opposed plate tectonics on the belief that the crust is too strong to deform. Sir Jeffreys was the mathematician-geophysicist who also had a major impact on Nixon's environment scientist, Gordon MacDonald, also unyieldingly opposed to plate tectonics theory.

The same year the 1972 Meyerhoff and Meyerhoff paper appeared in the petroleum geologists' bulletin, Paul Wesson, at St John's College, Cambridge, listed dozens of points he said were not satisfied by plate tectonics.[412] Enumerating all the way to point number 74, says Wesson, was a way of reminding readers of the serious failings of plate tectonics. Point number 5 claimed that some pieces are often omitted from supercontinent reconstructions to improve the fit; number 11 quoted Meyerhoff saying Ice Ages provide as much evidence against continental drift as for it; point 14 repeated Sir Jeffreys and concluded that the mantle is not able to flow; point 35 declared that aspects of seafloor spreading contradicts geochemistry data; and onwards to number 74.

Wesson concludes "the continents have almost certainly not moved with respect to each other" and that those positions derived by Vine, Matthews, Morley and others from remnant magnetism "are afflicted with an unknown cause of error." In a follow-up book published by the American Association of Petroleum Geologists two years later, Wesson adds, "Study of the modern theory of continental drift – including plate tectonics, convection, seafloor spreading, palaeomagnetism, and classical evidence – shows that there are more faults with the hypothesis of drift than reasonably can be accepted. It is suggested that seafloor spreading alone be regarded as valid, and that convection and palaeomagnetism be considered doubtful."[413]

Although he was later proved wrong about most of these assertions, Wesson argued good science. Some critics of plate tectonics are not so disciplined. Just as there continue to be people who believe the Earth is a flat disc and others who believe the planet is 6,000 years old, there are those who hold other fascinating beliefs about the planet's structure and evolution. One can not argue with those who are intent on dissenting for religious or philosophical reasons. In the end, science skeptics can simply ignore science-based evidence as planted by heretic forces intent upon testing one's faith or ideology. It is quite possible that the universe was created just a few thousand years ago. Or perhaps the universe was created a single moment ago, with all the trappings of a more lived-in space. If a deity can create a universe, surely creating it with fossils in place is not a problem. We can happily tolerate such philosophical or religious speculation if it occupies non-scientific realms of thought. It adds to the diversity and colour of our society. But, in this book at least, we have been attempting to determine how people have observed natural phenomena and have tried to explain the universe without requiring Greek gods (or any others) to be rolled upon the stage of science. Matters of faith are outside the realm of scientific inquiry, scientific methodology, and are incapable of proof, testing, or revision. They are, after all, matters of faith.

Among scientists, there continue to be bright geophysicists who question the mechanism of plate tectonics and unwaveringly reject crustal movement. But, for the foreseeable future, nothing explains the Earth – its sea basins and mountain ranges – as neatly as plate tectonics. Thoughtful scientists will address presumed failings in tectonics and tweak the theory. Perhaps one day evidence will require the rejection of the entire plate tectonics model. Meanwhile the new *status quo* (which favours plate movement) will fortify and protect its position, only reluctantly yielding space to enlightened contrarians. However, there are also rare scientists like Jason Morgan. We shall soon see that his work is the final bit of research that completed the drift revolution. The last dab of polish on the theory. Morgan became world-renowned for his mathematical description of plate motion. A colleague once asked Morgan what he could possibly do about plate tectonics to make an even greater name for himself. "I don't know. Prove it wrong, I guess."[414]

20

Moving On

    We have nearly unravelled the mystery of the mountains. We have found that plate tectonics makes continents collide and builds mountain ranges, But the seafloor is the really active component. This validates Professor Richard Field's 1936 comment that the problem with geology was its study had been confined to dry land for centuries: "you could not expect to have sensible views about the Earth if you studied only one third of its surface."[415] Within a generation, the seafloor was revealed, the great mysteries of the planet's structures laid bare. However, we still have one more piece of the puzzle to consider. Actually, several dozen pieces, but they are all cut from the same material: the plates. Plate tectonics would not be complete without plates, and until now we really haven't considered those crusty units as the ponderous pieces they are.

    An important and long-lasting criticism of plate tectonics was voiced by Sir Jeffreys, the Meyerhoffs, and Gordon MacDonald: Can the crust be weak enough to split yet strong enough to move without disintegrating? The answer lies in the rigid crustal units, the plates. Despite billions of years of pushing and shoving and squeezing and stretching, some portions of crust remain stubbornly cohesive.

    Billion-year-old chunks of continental shield in Canada, Brazil, Australia, the Baltics, Ethiopia, even Antarctica, are remnants of the planet's earliest cooled surface. It is among these shields that primordial Earth still exists. The most ancient of all, at about 4.4 billion years, is a bit of amphibolite that was found along Canada's Hudson Bay coast in Quebec by scientists from Montreal's McGill University. A few older crystals have also been discovered – zircon molecules in Australia, for example – but the Canadian amphibolite is the oldest extensive rock, the oldest piece of Earth that geologists can touch, hold, stand upon.[416] Or chip with hammer and chisel, if they must. This mottled and twisted rock, and others like it, has tenaciously held together through the world's violent episodes of heat, pressure, and erosion, all the while remaining in the upper crust. They formed shortly after the Moon's birth; life may have begun nearby. Theirs is a testament to the strength, longevity, and rigidity that seem to be evidence against plate tectonics. And yet these, too, support the story of the mobile continents.

The Mountain Mystery

Among the first scientists to explain how the crust can be both tenacious and malleable was Jason Morgan.[i] "The theory of plate tectonics he published in 1968 is one of the major milestones of US science in the 20th century," said Anthony Dahlen, Chair of the Princeton Department of Geosciences.[417] Morgan's significant paper with his milestone theory, "Rises, Trenches, Great Faults, and Crustal Blocks," was published in *Journal of Geophysical Research*, March 1968. Morgan carefully builds the case for plate tectonics through 24 comprehensively unified pages, but his emphasis was on crustal blocks, as that was the one remaining piece of plate tectonics that still needed to be resolved. In his thesis, the soft-spoken Georgian presents a geometric framework "to describe present day continental drift." Morgan proposes that "the Earth's surface is made of a number of rigid crustal blocks. It is assumed that there is no stretching, folding, or distortion of any kind within a given block."[418]

On a spherical surface such as the Earth, one broken rounded piece floating on the sphere can only move according to some rather old laws of physics. For the planet's plates, any floating motion over the mantle is controlled by a theorem first proposed by Leonhard Euler in 1776. Morgan dusted off the angular rotation idea of the Swiss mathematician, recognizing that Euler's Theorem could define the movement of tectonic plates, allow the reconstruction of ancient continents, and even predict future movements. The geophysicists loved Morgan's proposal – testable predictions and mathematics are the meat and potatoes of modern science. And it was useful to invoke the name of Euler, the creator of graph theory, infinitesimal calculus, the mathematical function, and the system used to compound the interest owed on your mortgage. The great Euler lent a measure of credibility to the new theory. Most significantly, Morgan's model of rigid rotating plates worked.

It was Jason Morgan's foresight that divided the crust into units that move as rigid blocks. He pointed out that Euler's Theorem states a block on a sphere can be moved to any other position by a single rotation about a properly chosen axis. It's not so different from spinning a baseball cap, keeping one point in contact with the head, whereby one can assume the rakishly charming look of a young gentleman with his sun bill pointing backwards. Morgan's combination of poles, angular velocity, and vectors define plate motion. This is exceedingly important. It relates the various plates to each other, allowing predictions of the planet's future look. Morgan gives dozens of plate boundaries in his paper. He concludes that "the evidence presented here favours the existence of large 'rigid' blocks of crust. That continental units have this rigidity has been implicit in the concept of continental drift. That large oceanic regions should also have this rigidity is perhaps unexpected."[419]

Morgan's landmark paper was published in 1968, but he first publicly aired his idea at a meeting of the American Geophysical Union a year earlier. He was

---

i   W. Jason Morgan (1935), American geophysicist.

Moving On

scheduled to talk about his PhD thesis on gravity measurements along the Puerto Rico Trench. But he switched topics, which is not uncommon for a speaker at such conferences. Typically, months pass from the time presenters register their intent to participate and the actual event. Months are a long bit of time when new science is unfolding. Nevertheless, it becomes important in this case because of a serious controversy that arose immediately following the geophysical conference.

A fast-rising star in the world of earth science also attended the meeting. Dan McKenzie[i] claims he did not hear Jason Morgan's talk about plate tectonics – he expected a gravity study of the Puerto Rico trench to be the topic. Puerto Rico did not interest him, so he walked out of the room just as Morgan began to speak. A few months later McKenzie published a paper very similar to Morgan's presentation. McKenzie said he himself had been working on the same motion theorem of rigid plates that Morgan had disclosed. There were immediate accusations that McKenzie was indeed at the presentation and had been persuaded by what he had heard. Perhaps, but it would not be unusual for two bright scientists to create the same theory independently. We saw it before, with the Canadian Lawrence Morley's failed attempts to publish his explanation of the magnetic stripes on the seafloor and in Robert Dietz's publication of the ocean's commotion, a paper that presented many of Harry Hess's thoughts. Dan McKenzie published a few months after Jason Morgan's talk – and a few months before Morgan's own paper was printed. And so again, the hugely important issue of priority of publication arose with supporters and detractors on both sides. Hugely important to the scientists involved, at least.

It remains a debated issue: the soft-spoken American, Morgan, certainly presented his idea before the visiting British scientist McKenzie published his work. And copies of Morgan's talk were widely circulated. But McKenzie may have been working on the same idea. McKenzie, from a high-achieving family, clearly wanted to create a name in science. His father was a gifted surgeon, but not as financially successful as his grandfather, also a doctor, who had a chauffeur who drove the elderly doctor to work every day from a posh Highgate estate. McKenzie's own father could only afford to rent a flat above some examination rooms. Dan McKenzie's mother came from a poor northern England family where her father earned his living grimly shovelling coal into the furnace of a power generator. But McKenzie's mother, Nan Fairbrother, benefited from a British scholarship program which awarded a free university education to the country's brightest 200 people each year. She won a scholarship, completed college, and made a brilliant career as a writer. Nan Fairbrother's books included *New Lives, New Landscapes*, which is considered a visionary look at land use in England. Dan McKenzie's own acceptance into university was also earned by merit – he says he faced an oral examination where the Cambridge interviewer asked him about Dostoyevsky and British wild orchids, on the strength of which he was

---

[i] Dan McKenzie (1942), British geophysicist.

admitted to study physics.[420]

As a 1966 graduate student of Sir Edward Bullard, McKenzie wrote a doctoral dissertation about the gravitational shape of the Earth, then worked on parts of the plate tectonics hypothesis for the next six years. Richard Fortey, in his excellent *Earth: An Intimate History*, writes of his encounters with McKenzie at Cambridge. He describes McKenzie as a rock star being flown to California even as a freshly minted PhD, so important was his work.[421]

McKenzie remembers it a bit differently. In an interview with Alan Macfarlane, in 2007, McKenzie recalls that he heard there was plenty of money for research in the States, and the American Navy was extremely interested in anything to do with deep sea oceanography. McKenzie relates that he knew nothing about how to go to the USA and get paid, so he went to the American Embassy in London and was handed an immigration visa. What McKenzie did not realize was that a valid US Immigration Visa made him immediately eligible for the American draft. This was at the height of the Vietnam war and he soon received draftee registration papers. Previously, in Britain, McKenzie had been politically active in what he described as the fairly violent left wing, "so I got on a plane and came home," he said.[422] He later returned, briefly, to attend the conference at which Jason Morgan spoke about the planet's plates and their relative motions. It is unfortunate Dan McKenzie missed the talk as Jason Morgan's presentation has been heralded as the final summary, or synthesis, of the entire plate tectonics theory.

Meanwhile, Dan McKenzie, with co-author Robert Ladislav Parker, published a version of plate motion in December 1967. With "The North Pacific: An example of tectonics on a sphere," they invoked Euler's Fixed Point Theorem and credited the earlier work of Cambridge's Bullard and Toronto's Tuzo Wilson, with "the essential additional hypothesis being that individual aseismic areas move as rigid plates on the surface of a sphere."[423] Matthews and Vine, Hess, and Sykes are also referenced – but not Morgan. McKenzie's paper was nevertheless good science. The authors limited the planet's tectonic plates to slabs where the insides are solid and not prone to earthquakes but the boundaries are seismically active. They also placed a limit on the total number of possible plates. As an interesting side-note, although McKenzie later claimed credit for the first published use of the phrase, his paper uses the words *plates* and *tectonics*, but not the phrase *plate tectonics*. This paper instead uses the lovely expression *paving stones* to describe crustal units, but that descriptive phrase never caught on.

Following publication of McKenzie and Parker's paving stones paper, the allegations began that McKenzie borrowed ideas from Morgan's spring 1967 presentation. After that pivotal conference, outlines with sketches and talking points were passed out by Morgan to various geophysicists, including a close associate of McKenzie's. But McKenzie, who began writing his paper a month after Morgan's talk, says he never saw the talk, the outline, nor the sketches.

Moving On

Xavier Le Pichon, a highly respected geophysicist who was also researching the way rigid plates slip around, felt is was "astonishing that McKenzie twice missed the opportunity to learn about Morgan's model."[424]

All of this created an obviously uncomfortable situation for the young British geophysicist. His paper was published in *Nature* in December 1967, while Morgan's virtually identical conclusion was printed in 1968.[425] Morgan seems to have taken the entire episode calmly, and even agreed to write a paper with McKenzie. Together they released a new study, the enticingly named "Evolution of Triple Junctions," which coordinated the two scientists' talents. In that joint 1969 paper, continental drift is referred by its modern name. They write, "the Earth's surface moves as rigid spherical caps and for this reason is often called '*plate tectonics*' . . . for the purposes of plate tectonics, the surface of the Earth is completely covered by a mosaic of interlocking plates in relative motion."[426]

In 2002, Jason Morgan was awarded America's National Medal of Science, the country's greatest science honour. According to President George W. Bush, who presented the citation at the White House, Jason Morgan was recognized "for development of the theories of plate tectonics and of deep mantle plumes which have revolutionized our understanding of the geological forces that control the Earth's crust and deep interior and consequently influence the evolution of the Earth's life and climate."[427]

Bush awards Morgan the National Medal of Science.

Princeton's Chair of Geosciences, Anthony Dahlen, credited Morgan's work further, saying, "Essentially all of the research in solid-earth geophysical sciences in the past 30 to 35 years has been firmly grounded upon Jason Morgan's plate tectonics theory. The scientific careers of a generation of geologists and geophysicists have been founded upon his landmark 1968 paper."[428]

Jason Morgan has continued to teach and to research the planet's crust, 50 years after his innovative use of Euler's math first solved the problem of rigid yet mobile plates. In October 2012, one of his most recent adventures took him to the

277

Indian Ocean where he observed an experiment to test the idea of plumes. Morgan had once taken Tuzo Wilson's original plume vision and given it further substance and character. His 1971 research paper described the plumes' manners and mechanics more fully, but in the intervening fifty years, plume existence had not been irrevocably confirmed. Sailing near Réunion Island with a group of scientists, the 77-year-old blogged about the efforts that go into team research. But he also expressed his continued sense of awe, ending his blog entry by writing "last night the sky was brilliant; a bright Milky Way and the bow of the ship pointing directly toward the Southern Cross."[429] He had spent his life unravelling the Earth's deep secrets, and like all men of science, he was fully enraptured by the planet's place in the cosmos.

Conversations with the Earth

Jason Morgan had established the model for plate tectonics and marked out many of the busy zones of plate contact. But another geophysicist, Xavier Le Pichon,[i] complemented Morgan's work by dividing the globe into its essential plates, drawing their borders, and producing the first true map of the continents and oceans as they really exist in their dozen or so major pieces. Le Pichon says that Morgan's 1967 talk at the American Geophysical Union in Washington, D.C., was the turning point in Le Pichon's career. Not that he understood much of what Morgan was talking about. "Jason has a special gift for disorienting his listeners and this gift was especially well displayed on that occasion. Apparently nobody, including myself, understood the importance of what he discussed then," wrote Le Pichon.[430] The real turning point from the meeting was the lengthy speaker's notes Morgan distributed. "As soon as I read it, I realized the importance of what he developed and dropped everything else I was doing."[431] Le Pichon immediately switched from physics to geophysics. He was 29. His life's work had started.

Le Pichon was born to French parents in Quy Nhon, a Vietnamese port city in French Indochina. His father managed a rubber plantation. At first it was an idealized childhood of tropical luxury. He said that as a child he first felt a passion to understand the Earth, the soil that grew the bamboo from which his home was thatched, and especially to know what was yet deeper, far below his feet. He later explained that he hoped, even as a child, he could somehow enter into a conversation with the planet, asking it questions and listening to its answers, as if on equal terms with a living ageless creature. It may have started as an idyllic childhood, but innocence ended abruptly. World War II began shortly after he was born. The Japanese would have invaded Vietnam, but France had surrendered to Germany, so French-controlled Vietnam was granted a truce with Japan. It was an uneasy peace which fell apart when Germany began to lose the war. The Japanese seized Indochina. Le Pichon's family, along with all French nationals, were

---

i   Xavier Le Pichon (1937), French geophysicist and humanitarian.

Moving On

suddenly enemies of Japan. The Japanese forced the French settlers into prison camps. Survivors described torture and starvation. Xavier Le Pichon was nine years old when released. After the war, the family moved to France.

At age 22, at the University of Caen in Lower Normandy, Le Pichon graduated with a physics degree. A few years later, he earned a doctorate trying to prove that the universal gravitational constant is not constant. An obsure, theoretical field, even for physicists. He shifted his studies towards the Earth, and attended the fortuitous meeting where Morgan presented his theories of the planet's plates. Le Pichon realized the huge significance of the new way the planet was being viewed. He wanted to be a part of that new science. After immersing himself in Morgan's work, Le Pichon divided the globe into six rigid plates, calculating their locations and rates of movement from Euler poles combined with palaeomagnetic and topographic data. It was a complicated amalgamation of extremely different types of data, but Le Pichon simplified the project and made a brave assertion: the rigid plates combine both continental and oceanic crust in some of the units. He drew a map – the first ever with plate boundaries – by boldly ignoring the ocean's water. His North American plate, for example, stretched from the centre of Iceland, across the western Atlantic Ocean, under New York City and Chicago, then nearly across the American continent. This was an important insight – geologists had just spent twenty years getting used to the extreme differences between oceanic and continental material, now it appeared these were sometimes fused together in an unexpected way. His map sparked the imagination of enthusiasts outside the field of earth physics and presented the public with a quick

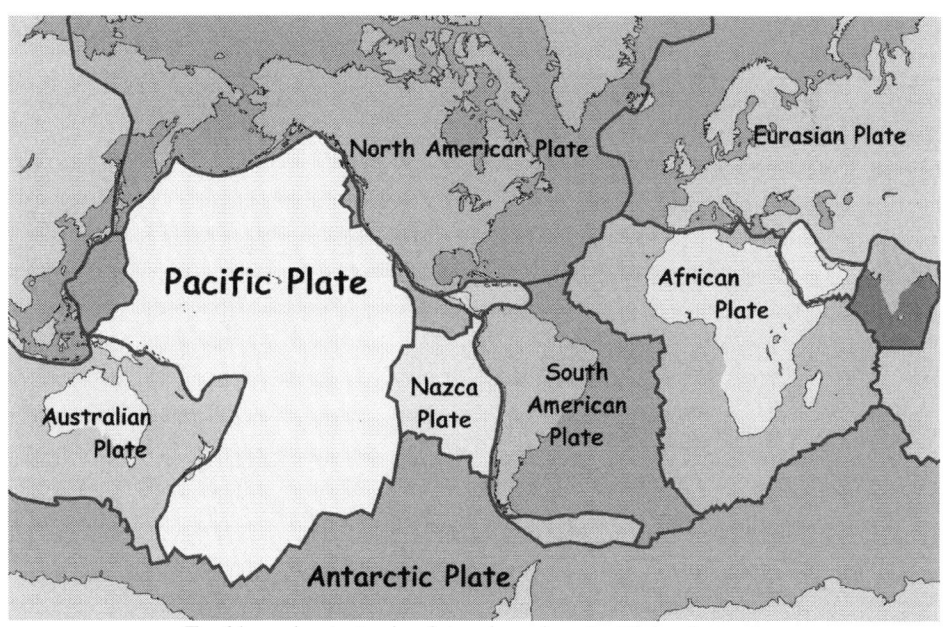

Earth's major tectonic plates, based on Le Pichon's work

summary of the new view of the world. Now in his thirties, Le Pichon was one of the most famous of all scientists, interviewed, toasted, respected. He and Earth were immersed in the intimate dialogue he had sought since childhood. Then, at age 36, Xavier Le Pichon did something truly unexpected.

Le Pichon abruptly resigned from all his positions on committees and at universities, and moved to Calcutta to work with Mother Teresa. He said he had found himself drawn deeply into understanding the planet, but at the same time, he felt unaware of the world of people. However, after six months in India, he was convinced by a friend, a priest, that although he should continue his commitment to help poor, suffering, and disabled people, he should also return to his work in France – as a geophysicist. His friend pointed out that he could help fragile and broken people but also continue his conversation with the fragile Earth. As a result of that advice, Le Pichon returned to France. For the next 40 years, he lived with his wife and six children in a large community house that integrated mentally challenged adults in a family home. His foster home was part of the L'Arche Program, which he helped found, and which now has over a hundred communities scattered around the world. Partly to explain his commitment to the disadvantaged, he told journalist Krista Tippett in a 2009 interview, "Life has an extreme diversity and in this diversity is its richness."[432]

Le Pichon is as passionate about his role in society as he is about his study of the evolution of the planet. His work remains at the heart of geophysics. Morgan showed how plates might move; Le Pichon showed us what those plates are – their extent, their territory. They are little nations, safe and solid at their cores, jagged and fiery at their edges where earthquakes and volcanoes mark boundaries with neighbouring plates. At first, the French geophysicist defined just six basic tectonic plates.[433] But Le Pichon pointed out that just as the complexity of human exchanges increases with understanding, more data about the Earth reveals more details. It's part of the conversation. Scientists eventually described 40 more plates. To reduce confusion, these are divided into primary, secondary, and tertiary groups. In addition to the massive primary units which include the North American and Pacific plates noted by Le Pichon, there are also minor slivers with names such as Nubian Plate, Anatolian Plate, and (not too surprisingly) the South Sandwich Plate. Defining all the individual solid discrete plates and using math theorems to anticipate how these should move were the final touches on the world-encompassing plate tectonics theory.

Where We Are

When the 1960s began, most geologists believed the surface of the Earth was solid and immobile. Complex but static. By the end of the 1960s, nearly all accepted that mantle convection currents displace broken bits of crust that are never at rest. The plates bounce, shake, jiggle, jostle, churn, rotate, and collide,

Moving On

forming mountains, plains, basins – all the impressive features we see on our planet. There was not much left to do, except to refine the general theory and find applications for the lofty new view of the Earth. J.J. Thomson, who was awarded the Nobel Prize for discovering the electron, said, "Applied science makes improvements; pure science makes revolutions."[434] The creative pure science was now secure; it was time to discover applications and make improvements. And, as scientists must, to highlight errors and inconsistencies in the great new model. These have been the centre of scientists' attention for the past 50 years.

Few geologists now doubt that continents have wanderlust. Along with the seismic noises, the magnetic striping, the matching rock outcrops and the ancient fossils, we can add maps drawn from radio interferometry and satellite data. In 1985, the proof that Alfred Wegener insisted would come by objectively measuring continental drift finally arrived. It had taken fifty years, but plate movement was finally observed when the distance between radio telescopes in Westford, Massachusetts, and Onsala, Sweden, increased a few millimetres.[435] The continents were measurably and most assuredly drifting apart. Dynamic Earth was reality; there could be no return to the stability of the past. Today, NASA's Jet Propulsion Laboratory, at the California Institute of Technology, operates 2,000 ground station receivers merging location data from 30 satellites. NASA plots the plates in motion.[436] This final proof of plate tectonics is complicated engineering, but not too different from the equipment that tells the world where your cell phone is at all times.

Development of the Global Positioning System (GPS) started in the 1960s and became fully operational in 1992. Because of technology limitations and military secrecy, accuracy with publicly-available devices before May 2000 was limited to 100 metres; now it approaches a centimetre when readings from multiple satellites are averaged. Compared to navigation just a hundred years ago, which relied on espying a nearby coastline and spotting celestial bodies with sextants, location information is now a million times more accurate. The new science includes pseudo-range and phase measurements, spectrum shifts, and complicated signal analysis. Some of the most stringent applications arise from meteorology (to track water vapour in the atmosphere and sea level rise) and geophysics (to measure creeping continental drift and deformation leading to earthquakes). Freeing the use of GPS data from military control to civilian use was an act of considerable, albeit belated, enlightenment. The enlightenment came by order of President Reagan a few months after a Soviet fighter destroyed a Korean passenger jet flying from New York City to Seoul in 1983. The commercial flight had drifted slightly into Russian airspace. It was spying, claimed the Russians. GPS data was not available to the Korean pilot. An Su-15 interceptor killed 269 travellers, blasting the airliner into the Sea of Japan. Reagan's aides quickly realized civilian use of GPS would have saved those lives.

Liberating this information for civilian use has allowed the development of the hugely effective mapping and navigation systems common today. GPS dog tags can report lost Fidos; trucking companies can dismiss drivers parked in front of pubs; male drivers can find destinations without asking for help. All of this contrasts sharply with the secrecy of maps and location information of the past. I have dealt with geophysical maps produced by Kazakh, Russian, and Bulgarian scientists during the 1960s and 70s. On those older maps, coordinates are often missing or encrypted – the idea of keeping Cold War data secure resulted in much of it becoming useless. I have spent dozens of hours repositioning such secretive maps by linking data to landmarks such as river bends and cemeteries, creatively georectifying the old maps, nudging them reluctantly into the twenty-first century. It becomes a much more progressive world when data are openly shared.

Today, hundreds of GPS stations track plate movements. Most of the drift is so slight as to be inconsequential in a human lifetime. Nearly all motion associated with the Atlantic Ocean occurs at about 2 centimetres (less than an inch) per year. Your hair probably grows faster. Since Leif Erikson's Vikings founded L'Anse aux Meadows, Newfoundland, there now exists an extra 20 metres of ocean separating Canada from Europe. If Vikings made a sentimental return journey from Gokstad on Norway's Sandefjord to their old settlement in Newfoundland, their 30-metre knarr ship would require the additional oarsman's energy of 1,200 kilocalories to recreate the trip. Coincidentally, that's the amount of energy in one can of Spam.[i] Thus, in the thousand years of human history which produced the Crusades, the Mona Lisa, the steam engine, electronics, and flights to the moon, the continents have parted scarcely a quarter of a soccer pitch. One thousand years, twenty metres. In the past million years, the Atlantic Ocean widened 20 kilometres, the distance you might drive in a quarter hour to see your grandmother. It has taken unfathomable time to shape this planet. Except in rare cases, we do not live long enough to experience the ride. But there are some exceptions.

Most of the major shapes on this planet are connected to the slow shifting of continents. Some plates, especially those abutting the Pacific, move much faster than the norm. The mid-sized Nazca Plate is forcing itself under South America at a rate exceeding 8 centimetres a year. But plate movements are not steady. These are not the progressions of plodding oxen, but rather the leaps of a kangaroo. Sometimes the abrupt jumps are astonishing. I narrowly missed experiencing a violent three-metre crustal leap during the 2010 Chilean earthquake. Near the epicentre, the town of Concepción moved the length of a car in a single lurch as one tectonic plate climbed over another.

Though many of us had been expecting this enormous Chilean earthquake, missing it by a week was not part of my travel plans. I had been teaching geophysics in Peru on a short project for the Canadian government. I extended my South American trip to include a few days visiting a Chilean friend, a man who

---

i Yes, this is an allusion to Monty Python. If you find this baffling, you can certainly find an internet video relating Vikings, Spam, and the flying circus.

Moving On

made a living with honey bees. Francisco Rey keeps several thousand colonies in a valley north of Santiago where his main work involves using bees to pollinate avocado trees. To make his work easier, and to keep vicious indigenous ants out of his hives, the beekeeper keeps his colonies perched atop posts. While visiting, I wondered aloud about those stilts that boosted the wooden hives a metre above the ground. Would the hives be broken and scattered in an earthquake? In 1960, the Chilean coast near Francisco's apiaries suffered the strongest earthquake ever recorded on Earth.[437] The week before the 2010 earthquake, I was in Chile's smoggy capital, up on the third-floor balcony of a yellowed, stucco-sided museum that had once been a government office. From that balcony, I felt earthquake tremors – slight, low-frequency vibrations under my feet. The sort of thing you might feel in an old building if a train lumbered past. But there were no trains, no big trucks. Just gently swaying palm trees in the plaza. It was a minor tremor, just a bit stronger than the type that occurred almost every day in Chile's great central valley. To describe it as a precursor, or warning, that a big earthquake was a few days away would be misleading. I left Chile just before the killer earthquake struck. In the early morning of February 27, 2010, the oceanic tectonic plate disappearing under South America took a deep dive. A thousand kilometre stretch of the Nazca Plate jerked into the subduction zone, pushing further under South America. Meanwhile, parts of the South American tectonic plate lurched west, up over that wedge of ocean crust.[438] The total energy of the 2010 Chilean earthquake was equal to 240 million tonnes of TNT. The energy of 20,000 Hiroshima bombs, released in a single minute.

Chile's second largest city, Concepción, was most severely shaken. The whole city, in one big piece, was conveyed along its thrust fault. The earthquake shifted a million people, their houses, cars, fire stations, schools, roads, trees, parks – in one violent motion. If you lived in Concepción, you'd think your neighbour's house was where it was the previous Friday evening. But you'd be wrong. According to GPS measurements conducted by geophysicists from Ohio State, the entire city had lurched a full three metres.[439] If you had the misfortune (and talent) to jump high into the air the moment the earthquake struck (and stayed aloft half a minute), you would have come down in your neighbour's yard. Well, actually, you would have dropped where you started, but the neighbourhood would have moved beneath your feet.

Meanwhile, three hundred kilometres away, Santiago's five million people were lifted and dropped about a third of a metre (one foot) west. Apartment buildings were split open, as were 125 million bottles of wine, their contents washed away – in some towns "the streets ran red" with wine.[440] At Magnitude 8.8, the 2010 Chilean event was the sixth largest ever recorded in the world. More than ten times stronger than that year's devastating Haitian earthquake, stronger even than a quake in China that killed over a half million in 1976. Chile lost about 600 people. Damage was nearly 10 billion dollars, not counting 40 billion dollars in lost production, wages, and uninsured losses.

All over the country, beehives were lifted and dropped from their pallets and perches. Neatly stacked tanks of honey were tossed and knocked on their sides – in one shop I'd visited, a million kilograms (over two million pounds) of Chilean honey were spilled on the tidy concrete floors. Francisco Rey's beehives were knocked to the ground. The beekeeper told me hundreds of his colonies were thrown from those little posts, the perches that made beekeeping easier and kept hungry ants at bay. On a branch in a nearby eucalyptus tree, disoriented worker bees were attracted by the pheromone of dislodged queens. Bees gathered by the hundreds of thousand, clinging to the tree above damaged boxes while Francisco and his beekeepers reassembled the carnage. Bees, honey barrels, wine bottles, apartment buildings, fallen bridges, and grieving families – the side-effects of the restless planet as the Earth's oceans widen. We have learned much about the planet's mobile crust – but earthquake predictions continue to elude us.

Hot Spots

In the same way advocates of the Flat Earth were finally defeated when humans took to space and could see the planet as a rotating orb, scientists could not refute satellite data showing the continents in motion. The continents move, and at rates and directions anticipated by geophysicists in the 1960s. However, much remains to be debated. Historically, scientists have clung to old theories, continually tweaking them to account for new observations that might unseat old notions. Finally, antiquated models such as an earth-centred universe become so cumbersome that a Copernican revolution is required to simplify the system. With earth science it was land bridges, contractions, and spontaneously regurgitated continents that could no longer explain faunal distribution and mountain building except with more and more tweaks which became less and less believable. Perhaps plate tectonics will one day be the cumbersome old dogma that needs to be replaced. Perhaps the speculated convection currents in the mantle will fail to exist. At some point, plate movement itself may require inconvenient adjuncts to sustain it. Facts may be unearthed that contradict tectonics as it is now described and an entirely new approach will be needed to understand the planet's physical evolution. Already there are many working on clever replacements.

A startling number of scientific papers reject the plate tectonics model. This may be an indication that nuances are wrong – or perhaps the entire theory. Likely not enough information has been found to explain some of the apparent irregularities. You may remember back to the very brilliant but very wrong Lord Kelvin who could not conceive of an Earth much over a few million years old – the Sun couldn't possibly last that long, it would burn out, he said. He fervently believed the planet's heat would have been long extinguished, the world merely a frozen sterile rock, if it were truly ancient. Kelvin also purportedly declared an end to the science of physics, believing everything important had been discovered. Scientists, he believed, would engineer their way to more and more precise measurements of all the existing particles and processes but nothing really new awaited discovery. If this remark is correctly attributed to him, as seems likely, it was presented with extremely poor timing. Within months of uttering this sentiment, nuclear energy explained the enduring solar heat, permitting a much more ancient Earth and Sun. Einstein demonstrated mass is energy, Heisenberg introduced randomness to quantum mechanics, and physics became a new world of unexpectedly creative excitement, no longer simply the realm of tinkerers measuring nuances.

Until the next earth science revolution, mainstream scientists will continue tweaking their ideas of moving plates, plumes, and hotspots. It is in plume theory that the history of continents may yet be rewritten. Formulated in 1963 in the imagination of Canadian geophysicist Tuzo Wilson while on his Hawaiian stop-over, plumes explained the Pacific island chain as the result of the slow drift of a tectonic plate above a fixed hotspot. A few years later, in 1971, Jason Morgan expanded on the idea, suggesting plumes, or narrow streams of hot mantle, rise from a 2,500 kilometre depth, near the core-mantle boundary. Some have pictured plumes as long-stemmed mushroom blobs, strings of molten rock with bulbous heads. The overwhelming heat of a plume – like a blowtorch under a sliding sheet of plastic – forces the ocean crust to bubble up as it slides over the fixed hotspot spewing its endless ribbon of super-hot mantle.

Morgan, at Princeton, expanded plume theory to include other hotspots far beyond Hawaii. For example, a mantle plume now under Réunion Island, east of Madagascar, probably began the cataclysmic creation of India's basaltic Deccan Plateau. This started when India was near Africa, before it began its long journey to Asia. Basalt flooded across one million square kilometres of the future subcontinent, flowing out of an enormous plume while the crust ambled across it. That protracted event – the explosive flood of basalt – killed everything within an area three times larger than France. Its occurrence 65 million years ago smoked the Earth – the event coincides with dinosaur extinction. Many palaeontologists surmise the darkened skies and poisoned gases from the erupting plume encouraged the dinosaur demise. Geologists have followed a 5,500 kilometre scar in the Earth across the Indian Ocean, retracing the subcontinent's path through those millions of years from its former location near east-coast Africa. However, within two years of Morgan's plume hypothesis, the existence of permanently fixed torches searing the underbelly of the crust was being questioned by Peter Molnar and Tanya Atwater. In particular, they concluded that the hotspots they examined – including Hawaii, the Indian Ocean's Réunion, Iceland, and Yellowstone – are not strictly fixed spots, but are themselves in motion.[441] Likely aspects of both ideas are correct, although the second notion has since become less fashionable.

The investigation of the ephemeral qualities of the world's dozens of suspected plumes continues. Recently, Qin Cao at Massachusetts Institute of Technology has been using seismic data to map the roots of Hawaii. Where Wilson theorized a localized hotspot, Morgan drew in a thin mantle plume, and Molnar/Atwater suggested a possibly drifting plume, Cao and her team think they have discovered a much more shallow heat source a thousand kilometres west of the islands. It may be feeding melted rock to the Hawaiian volcanoes along a relatively shallow pathway.[442] This could be similar to Arthur Meyerhoff's network of surge channels. By using earthquake-generated seismic waves that pass upwards towards the underside of mantle anomalies, Cao believes she can map different mineralization zones within the mantle. Areas with alternative mineral accumulations signal different heat and pressure regions. This may clarify her suggestion of an entirely new source for hotspot magma.

Other scientists propose hotspots are not over-heated, over-pressurized plumes but instead result from thin crustal zones where the seafloor has stretched and cracked, allowing magma to burst through, giving the illusion of an underlying hotspot. In this scenario, the Hawaiian-Emperor chain is formed from the stress of the Pacific Plate cooling. It contains nothing hotter than the usual scorching melted internal organs of our planet. In this theory those inorganic organs break through the surface because the crustal skin is locally weak.[443]

Decades after mantle plumes were first proposed, and long after they were claimed to explain Hawaii's islands and India's Deccan Plateau, we are still not altogether convinced plumes even exist. To finally learn if deeply-derived streams of hot rock actually rise from near the Earth's core, French and German researchers created the deliciously named RHUM-RUM – the Réunion Hotspot and Upper Mantle – Réunions Unterer Mantel Study. Scientists placed 57 strategically scattered seismometers on the floor of the Indian Ocean plus another 37 on land. They covered a vast area around Réunion Island's theorized mantle plume, the one that may have killed dinosaurs when it began flooding pre-India with basalt. The RHUM-RUM ocean-bottom seismometers were placed on the seafloor around the Mascarene islands and on Mauritius, La Réunion, and Madagascar. It's the largest experiment ever of this sort. All of the ocean-bottom sensors were retrieved in November 2013 and data analysis has begun.[444] The goal is to confirm the existence of the Deccan-Réunion plume. The results might also confirm the possible existence of other plumes elsewhere on the globe.

Karin Sigloch, a geophysicist working on RHUM-RUM, says it may take two years to interpret the data. She told me much of plate tectonics theory is being examined more closely. "We have yet to make the quantitative links to the convecting, viscous mantle that causes the plates to move and the mountains to form. The driving force is in the mantle, and much of it is still waiting to be figured out!"[445] Part of that solution may come from the data she is gathering over the Indian Ocean.

Moving On

Super Gravity Meters

Other researchers are using miniscule variations in gravity to attempt indirect detection of mantle plumes and convection currents. If less dense mantle is rising, it might be spotted as a gravity anomaly within its more dense neighbourhood. These are subtle changes in the density of material hundreds of kilometres below the planet's surface, approaching the Earth's core. It may be surprising that machines exist which can sense such nuances. I enjoyed a peek at a superconducting gravimeter when I played an exceedingly minor role in analyzing its data. An extraordinarily expensive (and extraordinarily accurate) machine, the Canadian Superconducting Gravimeter was employed to search for exactly such tiny gravity variations. The gravimeter was located in Quebec, just north of Ottawa, and shared by universities across Canada. From 1997 to 2006, my geophysics adviser, Jim Merriam, was Principal Investigator with the facility. When I saw the big blue device, I was mostly seeing a tank of liquid helium about the size of a rain barrel. Its insulating cage had been removed, and I could see the machine was resting on a concrete pier bolted to the planet, attached directly to raw Precambrian granite, part of the Canadian Shield underlying the gravimeter's building. If the crust of the Earth moved, the gravimeter would move. Otherwise, the machine was firmly fixed.

At the heart of the device, a hollow two-centimetre sphere of pure niobium floated in a permanent magnetic field. The temperature of the device was kept four degrees above absolute zero, near the point where all molecular motion freezes and all physics goes berserk. Such an icy disposition keeps the tool's magnetic field stable. Richard Warburton, whose firm makes the device, told me that an electric current flowing through two supercooled niobium wires produces a magnetic field that flows almost forever, hence it is a *super*-conducting system.[446] Observing the superconducting gravimeter as it sensed the inner Earth was slightly unnerving. The tool represents the zenith in technology, a pinnacle of human intellectual efforts. Yet, it is a rain barrel with tubes and wires. A speck of a machine taking the pulse of a planet a million trillion times larger.

However, this particular rain barrel is indeed a complicated amalgamation of engineering. The gravimeter's floating sphere of pure niobium, a rare grey-blue element, is extremely temperature stable, which means it doesn't shrink much, even in its frigid super-conducting environment. The niobium sphere floats inside a magnetic field that exactly balances the force of gravity. Approaching the sensor, my body-mass creates a tiny gravitational attraction on the little floating ball, drawing it towards me with an infinitesimal tug. Simultaneously, the magnetic field adjusts, recentres the orb, and the amount of adjusting energy is recorded by a computer. Meanwhile, a truck on the highway, half a kilometre away, rumbles past, tugging the ball towards it, and away from me. The gravimeter again recentres the sphere. Such is the sensitivity of a gadget that can measure changes in gravity of a single nanogal – one trillionth of the strength of the bit of gravity we experience every day on the Earth's surface. My role working with the superconducting gravimeter did not involve adjusting the sparse control knobs on the device. It was much less dramatic.

Funded by the National Research Council of Canada, I helped calculate the effects of the weight of passing storm clouds as they lumbered near the gravity meter. Using a mathematical trick called a Green's Function and applying three-dimensional calculus, the weight of the atmosphere could be estimated as it varied with changes in the weather. A passing high pressure system depresses the surface of the Earth about two centimetres because the air is heavy. So, the granite shield holding the gravimeter sinks, bringing the machine closer to the centre of the Earth. According to Newton's laws, the gravity signal should increase, but it doesn't. Instead, simultaneously, the heavy columns of air above the gravimeter are tugging upwards on the little niobium traveller with an even stronger pull. Scientists gathered measurements of atmospheric pressure within a thousand kilometres of the observatory, calculated its tug of gravity, then removed that value from the data, gleaning the hull from the kernel of gravity. Other scientists were responsible for calculating the gravitational pulls of the planets, the Sun, and especially the Moon and its tidal effects. All this noise – much more damaging to the data than a mere waving hand or a passing truck – is calculated and removed from the raw figures, making a filtered bunch of numbers that can tell researchers something more subtle than the weather or the moon's position. It is hoped superconducting gravimeter data will help us know if plumes exist or if perhaps they are simply a convenient conjecture. The gravity data also allows us to listen to the harmonic ring tone of the planet's fluid core: the Earth's bell.

It is a bell like no other. It tolls for every serious earthquake. Our planet's core rings like a giant chime for days after a strong tremor. Instruments have recorded deep inner ringing since 1960, when the fundamental tone – repeating every 54 minutes – was observed immediately following the disastrous 1960 Chilean earthquake. Since then, other modes have been recorded with higher frequency vibrations as part of the spectrum of oscillations occurring with each new major tremor. The ringing seems restricted to the fluid outer core. Called *core modes*, such vibrations tell us something of the composition of our planet, and each year they are observed with more and more detail as the resolution of gravity meters improves. Thus, following an earthquake, the planet shudders, and a world-wide array of superconducting gravimeters responds with little niobium balls dancing in controlled magnetic fields – dancing with a periodic rhythm captured in the form of gravity reverberations. From subtle nuances, irregularities in the boundaries between the inner and outer core and the mantle are inferred and properties of the inner Earth are surmised. As the ringing dampens, the interplay of earth elasticity, stratification, rotation, and gravity are revealed as those elements work to quell the chime.

Moving On

Rise of the Mega-Blobs

This discussion about plumes and pressures and inner core vibrations may seem trivial, but consider this. The most destructive thing yet to happen on this planet will be the explosive eruption of another major hotspot. Some scientists blame an adventurous plume for the demise of the dinosaurs. Others now speculate that Yellowstone National Park sits on the blunt end of a broad puffy plume that will eventually break through the crust in an unpleasant manner. Gravity and seismic data will someday provide the news that triggers a panicked evacuation of the park – and if predictions are correct, the rest of North America as well.

So it may be that plumes play a far greater role in the design of our planet's scenery than guessed by Wilson, Morgan, or their successors. Two Canadian scientists say basic ideas about continental movement may also need to be reconsidered. In *Nature*, Alessandro Forte of the University of Western Ontario and Jerry Mitrovica of the University of Toronto published their claim that huge plumes of hot rock are floating upward from the Earth's liquid core, having a huge unexpected influence on surface movement. Forte describes two immense rising hot plumes and two equally large sinking cold blobs. The ultimate lava lamp.

The two researchers monitored earthquake waves to map the mantle at depths of up to 3,000 kilometres. They noticed earthquake energy waves travelled more slowly in two large plume-shaped regions – one below the centre of the Pacific and the other under Africa. Meanwhile, earthquake shock waves propagated more quickly through two equally vast regions under the west and east margins of the Pacific Ocean.

For Mitrovica and Forte, the slow seismic areas seemed to indicate plumes that are rising while the faster seismic regions could be dense cooler materials sinking towards the core. The seismic data has been corroborated by variations spotted with superconducting gravimeters. The scientists also determined a zone of especially high viscosity nearly 2,000 kilometres below the Earth's surface. The discovery of this very sticky layer deep in the Earth's lower mantle may mean a great deal to understanding the elusive convection currents. The scientists' maps chart upwardly mobile mega-plumes in the lower mantle, under the Pacific and Africa.[447] These have the potential to split the crust, as may be happening already in Africa's Rift Valley.

From plumes, we have advanced to blobs and megablobs, as Forte and Mitrovica named their discovery. It sounds a bit Hollywood to refer to killer plumes capable of destroying civilization as megablobs, but as Michael Manga, a pony-tailed geophysicist in his mid-forties, comments in *Nature,* the team shows that distinct megablobs contribute to an internal ebb and flow; and Forte and Mitrovica verified that the entire mantle "appears to act as a single convective system driven primarily by thermal anomalies."[448]

The Mountain Mystery

This harkens back to the first convection sketches drawn by Arthur Holmes almost a hundred years ago. Holmes demonstrated that convection was powering Alfred Wegener's drifting continents.

It takes physicists conversant in fluid mechanics to make sense of it all. A Canadian from Ottawa, Michael Manga earned his Harvard PhD by tracing the behaviour of bubbles which he propelled through different sorts of fluids. Now at UCLA's Berkeley campus, the young geophysicist has been on a stellar path. It was probably his 2005 recognition as a MacArthur Fellow with its $500,000 "Genius Grant" that brought him to the attention of *People* magazine.

People placed Manga in their 2005 "Smart Guys" section as one of the world's sexiest men. As a busy professor, Manga has a full schedule. He had little time for the photo shoots that led to sharing a glossy page with Bono. But Manga explains, "I wanted to get information out to people who wouldn't normally hear or see anything about science."[449] He wanted people to know that "it is OK to be a scientist."[450]

Manga, a geophysicist on a stellar path
(Photo by permission of Bonnie Powell)

Manga is a gifted lecturer and carries a heavy teaching schedule, but he has helped write hundreds of papers on fluid-related geophysics focused on volcanoes, geodynamics, and more recently, evidence of past oceans on Mars, and present oceans on Jupiter's icy satellite, Europa. There is a strong expectation that such studies will give our best insight into the Earth's evolution.

The future of geophysics is evolving towards a deeper understanding of the mantle and its fluid motions. Therein may lie the the final secrets to our planet's mobile continents. If true, then Manga and his team may unravel the final mysteries of the mountains.

Moving On

So Far to Go

There is still much to know. Manga has a lengthy set of questions that he and his colleagues are trying to unravel: How do bubbles affect the flow of magma? What makes some volcanoes explosively eruptive, and others not so much? Why do earthquakes affect distant geysers? How have mantle plumes evolved with time? Are there distinct layers in the mantle, separated by mineral composition? How can liquid water bubble up onto the surface of an ice-moon like Europa?

Geophysicists can only gather indirect information and make inferences that seem to match observations. As Le Pichon said, one needs to have a conversation with the Earth. We can't ask the direct and obvious: *What is it?* Instead we play the children's game of twenty questions, honing our discovery with each answer. We unearth the hidden secrets indirectly. Humans have been posing their queries to the Earth for centuries. Each answer suggests more questions. We now realize we have a smaller piece of the complete story about our planet than Aristotle, Pliny, or Gilbert thought they had.

According to Thomas Herring of Massachusetts Institute of Technology, the theory of plate tectonics is "the equivalent of General Relativity and Quantum Theory in physics."[451] But Herring agrees there is much left to discover. If earthquakes are associated with plate boundaries, why do many of the great quakes occur in the middle of plates, thousands of kilometres from their rubbing edges? Are processes other than convection currents driving the system? Why are the rates of plate movement so variable? How do today's plate velocities compare with the past? Is everything slowing down? To Herring's list, we can add other loose ends we've encountered in this book. Do plumes actually exist? Is there a role for Earth contraction or expansion? Are the seafloors spreading because material is welling up, and shoving the halves apart, or because the plunging slabs disappearing into the deep dark trenches are tugging at the seafloor? In other words, are the plates pushed or pulled? Does inner-earth convection cause magnetic polar reversals? Will we be able to predict the next north-becomes-south polarity shift and prepare for it? Will we ever be able to predict earthquakes – even by a few minutes? How long will earthquakes remain expected but unpredictable surprises, striking while drivers are passing on bridges or pedestrians are walking along underpasses?

Answering these questions is a vexing exercise. The obvious problem with understanding the Earth's inner secrets is our lack of access. Most scientists – especially geologists – are hands-on folks. But we will never touch the Earth's core. We won't even get close. No one will ever explore the deep depths in a mantle-proof submarine. Much of our concept of inner-earth construction is inferred from remote measurements. Second-hand measurement can only allow suppositions of what we might find, if we could actually experience the inner Earth. Biology, physics, and chemistry are different. Those sciences permit experimentation – one can tinker with a formula, add more friction, burn chemicals at a higher temperature, feed a plant more nitrogen, and observe experimental results. But there is nothing an Earth scientist can do to affect the way the planet works. We can not change the global magnetic field, briefly stop a spreading seafloor, or speed up the planet's rotation. Experiments are impossible. Worse, we can not even isolate an interesting observation and separate it from all the variables that might have caused it. Earth science becomes a mind game, a puzzle solved through inferences. Models are constructed from observations and are true only until contradictory observations are made.

This book has been particularly harsh on scientists who allowed either inflated ego or philosophical dogma to supplant logic and reasoning. I have not been kind to people like the Chamberlins for their half-century's suffocation of geology from their perch at the University of Chicago and for their successful wounding of continental drift theory. Nor have some of the others been excused for allowing inertia to linger in their lecture halls. However, most scientists who created ideas we now dismiss as naive were brilliant beyond reckoning – Suess with his contracting world, and Carey with his expanding, jump immediately to mind. We have the advantage of hindsight, looking back from a smugly modern perspective of what we assume is knowledge. But we should have no illusion of mastery. One day, armed with still more elaborate data, our descendants will look back in delight at our own simplistic notions of the universe.

For now, the plate tectonics model with its spreading seafloor, plunging trenches, colliding plates, and convection currents is the best general explanation for ocean basins, islands, continents, and mountains. Every geologist accepts there will be modifications of plumes, channels, blobs, megablobs, and things yet undiscovered that will rewrite this story. However, as Marcia McNutt, past president of the American Geophysical Union recently said, "The development of plate-tectonic theory certainly warrants a Nobel Prize. There is no doubt that it ranks as one of the top ten scientific accomplishments of the second half of the 20th Century."[452]

The Nobel committee does not honour earth science. No one will ever get the prize for showing us how mountains have formed. But if they did, to whom should the trophy go? Alfred Wegener is recognized for continental displacement, but Arthur Holmes showed the power source for moving the continents. And he proved that the Earth is billions of years old, not millions, allowing time for processes to occur. Alexander du Toit in South Africa bravely heaped evidence upon continental mobility. Marie Tharp and Bruce Heezen discovered the ocean rifts, Harry Hess said the seafloor spreads from those rifts, and Morley, Matthews, and Vine saw the magnetic striping that proved it all. Isacks, Oliver, and Sykes pointed out how the ocean crust is subducted and recycled. Jason Morgan and Xavier Le Pichon carved up the plates and used Euler's laws to rotate them. Tuzo Wilson fixed a host of messy loose ends – finding plumes, transform faults, and

Moving On

cycles of ocean birth – and ocean death. It is our tendency to select a single figure as the symbol for progress and creativity, but none of these scientists worked in isolation. They all borrowed from Steno and Hutton and Lyell and Smith – who in turn built upon the ideas of their predecessors. There are discoveries worthy of a dozen Nobel Prizes.

Sometimes the advancement of science has been less a struggle against facts than egos. Repeatedly, prescient ideas have appeared, only to be beaten down by sponsors of older established traditions. We relish change; we cherish traditions. The most successful societies, when success is defined as the health, longevity, safety, and personal liberty of its citizens, have permitted unfettered development of arts and sciences. Expression of thought without fear of reprisal and persecution has led to the highest form of civilization. There will always be mysteries to solve, there will continue to be scientists questioning accepted answers and proposing new solutions. Each newly solved puzzle leads to more questions, assuring us that the mysteries shall continue.

It has not been an easy task, explaining those fish bones on the mountain slope where the sheep of Palaios once grazed. But now we think we know how the dead fish arrived on the living mountain. A vanishingly brief life; an eternity preserved in stone. Palaios's fish lived in a warm Miocene sea, ten million years before the Greeks began explaining the world. In death, it surrendered its body to mud which buried the creature. Bacteria and chemicals transformed and preserved its structure. The fish became a fossil, a permanent image, a crude stone copy of its animate self. It was buried under massive layers of mud and sand that arrived from the erosion of bygone hills. Within a few thousand years, the fish and mud had become stone. The African continent, arriving from somewhere far to the south, pressed under the sandstone and shale of the European rocks that held the fossil fish. Africa thrust the rocks and fossil up from the sea, lifting them with the enormous pressure of the southern continent's huge mass, relentlessly propelled by incessant currents of mantle below the surface. The rocks rose high above the sea, becoming lofty mountains hosting ancient tokens of bygone life. Finally, the fish fossil was exposed by the erosive power of baking sun, biting winds, and rare rainstorms. Nature unearthed the fossil in a rocky meadow where sheep grazed in the spring. And where a boy could pick up the stone fish and carry it home.

Acknowledgments and Thanks

As Giordano Bruno pointed out, it is our bonding, or connections, that make us functioning and potentially successful members of the human family. This book relied on the help of many, but any errors or omissions in this work are due to my own negligence.

Dozens of people helped me write this book. Some will never know my gratitude. These include the men and women who made the discoveries and lived the lives I have attempted to relate. I wish there would be a way to thank each one – Harry Hess and Sir William Gilbert and Saint Steno and Maria Tharp and Alfred Wegener and all the others. It is my humble hope that this book has paid proper tribute to them and has acknowledged their brilliance, their curiosity, their sacrifices, and their contributions to our society and our understanding of the beautiful world we inhabit.

I owe boundless thanks to the faculty of the University of Saskatchewan. I was an older student when I began my geophysics studies. The entire Earth Sciences department took a keen interest in my welfare and participated in my transition to the academic world. In particular, I am grateful to Professor Leslie Coleman, who tirelessly sought scholarships and funding for my studies – and who gave me a passing mark on his mineralogy and crystallography final laboratory exam. Also within the geology and geophysics faculty were the department's head, Hugh Hendry, who encouraged my enrollment and convinced me that good scientists could grow long beards and pick bluegrass music; Mel Stauffer, Zoli Hajnal, and Don Gendzwill endured unending questions and trouble from me and are also fondly appreciated. Among many others at the Saskatoon university who played a role in my success is Chary Rangacharyulu, who not only taught me quantum physics, but lent his house to my family during his sabbatical year abroad. But my biggest debt of gratitude goes to my mentor, Jim Merriam, the most brilliant geophysicist I have ever met. Jim guided me through the door of scientific research and encouraged me to participate in his work. I regret that I didn't stay at the university and complete a doctorate with him – it would have been fascinating work.

In my attempt to understand the varied facets of earth science, I have relied on many experts to explain their work to me. Among these are people who agreed to be interviewed, patiently answered my queries, commented on parts of the manuscript, and sent relevant papers and background materials. In particular, I would like to thank Karin Sigloch, a geophysicist specializing in tectonic motion and working with the plume-investigating RHUM-RUM study. She offered excellent feedback and suggested ideas to investigate. Other helpful correspondents include Kenneth Carpenter, Boris Behncke, George Hess, Dale Kaiser, Alice Payne, Ed Wiebe, and Richard Warburton.

Several people have thoughtfully read and edited this book. When I thought the book was somewhat readable, my son David presented me with roughly one hundred grammar errors, misspellings, and unclear statements – most of which I think I found and corrected. Among others whom I wish to heartily thank are Elaine Haggarty who read each word, inserted proper punctuation, corrected grammar, and suggested particular structural improvements. Praise and thanks go especially to my friend Gerhard Maier, a dinosaur expert and author of *African Dinosaurs Unearthed*. His thoughtful and meticulous editing and suggestions elevated *The Mountain Mystery* to a much more readable level and his fact checking prevented this writer from exposing some gaffes and lapses of logic. Another reader, my brother Larry, convinced me to stress that science does not have all the answers and that there are many ways to view the world. Finally, my wife, Eszter Miksha, offered invaluable advice that made this book more accessible to casual readers. Because of her, many obscure terms and obtuse arguments were struck from this book while others were refined or clarified in the hope that the reader may find the content meaningful.

Image Credits

Page 2: Fish Fossil, 1911, Brockhaus, *Kleines Konversations-Lexikon*

Page 3: Xenophanes, 1655, woodcut Stanley, *History of Philosophy*

Page 4: Cyclops, 1998, photographed by author, Munich, Germany

Page 6: Flood, 1850, Doré, *The Deluge*

Page 8: Alexander, 1866, engraving Laplante, *Education d'Alexandre par Aristotle*

Page 14: Map, 1570, Oertel, parted continents

Page 19: Compass, 1581, Norman, *The Newe Attractive*

Page 22: Bruno, 1578, Livre du recteu of the University of Geneva

Page 36: Mites, 1665, Hooke, *Micrographia*

Page 39: Earth transformed, 1681, Burnet, *The Sacred Theory of the Earth*

Page 49: Vesuvius, 1870, Duncanson, *Pompeii and Vesuvius Excavation*

Page 52: Mt Baker, 2012, photo used by permission of Edward C. Wiebe, Victoria

Page 68: Column, 1844, Phillips, *Memoirs of William Smith*

Page 70: Map, 1815, Smith, Geological Map of England

Page 73: Weald, 1892, Lubbock, *Beauties of Nature*

Page 75: Darwin, 1871, *The Hornet*

Page 83: Apple Earth, drawing by author

Page 85: Squeeze Box, 1875, BGS image P612832: NERC copyright

Page 86: Dana, 1904, *Annual Report* of the Smithsonian Institute

Page 96: Earthquake, 1885, Guyot, *Physical Geography*

Page 99: Dubrovnik, 2005, photographed by author, Dubrovnik, Croatia

Page 108: Earth Map, 1909, Mantovani, *Je m'instruis*

Page 112: Crater Lake, 1886, National Park Services Archives

Page 115: Seismogram, 1906, Göttingen recording, adapted by author

Page 116: Seismoscope, 1990, photographed by author, Dallas, Texas

Page 121: India, 1879, Index Map of The Great Trigonometric Survey of India

Page 125: Fissiparturition, drawing by author

Page 127: Mohorovičić, 2007 Seismological Research Letters, permission by Herak

Page 129: Time Curves, 1910, Mohorovičić, Notebooks

Page 132: Wildebeest, 2011, released to Public Domain by Yathin S Krishnappa

Page 132: Bison, Fish and Wildlife Service of US Government

Page 134: South America and Africa, drawing adapted by author

Page 135: Wegener, 1913, Bildarchiv Preussischer Kulturbesitz

Page 136: Map, 1927, Wegener, *Origin of Continents*

Page 137: Ice sheets, Public Domain from US Geological Survey

Page 138: Selected genera, Public Domain from US Geological Survey

Page 144: No Drift, 2013, adapted by author from 1920 Public Domain photograph

Page 152: Carboniferous, 1920, Wegener, *Die Entstehung der Kontinente und Ozeane*

Page 154: Holmes, 1912, unknown photographer

Page 156: Concordia, 2006, Pfunze Belt, released to Public Domain by Babakathy

Page 157: Convection currents, drawing by author

Page 158: Du Toit, 1896, unknown photographer

Page 159: Wagon, 1923, first published by Gever, 1949, *Life of Alex du Toit*

Page 168: Bowen series, drawing by author

Page 169: Turtle Mountain, 1903, photograph adapted by author

Page 173: Daly, 1899, Harvard Archives

Page 182: Ekati Mine, 2010, used by permission of photographer Jason Pineau

Page 186: Lucas, 1905, Texas Energy Museum

Page 189: Spindletop, 1902, Ostebee, *October 6, 1902: Spindletop*

Page 197: Hess, 1928, family photograph, used by permission of George Hess

Page 204: Heezen, 1947, used by permission of Lamont-Doherty Earth Observatory

Page 206: Seafloor, 1957, reprinted with permission of Alcatel-Lucent USA Inc

Page 207: Tharp, 1955, used by permission of Debbie Bartolotta, Marie Tharp Maps LLC.

Page 212: Carey, 1953, unknown photographer

Page 214: Earth cherry, illustration by author

Page 215: Hilgenberg, 1975, released to public domain by photographer Helge Hilgenberg

Page 218: Hess, 1962, Harvard Archives

Page 221: Joshua, 1885, Doré, *Joshua Commands the Sun*

Page 227: Isla, 1539, Magnus, detail of *Insula Magneta*

Page 229: Amundsen, 1909, National Oceanic and Atmospheric Administration

Page 231: An Urn, 1884, Van Gogh, detail of *Still Life with Three Bottles and Earthenware*

Page 234: Bullard, 1976, by permission Scripps Institution of Oceanography, U San Diego

Page 235: Blackett, 1950, unknown photographer

Page 237: Zebra Stripes, 1961, used by permission of Geological Society of America

Page 240: Morley, 1952, used by permission of the Canadian Geological Survey

Page 248: Seafloor spreading, drawing by author

Page 253: Goddess Pele, photograph and drawing by author

Page 257: Transform faults, drawing by author

Page 261: Aleutian arc, drawing by author

Page 262: Down-going slab, drawing by author

Page 270: Stop Plate Tectonics, 2012, used by permission of photographer Lisa Padilla

Page 277: Morgan, 2003, photograph from White House Archives

Page 279: The Plates, after Pichon, adapted by author

Page 290: Manga, 2005, used by permission of photographer Bonnie Powell

Bibliographic Notes and Selected Readings

For the reader seeking a deeper background than this brief history of geophysics has delivered, there are at least a dozen books that are particularly enlightening and rather enjoyable to peruse. I have listed some recommendations below.

As we have seen, geophysics is a dynamic, changing science. Discoveries are being unearthed daily. As the pace of science accelerates, faster communications – particularly on-line journals and blogs – help disseminate the latest conjectures, in turn hastening the rate of discoveries. Knowledge is nearly free and nearly ubiquitous. The challenge lies in acquiring the skill to discern science from pseudo-science, hype from understatement, wheat from chaff. Using cautious critical judgment, we can all avail ourselves of these evolving information resources and participate in the conversation.

Following the reading list is my collection of endnotes, the sources for much of *The Mountain Mystery*. Providing such references is one way this author has to express gratitude to the scientists and historians who made the fundamental discoveries about our planet and who documented the research and the researchers.

Selected Readings, Science:

Robert Norman: *The Newe Attractive, shewing the Nature, Propertie, and manifold Vertues of the Lodestone, with the Declination of the Needle*, 1581. This short book is a delight. Norman approaches science with unbridled curiosity and presents his message in an unexpectedly fresh manner for a sixteenth-century amateur scientist.

William Gilbert: *De Magnete, Magneticis que Corporibus, et de Magno magnete Tellure*, translation from Latin by P. Fleury Mottelay: *On the Lodestone and Magnetic Bodies*, 1600. This 400-year-old book is full of surprises – Gilbert names electricity, contrasts it with magnetism, and shows how he determined the Earth is a magnet. He also dispenses medical advice and promotes the Copernican solar system.

James Hutton: *Theory of the Earth, or an Investigation of the Laws observable in the Composition, Dissolution, and Restoration of Land upon the Globe, Transactions of the Royal Society of Edinburgh*, 1788. Written in English, but impossible to read, some nevertheless consider Hutton's book the foundation of modern geology.

The Mountain Mystery

Charles Lyell: *Principles of Geology: The Modern Changes of the Earth,* printed in three volumes 1830-1833. Lyell was Hutton's translator. As a friend of Darwin, a traveller, a thoughtful farmer, Lyell was probably the real founder of modern geology, but missed the accolade by being a generation younger than Hutton.

Alfred Wegener: *The Origin of Continents*, began as lecture notes in 1910 and expanded over the next twenty years of Wegener's life. The plate tectonics revolution earnestly began with the range of details and construction of the argument found in Wegener's book.

Arthur Holmes: *Principles of Physical Geology* is a text book, last updated by Holmes in 1940. It is a highly readable, entertaining, and prescient glimpse into the state of geological science just before the tectonics revolution.

Allan Cox: *Plate Tectonics and Geomagnetic Reversals* is a brilliant 1973 collection of the landmark papers published during the plate tectonics revolution. Papers on magnetic reversals are prominent, but the book yields ample space to papers on seafloor spreading and plates. These are original works published in *Science*, *Nature*, and various geophysics journals – that is not a caveat, the papers are highly readable.

Some fine contemporary books include Tuzo Wilson's *IGY: The Year of the New Moons (1961)*; Jack Oliver's *Shocks and Rocks: Seismology in the Plate Tectonics Revolution* (1966); John McPhee's *Annals of the Former World* (1995); Ron Redfern's *Origins: The Evolution of Continents, Oceans, and Life* (2001); Richard Fortey's *Earth: An Intimate History (2005)*; and Robert Hazen's *The Story of Earth* (2012). There are many others, but these books are a great introduction.

Selected Readings, History and Philosophy:

Thomas S. Kuhn: *The Structure of Scientific Revolutions*. This 1962 ground-breaking study of the interaction of science and culture in the context of significant (revolutionary) discoveries is usually noted as a paradigm shift in understanding how each influences the other. (*Paradigm shift*, incidentally is a phrase Kuhn introduced to the world in this book.) Rather than seeing scientific progress as linear, Kuhn advocated that advances are episodic and indeed revolutionary when they occur.

Anthony Hallam: *A Revolution in the Earth Sciences: From Continental Drift to Plate Tectonics*. Hallam, a geologist, was an eyewitness to the emergence of plate tectonics and as such was prescient in this well-regarded 1973 summary of the development of the new view of the geophysics of the Earth.

Ursula Marvin: *Continental Drift: The Evolution of a Concept*. Marvin is a noted planetary geologist. She worked at the Smithsonian Astrophysical Observatory and wrote a number of popular books on the geology of the Earth, including 1979's *Continental Drift*. Her book is similarly significant as Hallam's as she was also witness to the development of plate tectonics.

I. Bernard Cohen: *Revolution in Science*. I. B. Cohen, a Harvard science historian explored turning-points in scientific understanding as revolutions in thought which had fundamental cultural implications. First published in 1985, plate tectonics had just been accepted as geology's new paradigm. *Revolution in Science* sweeps across a wide arc of scientific revolutions – from Copernicus, Kepler, and Gilbert through Darwin, Freud, and Einstein. The saga ends with continental drift becoming plate tectonics. Despite the wide breadth of the material, detail relevant to the revolutionary consequences of the discoveries is convincingly provided.

Stephen J. Gould: *Time's Arrow, Time's Cycle*. Written in 1987. One of Gould's most readable books and an excellent introduction to the discovery of deep time and the development of geology.

S. Warren Carey: *Theories of the Earth and Universe: History of Dogma in Earth Sciences*. Carey was the unstoppable proponent of the Earth Expansion theory. His 1988 book is a sweeping view of the history of the nastiness of scientific rejection in general. Carey's manuscript is a bit of an attempt to scorn those who spent years ridiculing his ideas, but it is nevertheless an interesting point-of-view book.

David Oldroyd: *Thinking about the Earth: A History of Ideas in Geology*, 1996. A comprehensive study of the history of ideas as related to earth sciences.

Carl Sagan and Ann Druyan: *Demon-Haunted World: Science as a Candle in the Dark*, 1997. The ultimate argument for education in critical thinking. Witches and demons make their bids, but science wins.

Naomi Oreskes has at least two essential books about the development of plate tectonics theory. *Plate Tectonics: An Insider's History Of the Modern Theory Of the Earth* (2003), edited by Oreskes, is a collection of reminiscences of geophysicists – sometimes resulting in historical reconstructions which can be quite interesting. In 1999, Oreskes wrote *The Rejection of Continental Drift:*

Theory and Method in American Earth Science which develops her idea that the initial rejection of drift theory in America was largely based on cultural reasons.

Henry R. Frankel wrote exhaustive histories of the development of plate tectonics and interviewed many dozens of the scientists involved. His *Continental Drift Controversy: Wegener and the Early Debate*, written in 2012, is the latest in his *The Continental Drift Controversy* series. Frankel likely knows more about the history of the development of plate tectonics than anyone else on Earth. His massive books and perceptive insights prove it.

1. Herodotus (440 BCE). *Histories of Herodotus, Book 2, 2.75*: Translated by George Rawlinson, 1860.
2. Hamilton, W.R., P.J. Whybrow, H.A. McClure (1978). "Fauna of fossil mammals from the Miocene of Saudi Arabia," *Nature* Vol 274, No 5668, 20 July 1978, pp 248-249.
3. Isaak, Mark (2002). "Flood Stories from Around the World," http://www.talkorigins.org/ faqs/flood-myths.html#Yanomamo, retrieved February 19, 2013.
4. Babylonian Legend (c. 1250 BCE) *Epic of Gilgamesh, Tablet Eleven*.
5. Akkadian-Sumerian Legend, (c. 1750 BCE). *Epic of Atra-Hasis, Tablet 3*, in cuneiform at the British Museum, London.
6. Kuruvilla, Carol (2012). World feature in *New York Daily News*, December 14, 2012.
7. Anonymous (2012). "What Could Disappear," *The New York Times Sunday Review*, November 24, 2012.
8. Turney, Chris S.M., Heidi Brown (2007). "Catastrophic early Holocene sea level rise, human migration and Neolithic transition in Europe," *Quaternary Science Reviews*, Vol 26, Issues 17-18, pp 2036-2041.
9. Aristotle (350 BCE). *De Respiration*, Chapter 9.
10. Aristotle (c 340 BCE). *History of Animals*, 2.3.
11. Asara, John M., Schweitzer, Mary H., Freimark, Lisa M., Phillips, Matthew, Lewis C. Cantley (2007). "Protein Sequences from Mastodon and *Tyrannosaurus rex* Revealed by Mass Spectrometry," *Science,* Vol 316, No 5822, pp 280-285, 13 April 2007.
12. Daniel, Joseph, Karen Chin (2010). "The Role of Bacterially Mediated Precipitation in the Permineralization of Bone," *Palaios*, Vol 25, No 8, pp 507-516, August 2010.
13. Carpenter, Kenneth (2007). "How to Make a Fossil: Part 1 – Fossilizing Bone," *The Journal of Paleontological Sciences*: JPS.C.07.0001.
14. Schweitzer, Mary (2013). *Proceedings of the Royal Society*, November 26, 2013.
15. Seneca (c 62 AD). "Natural Questions," *Book 6.3*.
16. Crombie, Alistair (1990). *Science, optics, and music in medieval and early modern thought*. Continuum International Publishing Group, pp 108-109.
17. Toulmin, Stephen, June Goodfield (1965). *The Ancestry of Science: The Discovery of Time*, p 64, University of Chicago Press)
18. Gould, Stephen Jay (1998). *Leonardo's Mountain of Clams and the Diet of Worms*, p 43, Harvard University Press 2011 Edition.
19. Leonardo da Vinci (c 1500). Codex Leicester. Translated by Jean Paul Richter (1888). Sections 979-994.
20. Ibid.
21. Montgomery, David R. (2012). *Rocks Don't Lie: A Geologist Investigates Noah's Flood*, Chapter 3: Bones in the Mountains, WW Norton & Company, New York.
22. Vai, Gian Battista, W. Glen Caldwell (2006). *The Origins of Geology in Italy,* The Geological Society of America.
23. Friedman, Barry (2011). "Da Vinci Knew His Geology" *AAPG Explorer*, October 2011.
24. Wang Chong (c 80 AD). *Lun Heng*.
25. Étienne Du Trémolet de Lacheisserie (2003). *Magnetism: Fundamentals*, "Magnetism, from the dawn of civilization to today," p 5. Edited by Étienne Du Trémolet de Lacheisserie, Damien Gignoux, and Michel Schlenker. Kluwer Academic Publishers.
26. Dunlop, David, Özden Özdemi (2001). "A Brief History: Earth Magnetism," *Rock Magnetism: Fundamentals and Frontiers*.
27. Peregrinus, Petrus (1269). *The Letters of Petrus Peregrinus on the Magnet, A.D. 1269*, p12, Translated in 1904 by Brother Arnold, McGraw Hill Publishing Company, New

York.
28. Wasilewski, P., and Kletetschka, G. (1999). "Lodestone: Nature's only permanent magnet – What it is and how it gets charged," *Geophysical Research Letters,* Vol 26, No 15, pp 2275-78.
29. Norman, Robert (1581). *The Newe Attractive, shewing the Nature, Propertie, and manifold Vertues of the Loadstone, with the Declination of the Needle,"* written in 1581, reprinted 1720. All quotes attributed to Norman appear among its 43 pages.
30. Thompson, S.P. (1891). *Gilbert, of Colchester; an Elizabethan Magnetizer,* p 10, Chiswick Press, London.
31. Thompson, S.P. (1903). *"William Gilbert, and Terrestrial Magnetism in the Time of Queen Elizabeth,"* p 7.
32. Bruno, Giordano (1591). *De Immenso Book IV, x, pp 56-57.* Described by Paterson in the 1973 "Giordano Bruno's View on the Earth without a Moon," *Pensee* Vol. 3.
33. Mercati (1943). *Il sommario del processo di Giordano Bruno, con appendice di documenti sull'eresia e l'inquisizione a Modena nel secolo XVI.*
34. Del Col, Andrea (2010). *L'Inquisizione in Italia.* Oscar Mondadori, Milan. pp 779-780.
35. Sagan, Carl and Ann Druyan (1997). Demon-Haunted World: Science as a Candle in the Dark, Ballantine Books.
36. Berget, Alphonse (1915). *The Earth: Its Life and Death,* translated by E.W. Barlow. G.P. Putnam's Sons. p 237.
37. Thompson, S.P. (1903). *William Gilbert, and Terrestrial Magnetism in the Time of Queen Elizabeth*, p 4.
38. Gilbert, William (1600). *De Magnete, Magneticis que Corporibus, et de Magno magnete Tellure,* translation from Latin by P. Fleury Mottelay: *On the Lodestone and Magnetic Bodies*, p 55, London, 1893.
39. Ibid.
40. Thompson, S.P. (1891). *Gilbert, of Colchester; an Elizabethan Magnetizer,* p 34, Chiswick Press, London, 1891.
41. Gilbert, William (1600). *De Magnete, Magneticis que Corporibus, et de Magno magnete Tellure,* translation from Latin by P. Fleury Mottelay: *On the Lodestone and Magnetic Bodies*, p 69, London, 1893.
42. Thompson, S.P. (1891). *Gilbert, of Colchester; an Elizabethan Magnetizer,* p 15 Chiswick Press, London.
43. Thompson, S.P. (1891). *Gilbert, of Colchester; an Elizabethan Magnetizer,* p 34, Chiswick Press, London, 1891. Thompson writes about Gilbert's *de Magnete*: "The world was hardly prepared to accept a sober treatise based on simple facts in place of the wild and speculative treatises which had hitherto passed as philosophic."
44. Bacon, Sir Francis (1620). "Pan, or Nature: Explained of Natural Philosophy," from Chapter XIII of *Advancement of Learning.*
45. Bacon, Sir Francis (1620). *On the Dignity and Advancement of Learning,* Section 3.
46. Benjamin, Park (1898). *A History of Electricity (The Intellectual Rise in Electricity)*, J. Wiley & Sons. p 328.
47. Gilbert, William (1600). *De Magnete, Magneticis que Corporibus, et de Magno magnete Tellure,* translation from Latin by P. Fleury Mottelay: *On the Lodestone and Magnetic Bodies*, p 308, London, 1893.
48. Benjamin, Park (1898). *A History of Electricity (The Intellectual Rise in Electricity)*, J. Wiley & Sons. p 328.
49. Bacon, Sir Francis (1620). *Novum Organum, or True Suggestions for the Interpretation of Nature,* Section 64.
50. Ibid., Section 27.

51 Carey, S. Warren (1977). *"The Expanding Earth: An Essay"* p 7.
52 Gilbert, William (1600). *De Magnete, Magneticis que Corporibus, et de Magno magnete Tellure*, translation from Latin by P. Fleury Mottelay: *On the Lodestone and Magnetic Bodies*, p 6, London, 1893.
53 Ibid., p 14.
54 Thompson, S.P. (1891). *Gilbert, of Colchester; an Elizabethan Magnetizer*, p 45, Chiswick Press, London.
55 Brush, Stephen G. (1982). "Chemical History of the Earth's Core," *EOS*, Vol 63, No 47, pp 1185-1188. (23 November 1982).
56 Steno, Nicholas (1666). *Dissertation Concerning a Solid Body*.
57 Ibid.
58 Garrett Winter, John (1916). "Life of Steno," the introduction to Steno's *Dissertation Concerning a Solid Body Enclosed by Process on Nature within a Solid*, translated by Winter. MacMillan Company. pp 184-186.
59 Hooke, Robert (1662). *Micrographia,* pp 206-207.
60 Burnet, Thomas (1691). *The Sacred Theory of the Earth*, pp 14-15.
61 Ibid.
62 Anonymous (2013). Jewish Virtual Library: "La Peyrere, Isaac" http://www.jewishvirtuallibrary.org/jsource/judaica/ejud_0002_0012_0_11879.html, retrieved March 9, 2013.
63 Popkin, Richard H. (2003). *The History of Scepticism: From Savonarola to Bayle*. p 228. Oxford University Press.
64 Croft, Herbert (1685). *Some Animadversions upon a Book Intituled the Theory of the Earth*.
65 King, William (1776). *The Original Works*.
66 Gould, Stephen J. (1987). *Time's Arrow, Time's Cycle*, pp 21-59, Harvard University Press.
67 Gould, Stephen J. (1987). *Time's Arrow, Time's Cycle*, pp 21-59. Harvard University Press.
68 Hutton, James (1795). *Theory of the Earth with Proofs and Illustrations,* Volume I, p 271.
69 Lyell, Charles (1830). *Principles of Geology III,* p 37.
70 Ruse, Michael (1978). "What Kind of Revolution Occurred in Geology?" *PSA: Proceedings of the Biennial Meeting of the Philosophy of Science Association*, Vol 1978, Volume Two: Symposia and Invited Papers. p 261.
71 Geike, Archibald (1905). *The Founders of Geology,* p 66. Macmillan & Co.
72 Laurance, John (1835). *Geology in 1835,* p 4. Simpkin, Marshall & Co., London.
73 Price, David (1989). "John Woodward and a Surviving British Geological Collection from the Early Eighteenth Century," *Journal of the History of Collections* Vol 1, No 1, pp 79-95.
74 Woodward, John (1728). From John Woodward's will.
75 Strabo (c 20 A.D.) Geographica, Book V, Chapter 4, Part 8.
76 Guyot, Arnold (1885). *Physical Geography*, p 14. The American Book Company, New York.
77 Diodorus Siculus (c 40 B.C.). *Biblioteca historica*, Book IV, Chapter 21.
78 Pliny the Younger (AD 106). Pliny Letter LXVI to Tacitus.
79 Ibid.
80 Anonymous (1980). "Volcanologist Reported Missing," *Wilmington Morning Star*, p 12-C, 21 May 1980.
81 Anonymous (2013). "No more rising sun: The capital's last street-level view of Mount

Fuji is about to be obscured," *The Economist,* January 5, 2013, London.
82 Klemetti, Erik (2013). "The Right and Wrong Way to Die When You Fall into Lava," Denison University Geosciences Department *http://www.wired.com/wiredscience/2011/12/the-right-and-wrong-way-to-die-when-you-fall-into-lava/* retrieved January 12, 2013.
83 Behncke, Boris (2014). Personal communication, 19 January 2014.
84 Buffon, Georges-Louis Leclerc (1797). *Natural History: A Theory of the Earth,* Articles XVI, XVIII, XIX.
85 Ibid.
86 Cutler, Alan (2003). *The Seashell on the Mountaintop,* p 73. Dutton, New York.
87 Franklin, Benjamin (1784). From his December 22 1784 lecture, "Meteorological Imaginations and Conjectures," as recorded by Dr. Percival, London.
88 Franklin, Benjamin (1782). Letter to Abbé Jean-Louis Giraud Soulavie, August 12, 1782, Paris.
89 Halpern, Joel Martin (1951). "Thomas Jefferson and the Geological Sciences," *Rocks and Minerals,* Nov-Dec 1951, p 601.
90 De la Beche, Henry (1833). *A Geological Manual,* p 213. Charles Knight, London.
91 Ibid.
92 Ibid., p 214.
93 Hutton, James (1788). *Theory of the Earth, or an Investigation of the Laws observable in the Composition, Dissolution, and Restoration of Land upon the Globe, Transactions of the Royal Society of Edinburgh, Volume 1.*
94 Ibid.
95 Hutton, James (1753). From Hutton's letter to Sir John Hall.
96 Hutton, James (1788). *Theory of the Earth, or an Investigation of the Laws observable in the Composition, Dissolution, and Restoration of Land upon the Globe, Transactions of the Royal Society of Edinburgh, Volume 1.*
97 Ibid.
98 Lyell, Charles (1830). *Principles of Geology.*
99 Hutton, James (1788). *Theory of the Earth, or an Investigation of the Laws observable in the Composition, Dissolution, and Restoration of Land upon the Globe, Transactions of the Royal Society of Edinburgh, Volume 1.*
100 Porter, Roy (1976). "Charles Lyell and the Principles of the History of Geology," *The British Journal for the History of Science,* Vol 9, No 2, p 91.
101 McGowan, Christopher (2001). *The Dragon Seekers: How an Extraordinary Circle of Fossilists Discovered the Dinosaurs and Paved the Way for Darwin,* pp 203-204. Persus Publishing, Cambridge, Massachusetts.
102 Ibid., p 201.
103 Jablonski, Nina (2012). *Living Color: The Biological and Social Meaning of Skin Color,* University of California Press.
104 Swallow, Dallas (2003). "Genetics of Lactase Persistence and Lactose Intolerance," *Annual Review of Genetics,* Vol 37, pp 197-219.
105 Lyell, Charles (1827). From *The Life and Letters of Sir Charles Lyell, vol 1,* edited in 1881 by his sister-in-law, K. Lyell, p 168. Published by John Murray, London.
106 Darwin, Charles (1863). *The correspondence of Charles Darwin,* Vol 11, p 181. Edited by Frederick Burkhardt, et.al., published 1999. Cambridge.
107 Fourier, J.B.J. (1822). *Théorie Analytique de la Chaleur, (Analytic Theory of Heat).*
108 Fourier, J.B.J. (1827). "Mémoire sur les températures du globe terrestre et des espaces planétaires," *Mémoires de l'Académie Royale des Sciences de l'Institute de France,* Vol 7, pp 569-604.

109 Fourier, J.B.J. (1824). "Remarques Générales Sur Les Températures Du Globe Terrestre Et Des Espaces Planétaires," *Annales de Chimie et Physique,* Vol 27, p 137.

110 Fourier, J.B.J. (1827). "Mémoire sur les températures du globe terrestre et des espaces planétaires," *Mémoires de l'Académie Royale des Sciences de l'Institute de France*, Vol 7, p 603.

111 Fourier, J.B.J. (1824). "Remarques Générales Sur Les Températures Du Globe Terrestre Et Des Espaces Planétaires," *Annales de Chimie et Physique,* Vol 27, p 138.

112 Watson, C. (1969). *Some nineteenth century British scientists*, pp 96-153. Oxford.

113 Anonymous (1872). "Notices of New Books," *New Englander and Yale Review,* Vol 31, Article IX, p 373.

114 Lord Kelvin (1889). Annual Address to the Christian Evidence Society.

115 Lord Kelvin (1902). Interview in *Newark Advocate*, p 4. 26 April 1902.

116 Bailey, E.B. (1939). "Professor Albert Heim: 1849-1937,"*Obituary Notices of the Fellows of the Royal Society* Vol 2, No 7, pp 471-474.

117 Heim, Albert (1878). *Mechanismus der Gebirgsbildung.*

118 Dana, James Dwight (1896). *Manual of Geology*, pp 380-396. American Book Company, New York.

119 Ibid., p 380.

120 Dana, James Dwight (1856). *Science and the Bible*, p 81. Warren F. Draper, Andover.

121 Dana, James Dwight (1863). Dana's letter to Darwin, from New Haven, Connecticut, February 5, 1863.

122 Darwin, Charles (1862). Darwin's letter to Dana, from Bromley, Kent, February 20, 1863.

123 Ibid.

124 Dana, James Dwight (1896). *Manual of Geology*, pp 1028-1035.

125 Ibid.

126 Ibid.

127 Owen, Robert (1857). *Key to the Geology of the Globe*, p 14. A.S. Barnes, New York..

128 Ibid.

129 Anonymous (1884). "Guyot's View of Creation," *Science* Vol 3, No 67, May 16, 1884. An editorial book review of Arnold Guyot's *Creation, or on the Biblical Cosmogony in the Light of Modern Science.*

130 Guyot, Arnold (1884). *Creation, or the Biblical Cosmology in the Light of Modern Science*, p 32. Charles Scribner's Sons, New York.

131 Ibid., p 2.

132 Ibid., p 126.

133 Ibid., p 125.

134 Guyot, Arnold (1885). *Physical Geography*, p 13. American Book Company, New York.

135 Ibid., pp 16-20.

136 Ibid.

137 Ibid.

138 Perrey, Alexis (1863). *Propositions sur les Tremblement de Terre.*

139 Lyell, Charles (1883). *Principles of Geology: The Modern Changes of the Earth,* p 233.

140 Anonymous (2013). *Potres u Dubrovniku*, http://www.hrt.hr/arhiv/2001/04/06/NDD.html, retrieved April 4, 2013.

141 Guyot, Arnold (1885). *Physical Geography*, p 17. American Book Company, New York.

142 Ibid., pp 19-20.

143 Perry, John (1895). "On the Age of the Earth," *Nature*, Vol 51, p 227.
144 England, Philip, Peter Molnar, Frank Richter (2006). "John Perry's neglected critique of Kelvin's age for the Earth: A missed opportunity in geodynamics" *GSA Today*, January, 2007.
145 England, Philip, Peter Molnar, Frank Richter (2007). "Kelvin, Perry, and the Age of the Earth," *American Scientist*, Vol 95, pp 342-349.
146 Meijer, R.J., W. van Westrenen (2008). "The feasibility and implications of nuclear georeactors in Earth's core-mantle boundary region," *South African Journal of Science*, Vol 104, No 3-4, pp 111-118.
147 Johnston, Hamish (2011). "Radioactive decay accounts for half of Earth's heat," July 19, 2011, *PhysicsWorld.com*, http://physicsworld.com/cws/article/news/2011/jul/19/radioactive-decay-accounts-for-half-of-earths-heat, retrieved January 25, 2013.
148 Kuroda, P.K. (1956). "On the Nuclear Physical Stability of the Uranium Minerals," *Journal of Chemical Physics,* Vol 24, No 4, pp 781-782; 1295-1296.
149 Anonymous (1973). "Natural Nuclear Reaction Theory of Kuroda Proven Back in 1956," *Northwest Arkansas Times,* June 9, 1973, p 9.
150 Lyell, Charles (1883). *Principles of Geology* Vol 2, p 238.
151 Scalera, Giancarlo (2009). "Roberto Mantovani (1854-1933) and his ideas on the expanding Earth, as revealed by his correspondence and manuscripts," *Annals of Geophysics*, Vol 52, No 6, p 615.
152 Anonymous (2013). *Earthquake Preparedness: Earthquake Probabilites*, http://www.data.scec.org/earthquake/probabilities.html, retrieved March 14, 2013.
153 Koto, Bunjiro (1893). "On the cause of the great earthquake in central Japan in 1891," *Journal of the College of Science, Imperial University of Tokyo*, No 5, pp 296-353.
154 Anonymous (1906). *The Complete Story of the San Francisco Horror,* p 231.
155 Dutton, Clarence (1904). *Earthquakes: in the Light of the New Seismology*, pp 12-13. G.P. Putnam's Sons, New York.
156 Stegner, Wallace (1954). *Beyond the Hundredth Meridian*, University of Nebraska Press.
157 Dutton, Clarence (1882). *American Journal of Science*, No 136, p 288.
158 Dutton, Clarence (1904). *Earthquakes: in the Light of the New Seismology*, p iv. G.P. Putnam's Sons, New York.
159 Chilingar, G.V., and B. Endres (2005). "Environmental hazards posed by the Los Angeles urban oilfields," *Environmental Geology*, No 47, p 310.
160 Anonymous (2012). "Quake fallout takes many paths," January 2, 2012, *Vindicator: Mahoning Valley Newspaper.*
161 Anonymous (1935). "James Alfred Ewing," *Obituary Notices of Fellows of the Royal Society of London*, Vol 1, No 4, p 476.
162 Ibid.
163 Ewing, James (1883). *Earthquake measurement,* p 52.
164 Dutton, Clarence (1904). *Earthquakes: in the Light of the New Seismology*, pp 29-30. G.P. Putnam's Sons, New York.
165 Jackson, Patrick N.Wyse (1997). "Fluctuations in Fortune: Three Hundred Years of Irish Geology," in Foster J.W. & Chesney H.C.G. *Nature in Ireland: A Scientific and Cultural History.* McGill-Queens. p 101.
166 Dutton, Clarence (1904). *Earthquakes: in the Light of the New Seismology*, p 30. G.P. Putnam's Sons, New York.
167 Dutton, Clarence (1904). *Earthquakes: in the Light of the New Seismology*, p 33. G.P. Putnam's Sons, New York.
168 Anonymous (1911). "Interesting Theory Advanced by Dr. Howard B. Baker, a Detroit

Student of the Planetary System," *Detroit Free Press, E7*, April 23, 1911.
169 Oreskes, Naomi (1994). "Weighing the Earth from a Submarine," *History of Geophysics Volume 5*, edited by Gregory Good, p 56. American Geophysical Union, Washington.
170 Mohorovičić, Andrija (1897). "Klima grada Zagreba," *Rad JAZU*, Vol 131, pp 72-111.
171 Mohorovičić, Andrija (1909). Laboratory notebooks.
172 Mohorovičić, Andrija (1910). "The Earthquake of October 8, 1909," *Godisnje Izvjesce Zagrebackog Meteoroloskog Opsevervatorija*, Zagreb.
173 Mohorovičić, Andrija (1909). Yearbook of the Meteorological Observatory, Zagreb.
174 Skoko, Dragutin (1982). *Mohorovičić*, Školska Knjia, Zagreb, Croatia, p 7.
175 Wegener, Alfred (1910). Letter to Else Köppen.
176 Wegener, Alfred (1912) *Die Entstehung der Kontinente, Geologische Rundschau*, Vol 3, pp 276-292.
177 Green, W.L. (1875). *Vestiges of the Molten Globe Volume 2*, pp 116-119.
178 Matthew, William Diller (1915). "Climate and Evolution," *Annals of the New York Academy of Sciences*, Vol 24, Issue 1, pp 171-318.
179 Anonymous (1932). "Biographic Memoir of William Diller Matthew," *Biographical Memoirs of the Fellows of the Royal Society* Vol 1, No 1, p 71.
180 Matthew, William Diller (1915). "Climate and Evolution," *Annals of the New York Academy of Sciences*, Vol 24, Issue 1, pp 172-178.
181 Wegener, Alfred (1912). "The origins of continents" lecture at Geologische Vereinigung in Frankfurt, January 6, 1912.
182 Termier, Pierre-Marie (1925). "The Drifting of the Continents," *1924 Annual Report of the Smithsonian Institute*, p 236.
183 Bullard, Edward (1975). "The Emergence of Plate Tectonics: A Personal View," *Annual Review of Earth and Planetary Science*, Vol 3, p 5.
184 Frankel, Henry R. (2012). *Continental Drift Controversy: Wegener and the Early Debate*, p 433. Cambridge University Press.
185 Lake, Philip (1923). The quote was from a spoken presentation at the Royal Geographical Society, January 22, 1923. An abridged version is found in Lake's 1923 paper, "Wegener's hypothesis of continental drift," *Nature* Vol 111, pp 226-228.
186 Dott, Robert H., and Donald R. Prothero (1994). *Evolution of the Earth*, McGraw-Hill, New York.
187 Hallam, A. (1973). *A Revolution in the Earth Sciences*. Clarendon Press, Oxford.
188 Oreskes, Naomi (1999). *The Rejection of Continental Drift: Theory and Method in American Earth Science*, p 126. Oxford University Press.
189 Chaney, Ralph W. (1940). "Tertiary Forests and Continental History," *Bulletin of the Geological Society of America*, v.51, No 3, pp 469-488. March 1, 1940.
190 Krill, Allan (2008). "Mobile continents and fixed published opinions," presented at International Geological Congress, Oslo, Norway.
191 Simpson, George Gaylord (1943). "Mammals and the Nature of Continents," *American Journal of Science* Vol 241, pp 1-31.
192 Wegener, Kurt (1930). *The Origin of Continents and Oceans*, Forward Section. Dover Publications, 1966 reprint.
193 Chamberlin, Rollin T. (1928). "Some of the Objections to Wegener's Theory," *Theory of Continental Drift: A Symposium*, ed. Van der Gracht, p. 83.
194 Bullard, Edward (1975). "The Emergence of Plate Tectonics: A Personal View," *Annual Review of Earth and Planetary Science*, Vol 3, p 5.
195 Chamberlin, Thomas C. and Rollin D. Salisbury (1907). *Geology Volume 3*, p 536. Henry Holt and Company, New York.

196 Chamberlin, Thomas C. (1899). "Lord Kelvin's address on the age of the Earth as an abode fitted for life": *Science*, Vol 9, pp 889-901.
197 Pettijohn, F.J. (1970). "Rollin Thomas Chamberlin," *Biographical Memoirs of the National Academy of Sciences*, p 101.
198 Ibid., p 96.
199 Chamberlin, Rollin T. (1928). "Some of the Objections to Wegener's Theory," *Theory of Continental Drift: A Symposium*, ed. Van der Gracht, p. 83.
200 Muir-Wood, Robert (1985). *The Dark Side of the Earth: The Battle for the Earth Sciences, 1800-1980*, p 83. Unwin Hyman Publishers.
201 Melton, F.A. (1930). "Review of Theory of Continental Drift – A Symposium," *The Journal of Geology*, Vol 38, No 3 (April-May, 1930), p 284.
202 Berry, Edward W. (1928). *Theory of Continental Drift – A Symposium*, p 195. American Association of Petroleum Geologists, Tulsa, Oklahoma
203 Bullard, Edward (1975). "The Emergence of Plate Tectonics: A Personal View," *Annual Review of Earth and Planetary Science*, Vol 3, p 6.
204 Ibid.
205 Cohen, I. Bernard (1985). *Revolution in Science*, p 453.
206 Andrews, E.C. (1938). "Some major problems in structural geology," *Proceedings of the Linnean Society of New South Wales*, Vol 63, p v.
207 Krill, Allan (2008). "Mobile continents and fixed published opinions," International Geological Congress in Oslo. http://www.cprm.gov.br/33IGC/1260319.html.
208 Bullard, Edward (1975). "The Emergence of Plate Tectonics: A Personal View," *Annal Review of Earth and Planetary Science*, Vol 3, p 5.
209 McKie, Robin (2012). "David Attenborough: force of nature". *The Observer*, website: http://www.guardian.co.uk/tv-and-radio/2012/oct/26/richard-attenborough-climate-global-arctic-environment, retrieved 29 October 2012.
210 Bullard, Edward (1975). "The Emergence of Plate Tectonics: A Personal View," *Annual Review of Earth and Planetary Science*, Vol 3, p 3.
211 Chamberlin, Thomas C. (1890). "The Method of Multiple Working Hypotheses," *Science*, February 7, 1890.
212 Oreskes, Naomi (2001). "From continental drift to plate tectonics," *Plate Tectonics: An Insider's History*, pp 10-11.
213 Longwell, Chester (1944). "Some Thoughts on the Evidence for Continental Drift," *American Journal of Science*, Vol 242, p 231.
214 Rodger, John (1976). "Biographical Memoir: Chester Ray Longwell," Vol 53, p 249.
215 Hofer, John (1978). *The History of the Hutterites*, pp 62-63. Published by the Mulitcultural Program of the Government of Canada.
216 Wegener, Kurt (1930). *The Origin of Continents and Oceans*, by A. Wegener. Preface by Kurt Wegener. Dover Publications, 1966 reprint.
217 Carlton, Ian Clark (1974). *An account of the Great Floods in the River's Tyne, Wear and Tees*, p 114. Gateshead Corporation.
218 Lewis, Cherry L.E. (2002). "Arthur Holmes: An Ingenious Geoscientist," *GSA Today*, p 16. The Geological Society of America, March 2002.
219 Holmes, Arthur (1911). "The Association of Lead with Uranium in Rock-Minerals and its Application to the Measurement of Geological Time," *Proceedings of the Royal Society A: Mathematical, Physical, and Engineering Sciences*, vol 85 no 578, pp 248-256.
220 Holmes, Arthur (1913). *The Age of the Earth*, Preface (ix). Harper Brothers, London.
221 Ibid., p 18.
222 Ibid., p 15.

223 Holmes, Arthur (1931). "Radioactivity and Earth Movements," *Transactions of the Geological Society of Glasgow* 18: 559-606.
224 Holmes, Arthur (1944). *Principles of Physical Geology*, p 306.
225 Ibid., Preface.
226 Haughton, S.H. (1949). "Alexander Logie du Toit," *Obituary Notices of Fellows of the Royal Society,* Vol. 6, No 18. pp 385-395. Published by The Royal Society, London.
227 Du Toit, A.L. (1937). *Our Wandering Continents,* Edinburgh, Oliver & Boyd.
228 Du Toit, A.L. (1927). *A Geological Comparison of South America with Africa.*
229 Brenner, Sydney (1994). "A Life in Science Told to Lewis Wolpert," http://www.webofstories.com/play/sydney.brenner/32, retrieved 19 April 2013.
230 Chamberlin, R.T. (1938). Review of du Toit's *Our Wandering Continents, Journal of Geology,* vol 46, p 791.
231 Du Toit, A.L. (1937). *Our Wandering Continents: An Hypothesis of Continental Drifting*, preface.
232 Ibid., p 3.
233 Gevers, T.W. (1950). "Life and work of Dr. Alex L. Du Toit," Transactions of the Geological Society of South Africa, Vol 52 Supplemental, pp 1-109.
234 Bullard, Edward (1975). "The Emergence of Plate Tectonics: A Personal View," *Earth and Planetary Science* Vol 3, pp 1-31.
235 Williams, Mark, Michael P. Clough, Jane Pedrick Dawson, Cinzia Cervato (2010). "Data Do Not Speak: The Development of a Mechanism for Continental Drift," National Science Foundation: http://www.storybehindthescience.org/pdf/drift.pdf
236 Hapgood, Charles (1958). *Earth's Shifting Crust*, p 14, Pantheon Books, New York.
237 Ibid.
238 Lyell, Charles (1883). *Principles of Geology* Vol 2, p 208.
239 Hapgood, Charles (1958). *Earth's Shifting Crust*, p 28, Pantheon Books, New York.
240 Ibid., pp 28-29.
241 Einstein, Albert (1955). Foreword to Charles Hapgood's *Earth's Shifting Crust*, Pantheon Books, New York.
242 Pezzati, Alex (2005). "Mystery at Acámbaro, Mexico," *Expedition Magazine*, Vol 47 No 3, pp 7-8. University of Pennsylvania Museum.
243 Noone, Richard W. (1997) *5/5/2000, ICE: The Ultimate Disaster.* New York, NY: Three Rivers Press.
244 Oreskes, Naomi (1999). *The Rejection of Continental Drift: Theory and Method in American Earth Science*, p 123. Oxford University Press.
245 Birch, Francis (1960). "Reginald Aldworthy Daly," *Biographical Memoirs of the National Academy of Sciences,* p 32.
246 Ibid.
247 Daly, Reginald Aldworth (1914). *Igneous Rocks and their Origins,*McGraw-Hill, New York.
248 Birch, Francis (1960). "Reginald Aldworthy Daly," *Biographical Memoirs of the National Academy of Sciences,* p 34.
249 Anonymous (1899). *Butte Weekly Miner,* March 23, 1899.
250 Anonymous (1908). "Rich Man Insane," *Morning Oregonian*, p 4. 17 June 1908.
251 Anonymous (1908). "H.L. Frank Demented," *Fernie Free Press.* 26 June 1908.
252 Cruden, D.M., C.D. Martin (2004). "A century of risk management at the Frank Slide, Canada," IAEG2006, p 772, *Geological Society of London.*
253 Birch, Francis (1960). "Reginald Aldworthy Daly," *Biographical Memoirs of the National Academy of Sciences,* p 34.

254 Birch, Francis (1960). "Reginald Aldworthy Daly," *Biographical Memoirs of the National Academy of Sciences,* p 42.
255 Sinclair, Andrew (1962). *Prohibition: The Era of Excess,* p 47. Little, Brown, Boston.
256 Birch, Francis (1960). "Reginald Aldworthy Daly," *Biographical Memoirs of the National Academy of Sciences,* p 42.
257 Hodgson, Ernest A. (1927). "Review of Daly's *Our Mobile Earth,*" *Journal of the Royal Astronomical Society of Canada,* Vol 21, p 113.
258 Gevers, T.W. (1949). "The life and work of Dr Alex du Toit," *Proceedings of the Geological Society of South Africa,* p 8.
259 Daly, Reginald Aldworth (1927). "Review of *The Geology of South Africa,*" *Journal of Geology,* Vol 35, pp 671-672.
260 Birch, Francis (1960). "Reginald Aldworthy Daly," *Biographical Memoirs of the National Academy of Sciences,* p 39.
261 Daly, Reginald Aldworth (1914). *Igneous Rocks and their Origin,* p 1.
262 Daly, Reginald Aldworth (1926). *Our Mobile Earth,* p 320. New York.
263 Ibid., Preface.
264 Hodgson, Ernest A. (1927). "Review of Daly's *Our Mobile Earth,*" *Journal of the Royal Astronomical Society of Canada,* Vol 21, p 113.
265 Daly, Reginald Aldworth (1926). "Origin of Mountain Ranges," *Our Mobile Earth* p 311. New York.
266 Ibid., pp 251-291.
267 Anonymous (1957). "Reginald A. Daly Geologist, 86, Dies: Harvard Professor Emeritus Received Honors for Study of the Earth's Interior," *The New York Times.* September 20, 1957.
268 Somers, Ransom (1927). "The Destructive Earth: Earthquakes," *Pamphlets on Biology: Kofoid Collection,* Vol 1061, p 47.
269 Daly, Reginald Aldworth (1926). *Our Mobile Earth,* pp 1-44, New York.
270 Hodgson, Ernest A. (1927). "Review of Daly's *Our Mobile Earth,*" *Journal of the Royal Astronomical Society of Canada,* Vol 21, p 118.
271 Daly, Reginald Aldworth (1926). *Our Mobile Earth,* p 291, New York.
272 Anonymous (1932). "Earth's Core Held Eternal Mystery," *Special to The New York Times,* p 12. December 28, 1932.
273 Anonymous (1926). "Says Earth Shakes to Gain Symmetry: Prof Daly of Harvard Offers a New Explanation of Quakes and Volcanoes," *The New York Times,* September 10, 1926.
274 Binder, A.B. (1974). "On the origin of the Moon by rotational fission," *The Moon,* Vol 11, No 2, pp 53-76.
275 Daly, R.A. (1946). "Origin of the Moon and Its Topography," *Proceedings of the American Philosophical Society,* Vol 90, No 2, pp 104-119.
276 Jutzi, M., E. Asphaug (2011). "Forming the lunar farside highlands by accretion of a companion moon," *Nature,* Vol 476, pp 69-72.
277 Simkin, Tom, Lee Siebert (2000). "Earth's Volcanoes and Eruptions: An Overview," *Encyclopedia of Volcanoes,* edited by Haraldur Sigurdsson, p 253.
278 Berger, Zeev (2013). Personal communication re: Jacob Agassi.
279 Bullard, Edward (1975). "The Emergence of Plate Tectonics: A Personal View," *Annual Review of Earth and Planetary Science,* Vol 3, p 8.
280 Anonymous (1925). "Think Land Linked Europe to America," *The New York Times,* Science Section, November 2, 1925.
281 Bullard, Edward (1975). "The Emergence of Plate Tectonics: A Personal View," *Annual Review of Earth and Planetary Science,* Vol 3, p 9.

282 Wilson, J. Tuzo (1981). "*Movements in Earth Science*," New Scientist, p 614, November 26, 1981.
283 Bullard, Edward (1975). "William Maurice Ewing," *National Academy of Sciences Biography Series*, p 119.
284 Etienne-Gray, Tracé (2013). "Higgins, Pattillo," *The Handbook of Texas Online*, Texas State Historical Association, http://www.tshaonline.org/handbook/online/articles/fhi07, retrieved February 6, 2013.
285 Bourne. Edward Gaylord (1904). *Narratives of De Soto*, translated by Buckingham Smith, Vol I, p 209.
286 Halbouty, Michel (1997). "Spindletop: The Original Salt Dome," *World Energy*, Vol 3, No 2, pp 108-112.
287 Ewing, Maurice (1922). In a letter to his parents.
288 Bullard, Edward (1975). "William Maurice Ewing," *Biographical Memoirs of the National Academy of Sciences*, Vol 21, p 119.
289 Worzel, Lamar J. (2001). *Autobiography*, pp 15-17.
290 Bullard, Edward (1975). "William Maurice Ewing," *Biographical Memoirs of the National Academy of Sciences*, Vol 21, p 125.
291 Ibid., p 126.
292 Wolf, T.H. (1935). "Professor Richard Field Hopes to Use Geophone to Rewrite Earth's History," *The Daily Princetonian*, p 1, November 12, 1935.
293 *Princeton Alumni Weekly* (1923). Vol XXIV, No 5, p 1, October 31, 1923.
294 Anonymous (1925). "Think Land Linked Europe to America," *The New York Times*, Science Section, November 2, 1925.
295 Leet, L. Don (1937). "Review of Geophysical Investigations in the Emerged and Submerged Atlantic Coastal Plain by Maurice Ewing," *Bulletin of Seismological Society of America*, Vol 27, pp 353-354.
296 Bullard, Edward (1975). "William Maurice Ewing," *Biographical Memoirs of the National Academy of Sciences*, Vol 21, p 129.
297 Bullard, Edward (1975). "Emergence of Plate Tectonics: A Personal View," *Annual Review of Earth and Planetary Sciences 1975,* Vol 3, p 26.
298 Hamilton, Andrew (1949). "SOFAR – the navy's lost and found," *Popular Mechanics*, Vol 92, No 6, pp 166-169, December 1949.
299 Anonymous (1907). "The City's Population," *The New York Times,* March 31, 1907.
300 James, Harold L. (1973). "Harry Hammond Hess," *Biographical Memoirs of the National Academy of Sciences*, Vol 43, p 4.
301 Ibid.
302 Anonymous (1969). "Dr. Harry H. Hess, Princeton Geologist, Dies at 63," *The New York Times,* August 26, 1969.
303 *The New York Times* (1957). "Geologists Spend Vacation in Work," August 3, 1957.
304 Wilson, Tuzon (1972). "Continental Drift, Sea-Floor Spreading, and Plate Tectonics: Introduction," special issue: *Readings from Scientific American,* p 38. W. H. Freeman and Company, San Francisco.
305 Cohn, David L. (1940). "Pathfinder of the Seas," *The Nautical Gazette*, May 1940.
306 Maury, Matthew F. (1855). *The Physical Geography of the Sea.*
307 Tharp, Marie (1999). "Connect the Dots: Mapping the Seafloor and Discovering the Mid-Ocean Ridge," from *Lamont-Doherty Earth Observatory of Columbia: Twelve Perspectives on the First Fifty Years 1949-1999,* Woods Hole Oceanographic Institute.
308 Hess, Harry (1957). Spoken to Bruce Heezen, March 26, 1957, at a Princeton symposium. Reported by Walter Sullivan (1974) *Continents in Motion*, p 57.
309 Freeman, Ira Henry (1957). "Crack in World is Found at Sea," *The New York Times*,

February 1, 1957.
310 Tharp, Marie (1999). "Connect the Dots: Mapping the Seafloor and Discovering the Mid-Ocean Ridge," from *Lamont-Doherty Earth Observatory of Columbia: Twelve Perspectives on the First Fifty Years 1949-1999,* Woods Hole Oceanographic Institute.
311 Longden, Tom (2013). Des Moines Register, "Bruce Heezen," retrieved from http://data.desmoinesregister.com/dmr/famous-iowans/bruce-heezen on July 18 2013.
312 Joly, John (1915). *The Birth-time of the World,* p 133, Dutton & Company, New York.
313 Joly, John (1908). "Uranium and Geology," *Science*, New Series, Vol 28, No 726 (November 27, 1908), pp 737-743.
314 Egyed, László (1959). "Zsugorodás, tágulás, vagy magmaáramlások?" [Shrinking, expanding, or magmatic currents?], *Földrajzi Közlemények*, Vol 83, No 1, pp 1-20.
315 Scalera, G., T. Braun (2003). "Ott Christoph Hilgenberg in twentieth-century geophysics," *Why an Expanding Earth? A Book in Honour of O.C. Hilgenberg,* pp 30-32.
316 Koziar, Jan (1991). *A New Reconstruction of Gondwana on the Expanding Earth,* Wroclaw Geotectonic Laboratory, 1991, revised 2012.
317 Maxlow, James (2001). "Quantification of an Archaean to Recent Earth Expansion Process Using Global Geological and Geophysical Data Sets." PhD thesis at the Department of Applied Geology, Curtin University of Technology, Perth, Australia.
318 Hess, H.H. (1962). "History of Ocean Basins," *Petrologic Studies: A Volume to Honor A.F. Buddington*, p 618.
319 Ibid., p 599.
320 Sullivan, Walter (1966). "Science: The Velikovsky Affair" *The New York Times*, October 2 1966.
321 Sullivan, Walter (1966). "Science: The Velikovsky Affair," *The New York Times*, October 2 1966.
322 Velikovsky, Immanuel (1983), *Stargazers and Gravediggers*, p 313. William Morrow & Company, New York.
323 Velikovsky, Immanuel (1972). "H. H. Hess and My Memoranda," *Pensée* Vol 2, No 3
324 Nason, R.D, W.H.K. Lee (1962). "Preliminary Heat-Flow across the Atlantic," *Nature,* Vol 196, p 975.
325 Von Heezen, R.P., A.E. Maxwell (1969) "Sea Floor Spreading," *Sierra Leone Rise: Summary and Conclusion.* p 459. Research results of the Deep Sea Drilling Project.
326 Laj, C., J.E.T. Channell (2007). "Geomagnetic Excursions," *Treatise on Geophysics*, vol 5., *Geomagnetism*, edited by M. Kono. p 373, Elsevier Press, Amsterdam.
327 Mercator, Gerardus (1577). Mercator quoted from a now-lost book, *The Travels of Jacobus Cnoyen*, based on another now-lost book *Inventio Fortunata*. In both, English Friar Jacobus Cnoyen was claimed to have ventured north to the pole in 1360 and discovered a passage to the inside of the hollow Earth. Mercator's magnetic island came from the monk's writings and Mercator's description of it is preserved in Mercator's 1577 letter to John Dee, Queen Elizabeth's official astrologer.
328 Madill, R.G. (1949). "The Search for the North Magnetic Pole," p 8. Mines, Forest, and Scientific Services Branch, Department of Mines and Resources, Government of Canada.
329 Ibid., p 14.
330 Thellier, Emile (1981). "Sur la direction du champ magnétique terrestre, en France, durant les deux derniers millénaires," *Physics of the Earth and Planetary Interiors*, Vol 24, pp 89-132.
331 Leonhardt, R., K. Fabian, M. Winklhofer, A. Ferk, C. Laj, C. Kisse (2009). "Geomagnetic field evolution during the Laschamp excursion," *Earth and Planetary Science Letters,* Vol 278, No 1-2, pp 87-95.

332 McKenzie, D.P. (1980). "Edward Crisp Bullard," *Biographical Memoirs of Fellows of the Royal Society,* Vol 33, pp 66-98.

333 Blackett, P.M.S. (1948). *Fear, War, and the Bomb: Military and Political Consequences of Atomic Energy.* New York: McGraw-Hill.

334 Blackett, P.M.S. (1947). "The Magnetic Field of Massive Rotating Bodies," *Nature,* Vol 159 Issue 4046, pp 658-666.

335 Mason, Ron (2001). "Stripes on the Sea Floor," *Plate Tectonics: An Insider's History of the Modern Theory of the Earth,* edited by Naomi Oreskes. p 31. Westview Press, Cambridge, Massachusetts.

336 Lyman, John (1972). "Memo for History of Oceanography: R.G. Mason's geomagnetic surveys in Pioneer to Ben J. Korgen," - a letter from Lyman to Korgen written in 1972, published in 1995 in *Oceanography,* Vol 8, No 1, p 20.

337 Mason, Ron (2001). "Stripes on the Sea Floor," *Plate Tectonics: An Insider's History of the Modern Theory of the Earth,* edited by Naomi Oreskes. p 36. Westview Press, Cambridge, Massachusetts.

338 Lyman, John (1972). "Memo for History of Oceanography: R.G. Mason's geomagnetic surveys in Pioneer to Ben J. Korgen," - a letter from Lyman to Korgen written in 1972, published in 1995 in *Oceanography,* Vol 8, No 1, p 20.

339 Mason, Ron G. (1958). "A magnetic survey off the west coast of the United States," *Geophysical Journal of the Royal Astronomical Society,* Vol 1, pp 320-329.

340 Morley, Lawrence (2001). "The Zebra Pattern," *Plate Tectonics: An Insider's History of the Modern Theory of the Earth,* edited by Naomi Oreskes. p 68. Westview Press, Cambridge, Massachusetts.

341 Morley, Lawrence (2011). "The Lunatic Fringe," *Friends of GSC History, Series A - Historical Contributions* No GSCHIS-A016, p 1.

342 Anonymous (2013). "Resources Section," *Ottawa Citizen,* April 26, 2013.

343 Morley, Lawrence (2001). "The Zebra Pattern," *Plate Tectonics: An Insider's History of the Modern Theory of the Earth,* edited by Naomi Oreskes. p 70. Westview Press, Cambridge, Massachusetts.

344 Frankel, Henry R. (1987). *The Continental Drift Controversy: Evolution into Plate Tectonics,* Vol 4, p 126. Cambridge University Press.

345 Ibid.

346 Morley, Lawrence (2001). "The Zebra Pattern," *Plate Tectonics: An Insider's History of the Modern Theory of the Earth,* edited by Naomi Oreskes. p 71. Westview Press, Cambridge, Massachusetts.

347 Morley, Lawrence (1963). Morley's "Letter" from John Lear's article, "Canada's unappreciated role as scientific innovator," *Saturday Review* (2 Sep 1967), p 47.

348 Morley, Lawrence (2001). "The Zebra Pattern," *Plate Tectonics: An Insider's History of the Modern Theory of the Earth,* edited by Naomi Oreskes. p 84. Westview Press, Cambridge, Massachusetts.

349 Ibid.

350 Vine, Frederick (2001). "Reversals of Fortune," *Plate Tectonics: An Insider's History of the Modern Theory of the Earth,* edited by Naomi Oreskes. p 49. Westview Press, Cambridge, Massachusetts.

351 Ibid., p 52.

352 Ibid., pp 54-55.

353 Vine, F.J., D.H. Matthews (1963). "Magnetic Anomalies over Oceanic Ridges," *Nature,* Vol 199, No 4897, pp 947-949.

354 Vine, Frederick (2001). "Reversals of Fortune," *Plate Tectonics: An Insider's History of the Modern Theory of the Earth,* edited by Naomi Oreskes. p 57. Westview Press,

Cambridge, Massachusetts.
355 Vine, Frederick (2007). Spoken at the Geological Society Awards ceremony upon receipt of the Prestwich Medal.
356 Vine, Frederick (2001). "Reversals of Fortune," *Plate Tectonics: An Insider's History of the Modern Theory of the Earth*, edited by Naomi Oreskes. p 58. Westview Press, Cambridge, Massachusetts.
357 Sclater, John G. (1992). "The Development of Plate Tectonics: A personal Perspective; A reply to a series of questions from Professor H. Frankel," p 34. *SIO Reference Series 92-30* Scripps Institution of Oceanography, La Jolla, California.
358 Vine, Frederick (2001). "Reversals of Fortune," *Plate Tectonics: An Insider's History of the Modern Theory of the Earth*, edited by Naomi Oreskes. p 59. Westview Press, Cambridge, Massachusetts.
359 Vine, Frederick (2003). "Ophiolites, ocean crust formation, and magnetic studies: A personal view," Geological Society of America Special Paper 373, from *Ophiolite Concept and the Evolution of Geological Thought*, edited by Yildirim Dilek and Sally Newcomb, p 74. Published by Geological Society of America, Boulder, Colorado.
360 Vine, Frederick (2001). "Reversals of Fortune," *Plate Tectonics: An Insider's History of the Modern Theory of the Earth*, edited by Naomi Oreskes. p 66. Westview Press, Cambridge, Massachusetts.
361 Vine, F.J., J.T. Wilson (1965). "Magnetic anomalies over a young oceanic ridge off Vancouver Island" *Science,* Vol 150, pp 485-489.
362 Day, Deborah (1999). "Robert Sinclair Dietz Biography," UCSD Libraries, La Jolla, CA.
363 Dietz, Robert (1961). "Continent and Ocean Basin Evolution by Spreading of the Sea Floor," *Nature,* Vol 190, No 4779, pp 854-857.
364 Dietz, Robert (1963). "Continent and Ocean Basin Evolution by Sea Floor Spreading," *Tulsa Geological Society Digest,* Vol 31, pp 241-242.
365 Bell, Sebastian (2006), "Reminiscences of Harry Hess," *Harry Hess Centennial, The Smilodon, The Princeton Geosciences Newsletter* – Spring 2006, Vol 47, No 1, pp 11-14.
366 Ibid.
367 Dietz, Robert S. (1968). "Reply to: Arthur Holmes – Originator of spreading ocean floor hypothesis, by A.A. Meyerhoff (1968)" *Journal of Geophysical Research,* Vol 73, No 20, p 6567.
368 Meyerhoff, A.A. (1968). "Arthur Holmes – Originator of spreading ocean floor hypothesis," *Journal of Geophysical Research,* Vol 73, No 20, pp 6563-6565.
369 Holmes, Arthur (1945). *Principles of Physical Geology,* p 508. The Ronald Press Company, New York.
370 Plumstead, Edna P., R. Krausel (1962). "Fossil floras of Antarctica," London: Trans-Antarctic Expedition Committee.
371 Payne, Alice (2001). *Quin Kola: Tom Payne's Search for Gold*, Crossfield Publishing.
372 McNicholl, Martin K. (2013). "John Tuzo Wilson," *The Canadian Encyclopedia.* Retrieved 11 May 2013 from http://www.thecanadianencyclopedia.com/articles/john-tuzo-wilson.
373 Abbott, Patrick (2009). *Natural Disasters,* p 54. McGraw-Hill, Boston.
374 York, Derek (2001). "J. Tuzo Wilson," *Rock Stars in GSA Today,* pp 24-25.
375 Wilson, J. Tuzo (1963). "A Possible Origin of the Hawaiian Islands," *Canadian Journal of Physics* Vol 41, No 6, pp 863-870.
376 United States Geological Survey (2012). "Summary of Pu'u'o - Kupaianaha Eruption, Kilauea Volcano, Hawai`i," Hawaiian Volcano Observatory-United States Geological

Survey. http://hvo.wr.usgs.gov/kilauea/summary/ retrieved March 10, 2013.
377 Wilson, J. Tuzo (1966). "Did the Atlantic Close and the Re-Open?" *Nature,* Vol 211, pp 676-681. (13 August 1966).
378 Duarte, João (2013). "Are subduction zones invading the Atlantic? Evidence from the southwestern Iberia margin," *Geology,* Vol 41, No 8, pp 829-842. (6 June 2013).
379 Wilson, J. Tuzo (1965). "A New Class of Faults and their Bearing on Continental Drift," *Nature*, 207, pp 343-347.
380 Bevis, Mike (2011). From a tribute to Jack Oliver, "http://www.eas.cornell.edu/people/honorary/oliver.cfm," retrieved June 2, 2013.
381 Bevis, Michael, F.W. Taylor, et al. (2002). "Geodetic observations of very rapid convergence and back-arc extension at the Tonga arc," *Nature,* Vol 374, No 6519, pp 249-251. (16 March 2002).
382 Oliver, Jack (2001). "Earthquake Seismology in the Plate Tectonics Revolution," *Plate Tectonics: An Insider's History of the Modern Theory of the Earth*, edited by Naomi Oreskes. pp 162. Westview Press, Cambridge, Massachusetts.
383 Oliver, Jack (1996). *Shocks and Rocks: Seismology in the Plate Tectonics Revolution*, p 107, American Geophysical Union, Washington, D.C.
384 Isacks, Bryan, Oliver, Jack, and Lynn Sykes (1968). "Seismology and the New Global Tectonics," *Journal of Geophysical Research,* Vol 73, No 18, pp 5855-5899.
385 Oliver, Jack (2001). "Earthquake Seismology in the Plate Tectonics Revolution," *Plate Tectonics: An Insider's History of the Modern Theory of the Earth*, edited by Naomi Oreskes. p 159. Westview Press, Cambridge, Massachusetts.
386 Oliver, Jack (1979). "This Week's Citation Classic: Deep earthquake zones, anomalous structures in the upper mantle, and the lithosphere." Retrieved from www.citationclassics.org March 28, 2013.
387 Wilson, J. Tuzo (1975). *The Planet of Man*, Toronto, Canada.
388 Muller, Richard A., and Gordon J. MacDonald (2002). *Ice Ages and Astronomical Causes*. Springer-Verlag, London.
389 MacDonald, Gordon (2001). "How Mobile is the Earth?" *Plate Tectonics: An Insider's History of the Modern Theory of the Earth*, edited by Naomi Oreskes, p 125. Westview Press, Cambridge, Massachusetts.
390 MacDonald, Gordon (2002). "Gordon James Fraser MacDonald," *Biographical Memoirs of the National Academy of Sciences,* Vol 84, pp 232-234. Published in 2004, after MacDonald's death.
391 MacDonald, Gordon (1994). Interviewed by James Fleming in San Diego, California, March 21, 1994, for the Center for History of Physics of the American Institute of Physics, Oral History Project.
392 U.S. Department of Energy's Carbon Dioxide Information Analysis Center (CDIAC) "Global CO2 Emissions," http://cdiac.ornl.gov/ftp/ndp030/global.1751_2009.ems, retrieved July 17, 2013.
393 Lewis, Cherry (2000). *The Dating Game: One Man's Search for the Age of the Earth,* p 159. Cambridge University Press, U.K.
394 Jeffreys, Harold (1982). "Tidal fiction; the core; mountain and continent formation," *Geophysical Journal of the Royal Astronomical Society*, Vol 71, No 3, pp 555-556.
395 MacDonald, Gordon (2001). "How Mobile is the Earth?" *Plate Tectonics: An Insider's History of the Modern Theory of the Earth*, edited by Naomi Oreskes, p 113. Westview Press, Cambridge, Massachusetts.
396 MacDonald, Gordon (2002). "Gordon James Fraser MacDonald," *Biographical Memoirs of the National Academy of Sciences,* Vol 84, pp 228. Published in 2004, after MacDonald's death.

397 MacDonald, Gordon (2001). "How Mobile is the Earth?" *Plate Tectonics: An Insider's History of the Modern Theory of the Earth*, edited by Naomi Oreskes, p 124. Westview Press, Cambridge, Massachusetts.

398 Ibid., p 127.

399 Cox, Allan (1973). *Plate Tectonics and Geomagnetic Reversals,* pp 535-536. W.H. Freeman & Company, San Francisco.

400 Chang, Keneth (2011). "Quake, Tectonic and Theoretical," *The New York Times: Week in Review,* January 15, 2011.

401 Bullard, Edward (1975). "William Maurice Ewing," *Biographical Memoirs of the National Academy of Sciences*, Vol 21, p 154.

402 McKenzie, Dan P., W. Jason Morgan (1969). "Evolution of Triple Junctions," *Nature,* Vol 224, p 125.

403 McKenzie, Dan (2001). "Plate Tectonics: A Surprising Way to Start a Scientific Career," *Plate Tectonics: An Insider's History of the Modern Theory of the Earth*, edited by Naomi Oreskes. p 184. Westview Press, Cambridge, Massachusetts.

404 Brook, Anthony (2011). "On the first usage of 'Plate Tectonics'," *HOGG: Newsletter of the History of Geology Group of the Geological Society of London,* No 43.

405 Schopf, James (1969). "Ellsworth Mountains: Position in West Antarctica due to Sea-Floor Spreading," *Science,* Vol 164, pp 63-66.

406 Anonymous (1970). "Theory is Upheld on Earth Plates," *The New York Times,* January 4, 1970.

407 Meyerhoff, Howard (1936). "Floods and Dust Storms," *Science,* Vol 83, No 2165, p 622.

408 Coates, Donald R. (1963). *Geology of South Central New York*, p 7. New York State Geological Association.

409 Meyerhoff, Arthur, Howard A. Meyerhoff (1972). "The new global tectonics: Major inconsistencies" *American Association Petroleum Geologists Bulletin*, Vol 56, p 297.

410 Ibid., pp 269-336.

411 Hull, Donna Meyerhoff (1996). Editor's notes in *Surge Tectonics: A New Hypothesis of Global Geodynamics*. Springer Publishing.

412 Wesson, Paul S. (1972). "Objections to Continental Drift and Plate Tectonics," *Journal of Geology,* Vol 80, pp 185-197.

413 Wesson, Paul S. (1974). "Problems of Plate Tectonics and Continental Drift," pp 146-154, in the compendium *Plate Tectonics – Assessments and Reassessments: Special Volume A154,* published by the American Association of Petroleum Geologists.

414 McPhee, John (1998). Ken Deffeyes, quoted in *Annals of the Former World,* Farrar, Strauss & Giroux.

415 Bullard, Edward (1975). "The Emergence of Plate Tectonics: A Personal View," *Annual Review of Earth and Planetary Science*, Vol 3, p 8.

416 O'Neil, Jonathan, Richard W. Carlson, Jean-Louis Paquetter, Don Francis (2012). "Formation age and metamorphic history of the Nuvvuagittuq Greenstone Belt," *Precambrian Research,* Vol 220-221, November 2012, pp 23-44.

417 *Princeton Press* (2003). "Princeton geophysicist to receive National Medal of Science: Morgan honored for discoveries underlying modern studies of earthquakes, volcanoes," October 22, 2003, News Release, Princeton's Office of Communications.

418 Morgan, W. Jason (1968). "Rises, Trenches, Great Faults, and Crustal Blocks," *Journal of Geophysical Research*, Vol 73, No 6, p 1980.

419 Ibid.

420 McKenzie, Dan (2001). "Plate Tectonics: A Surprising Way to Start a Scientific Career," *Plate Tectonics: An Insider's History of the Modern Theory of the Earth*,

edited by Naomi Oreskes. pp 171-172. Westview Press, Cambridge, Massachusetts.
421 Fortey, Richard (2005). *Earth: An Intimate History*, p 158. Vintage Publishing.
422 McKenzie, Dan (2007). Interviewed by Alan Macfarlane, May 11, 2007, Cambridge University, http://www.dspace.cam.ac.uk, retrieved March 15, 2013.
423 McKenzie, Dan P. and Robert L. Parker (1967). "The North Pacific: An Example of Tectonics on a Sphere," *Nature*, Vol 216, pp 1276-1280.
424 Le Pichon, Xavier (1991). "Perspectives and Retrospectives: Introduction to the publication of the extended outline of Jason Morgan's April 17, 1967 American Geophysical Union Paper," *Tectonophysics,* Vol 187, pp 1-22.
425 McKenzie, Dan (2007). Interviewed by Alan Macfarlane, May 11, 2007, Cambridge University, http://www.dspace.cam.ac.uk/, retrieved March 15, 2013
426 McKenzie, Dan, Jason Morgan (1969). "Evolution of Triple Junctions, *Nature,* Vol 224, pp 125-127.
427 National Science Foundation (2013). "Vital Information – Laureates of the 2002 National Medal of Science," http://www.nsf.gov/news/news_summ.jsp?cntn_id=103051, retrieved April 17, 2013.
428 Princeton Office of Communications (2003). "Princeton geophysicist to receive National Medal of Science," http://www.princeton.edu/pr/news/03/q4/1022-morgan.htm, retrieved April 15, 2013
429 Morgan, W. Jason (2012). "Floating Seminar," blog post on RHUM-RUM website, http://www.rhum-rum.net/en/blog/item/floating-seminar, retrieved January 18, 2014.
430 Le Pichon, Xavier (2003). Side notes to "Asymmetry in elastic properties and the evolution of large faults," *Plates and Plumes: A Celebration of the Contributions of W. Jason Morgan*, Princeton Department of Geosciences.
431 Ibid.
432 Le Pichon, Xavier (2009). In interview with Krista Tippett, "Fragility and the evolution of our humanity," from the *On Being* series.
433 Le Pichon, Xavier (1968). "Sea-floor spreading and continental drift," *Journal of Geophysical Research*, Vol 73, pp 3661-3697.
434 Blackett, Patrick (1948). Lord Blackett, at his Nobel Acceptance Banquet Speech, attributed this quote to J.J. Thomson.
435 Olofsson, Hans (2012). "Onsala Space Observatory Strategic Plan 2012-2016," p 65. Published by Onsala Space Observatory, Sweden.
436 NASA Jet Propulsion Laboratory (2013). Site: "GPS Time Series" http://sideshow.jpl.nasa.gov/post/series.html, retrieved February 12, 2013.
437 Fowler, C.M.R. (1990). *Solid Earth Geophysics*: Earthquake Seismology, p 91. Cambridge University Press, 1990.
438 Cereceda, Errazuriz, and Lagos (2011). *Terremotos y Tsunamis en Chile*, pp 14-27. Origo Ediciones, 2011.
439 Research News of Ohio State University (2010). "Researchers Show How Far South American Cities Moved In Quake," http://researchnews.osu.edu/archive/chilemoves.htm, retrieved May 20, 2013.
440 Morales, Pamela (2010). *"Wine Industry Bouncing Back from Earthquake,"* Santiago Times, April 13, 2010.
441 Molnar, Peter, Tanya Atwater (1973). "Relative Motion of Hot Spots in the Mantle," *Nature,* Vol 246, No 5431, pp 288-291.
442 Cao, Q., R.D. Van der Hilst, M.V. De Hoop, S.H. Shim (2011). "Seismic Imaging of Transition Zone Discontinuities Suggests Hot Mantle West of Hawaii," *Science,* Vol 332, No 6033, pp 1068-1071.
443 Stuart, William D., Gillian R. Foulger, and Michael Barall (2007) "Propagation of the

Hawaiian-Emperor volcano chain by Pacific plate cooling stress," *Geological Society of America Special Papers*, Vol 430, pp 497-506.
444 Sigloch, Karin (2012). "RHUM-RUM investigates La Réunion mantle plume from crust to core" Retrieved June 12, 2013 from http://www.rhum-rum.net/en/project/7-rhum-rum-investigates-la-reunion-mantle-plume-from-crust-to-core
445 Sigloch, Karin (2014). Personal communication.
446 Warburton, Richard (2013). Personal communications, June 2013.
447 Alessandro M Forte and Jerry X Mitrovica (2001). "Deep-mantle high viscosity flow and thermochemical structure inferred from seismic and geodynamic data," *Nature*, Vol 410, pp 1049-1056.
448 Manga, Michael (2001). "Shaken, not stirred," *Nature*, Vol 410, pp 1041-1042.
449 Edelstein, Wendy (2005). "Hesitant Hottie," *UC Berkley News*, December 2, 2005.
450 Gyllenhaal, Jake (2005). *The Times of India*, December 26, 2005, World Section.
451 Herring, Thomas (2002). "Plate Tectonics Lecture," retrieved from www.gpsg.mit.edu/~tah/ July 23, 2013.
452 Sternlof, Kurt (2000). "Morgan, Pitman, and Sykes Win Vetlesen Prize for Earth Science Achievement," Columbia University News. January 21, 2000.

Manufactured by Amazon.ca
Bolton, ON